*Entertaining Elephants*

ANIMALS, HISTORY, CULTURE

Harriet Ritvo, Series Editor

# Entertaining Elephants

## Animal Agency and the
## Business of the American Circus

SUSAN NANCE

The Johns Hopkins University Press
*Baltimore*

© 2013 The Johns Hopkins University Press
All rights reserved. Published 2013
Printed in the United States of America on acid-free paper
2  4  6  8  9  7  5  3  1

The Johns Hopkins University Press
2715 North Charles Street
Baltimore, Maryland 21218-4363
www.press.jhu.edu

Library of Congress Cataloging-in-Publication Data

Nance, Susan.
Entertaining elephants : animal agency and the business of the American
circus / Susan Nance.
pages cm. — (Animals, history, culture)
Includes bibliographical references and index.
ISBN 978-1-4214-0829-3 (hdbk. : acid-free paper) —
ISBN 1-4214-0829-5 (hdbk. : acid-free paper) —
ISBN 978-1-4214-0873-6 (electronic) — ISBN 1-4214-0873-2 (electronic)
1. Elephants—United States—History—19th century. 2. Animal welfare—
United States—History—19th century. 3. Circus animals—United States—
History—19th century. 4. Circus—United States—History—
19th century. I. Title.
QL737.P98N36 2913
636.088'8—dc23    2012023932

A catalog record for this book is available from the British Library.

*Special discounts are available for bulk purchases of this book. For more information,*
*please contact Special Sales at 410-516-6936 or specialsales@press.jhu.edu.*

The Johns Hopkins University Press uses environmentally friendly book
materials, including recycled text paper that is composed of at least 30 percent
post-consumer waste, whenever possible.

# CONTENTS

*For Tyke*

ACKNOWLEDGMENTS

As I was working on this project, my Mum would occasionally ask me about my progress by inquiring, "Are the elephants out of the barn yet?" For six years I could answer only that I was *trying* to get the elephants out of the barn but that it was proving slow and difficult. Now that the elephants are out (since the book is finished), I am very glad to thank the many beings who helped me get it done. I figured out how to inquire about the lives of historical elephants through my interaction with the supportive and wise group of scholars in the fields of animal studies and animal history. For their advice, criticisms, humor, and passion for their work, I would like to tip my hat to all of them, but especially to Brett Mizelle, Sheryl Vint, Helena Pycior, Susan Pearson, Clay McShane, Nigel Rothfels, Victoria Anderson, Alice Hovorka, and Georgina Montgomery. For various enlightening talks about elephants and others and for reminding me to take it for granted that animals have a history and that it should be written, I am grateful to Georgia Mason, Ian Duncan, Derek Haley, Kim Sheppard, Patricia Turner, Karol Matthews, Tina Widowski, and the students likewise affiliated with the Campbell Centre for the Study of Animal Welfare at the University of Guelph.

Particularly crucial to this project, Denna Benn and the staff of Animal Care Services at the University of Guelph impressed upon me the importance of animal experience and ways of distinguishing it from human needs and impressions. Additionally, although they may not have realized it, I have learned a further multitude of lessons about the workaday dilemmas of human-animal relationships from the various cattle, donkeys, sheep, chickens, salamanders, rabbits, pigs, shrews, horses, mice, snakes, rats, fish, dogs, toads, cats, scientists, veterinarians, animal techs, graduate students, and facility managers I have met at the University of Guelph. For the same reasons I am very appreciative of the circus performers, animal managers, and other anonymous individuals who told me off the record of their experiences in show business.

A special thanks goes to my colleagues in the History Department and the College of Arts for regularly inquiring about and showing interest in my unconventional research activities, especially the members of the departmental research reading group who critiqued an experimental early draft of chapter 1. Mostly, though, I need to sing the praises of Bob Brugger, Harriet Ritvo, and the anonymous reviewer who helped me shape this book, as well as Barbara Lamb, Kara Reiter, and the rest of the patient staff at the Johns Hopkins University Press, who made it material.

This project received crucial funding at early stages from the generous folks at the Princeton Rare Books Library, the German Historical Institute, the University of Delaware, the Hagley Museum and Library, and the University of Guelph College of Arts. On the research front, a round of applause is also due to Chris Quinn, for helping me wade through industry trade journals and track down some of the biographical details of various elephants; the students of my fall 2008 Animals in History junior seminar, whose questions and insights helped me think through all the big issues concerning human-animal relations in the past; the kind librarians at the Robert L. Parkinson Library at the Circus World Museum in Baraboo, Wisconsin (who inexplicably lent me books with no guarantee I'd ever return them!—I did, by the way); the patient and diligent souls who run the University of Guelph Interlibrary Loan Office (who, in spite of my unreasonable stream of requests since 2004, have never cut me off); the helpful staffs at the New-York Historical Society, the New York Public Library, and the John and Mabel Ringling Museum of Art.

Finally, I would like to thank for their support and friendship Wes Macdonald, Margit Nance, Penny Gallagher, and everyone in Edmonton. My gratitude also goes to Lilly, Trixie, and Okie, who are still waiting to be paid, and old Priscilla, who never worked a day in her life.

*Entertaining Elephants*

# Turning the Circus Inside Out

Consider the career of an enduring if controversial icon of American entertainment: the genial circus elephant. A being beautiful and carnivalesque in appearance, she is physically powerful but judicious and kind to humans. In a whimsical costume with trunk raised in a salute to the viewer, she is so glad to entertain her human audience that she even appears to be smiling. Beginning in the early republic, generations of Americans came to know the genial circus elephant as a show business ambassador who cried out: "Look here for fun and novelty!" She was so comprehensible that she inhabited not only the circus ring and street parade but also handbills found at the local tavern, fences coated with circus bills, show programs and trade cards, political cartoons and newspaper advertising, movie backdrops, book plots, toy catalogs, and candy packaging. In 1993 the genial circus elephant even graced a U.S. postage stamp. The philatelic issue was tacit state endorsement for an American tradition then faltering in the face of a changing entertainment industry and assertive animal rights advocates.[1]

While she has been especially troubled over the last one hundred years, the genial circus elephant exerted considerable influence on American consumers in the nineteenth century. This was the golden age of the circus. Traveling shows were the nation's preeminent mass entertainment, and they catered to customers from every walk of life and every region. Living elephants provided these ventures with physical and cultural labor that made the sign of the genial circus elephant possible. Initially, the trade began as experimental animal exhibitions in the early republican northeast. Later, antebellum menagerie wagon shows offered caged exotic creatures, trained dogs, horses and monkeys, and perhaps a single elephant. As the nation's borders moved outward, the genial elephant accompanied the circus into each new territory and state. Following the Civil War, highly capitalized modern circuses merged the antebellum menageries and horse circuses into unified productions. They employed huge workforces of people and animals, modern technologies, and management schemes to serve consumers by wagon trail, riverboat, and railway.

In that century, ticket buyers and showmen alike believed that any entertainment called "circus" *required* elephants. Only living elephants could be trans-

A late-twentieth-century iteration of the genial circus elephant. Stamp block celebrating "200 Years of the Circus in America." United States Postal Service, 1993.

formed by the right costuming, patter, advertising, and human performers to produce the icon of the genial circus elephant. Only elephants were strong enough to raise the masts required by the largest circus tents. Only they had the power to shove heavy circus wagons through mud that bogged down exhausted men and horses. Eventually appearing in herds of up to twenty on a single production, the uniquely massive and powerful elephant alone could advertise the ambition of circus impresarios like Adam Forepaugh, P. T. Barnum, or James Bailey. By 1900, the biggest American circuses were branded commercial entities that showed on multiple continents. They were the largest entertainment companies people had ever seen and a rags-to-riches story in the classic American style. In spite of the form's ancient roots in Europe and Asia, to create such enormous circus shows by the systematic use of captive elephants was a uniquely American innovation. Thus, the long nineteenth century of the American circus was equally the elephant's long nineteenth century in America. It began with the arrival of the first elephant in 1796 and lasted until the 1907 destruction of the first elephant born in the United States, right at the peak of circus business history.

The genial circus elephant, however, carries more significance than this, for she ushered in sophisticated mass markets for consumer experience in the United States before there were comparable markets for products or services. She preceded many American characters taken as a sign of the growth of market life: Zip Coon and the plantation "Darky," Brother Jonathan and Davy Crockett, Barnum's "What is it?," Jenny Lind, and the Feejee Mermaid. In the 135 years after independence, the country underwent vast changes in geographic range, cultural diversification, and Anglo-American power consolidation. Once an unstable rebel state, by 1907 an imperial, if indebted, world power, the nation was traversed by new roads, canals, and railways. That infrastructure was made useful by enormous pools of equine and human labor by which people made what they could of the continent's abundant natural resources. In this, Americans were part of global efforts to contain, control, and extract value from plants, animals, and the environment, which were crucial to industrialization where it occurred in the nineteenth century.[2] The United States was also the world's preeminent consumer economy and, despite widespread poverty, held lucrative national markets for live entertainment, advertising, and publishing. Citizens learned how to find personal identity in the ethical and economic systems associated with these changes by spending on entertainment. That is, people worked out how to define themselves as individuals by paying to see things just as they did by paying to own and display things.[3]

Animal exhibitions, menageries, and circuses reshaped American markets for experience by employing the jovial circus elephant icon to show Americans how to become modern consumers of animal celebrity. Appearing on the cultural landscape in the 1810s, animal celebrity constituted an abstract way of conceiving wild animals as public, named, and commercially engaged individuals. The resulting animal figure was a product of people who decided they could make a living by creating and selling portrayals of noted animals that flattered audience humor, visual interest, or morality. Those portrayals morphed into the icon of the genial circus elephant and worked so well that show business people came to believe that the full nature of the living animals they used to produce her was a behind-the-scenes secret best left off the show program.

The practices of marketing and consuming nonhumans in these ways were as central to the development of American culture as blackface minstrelsy, although, unlike minstrelsy, such practices are still alive and well today. Like burnt-cork performance, animal celebrity was a flexible form that saw human performers and their audiences collaborate to create satire, protest, accommodation, and oppression grounded in carefully crafted clichés and misrepresentations. As the blackface form helped people to define race in the United States, Americans employed animal celebrity to complicate or to clearly mark out the imagined lines distinguishing humans from nonhumans. The two forms grew in parallel, and many Americans throughout the nineteenth century would compare animals and African Americans in order to talk about sympathy, loyalty, and suffering, to argue about exploitation under capitalism, or to demonize one or both as bestial.[4] At the end of the century, both African Americans and elephants would share a historic nadir, which saw many African Americans lynched or otherwise punished for challenging or appearing to challenge the people who sought to dominate them. Unlike the many African Americans and other Americans who clearly understood and often criticized the broad tradition of blackface and the network of power and terror it facilitated, nonhumans had no abstract understanding of animal celebrity, no opportunity to argue back or publicly denounce those Americans who accepted it. Elephants and other animals only inadvertently subverted animal celebrity and its articulations of human supremacy when they defied their human keepers and the conditions of their experience.

Circuses did not invent human practices of imagining animals as totems or symbols, of course, but American animal shows pioneered ways of adapting those habits to commercial purposes for a broad consumer audience. The United States offered one of the earliest mass entertainment markets to the people who mapped out these practices by taking their shows down bumpy wagon trails and along railways to the Pacific. Over the nineteenth century these showmen developed the selling of animal celebrity into a highly mediated process, one that would proliferate during the twentieth century by expanding beyond the elephant to include many other species and media, from the animal amusement park and zoo to the cinema and product advertising.[5]

Today we take for granted that animal characters—both the purely fictional and those developed from actual animals—are ubiquitous in mature consumer societies.[6] But this was not inevitable. Someone needed to teach people how to accept animals framed as zoo pets or movie stars or reality TV characters by seeing that one might gain power as a human in a consumer capitalist society by doing so. The genial circus elephant icon suggested to viewers that they could come to understand animals by paying to be entertained by them. She assured people that they could believe that captive animals shared human goals and sought to cooperate with the public. She encouraged Americans to take on a custodial and ultimately hubristic attitude toward animals, particularly nonnative wild creatures like elephants. The jovial elephant figure told audiences that, no matter how exotic or fragile or rare wild animals were, circuses could be trusted to package and deliver them to a convenient location at a low price and that the resulting privatization of nature represented an educational and patriotic sign of Americans' power in the world.[7]

The icon of the jovial circus elephant, therefore, introduced Americans to a mode of capitalism linked to expanding human supremacy over other animals. Premised on mass production and mass consumption of animal bodies, this was a consumer capitalism facilitated by images of animals, like the circus elephant, that obscured the backstage realities of their production.[8] Or rather, chicken-and-egg style, sometimes it was audience enjoyment of a given animal celebrity sign that drove increasing for-profit uses of actual animal bodies.

Providing the seminal fantasy of gregarious elephant performer at liberty in America was exhausting and contentious work for the people and the actual living elephants involved. The consumer-friendly animals that Anna Sewell produced with text, or that Disney would begin creating in the early twentieth century with animators or carefully edited footage of wild animals, or that SeaWorld or the National Zoo still provide in stationary facilities, circuses delivered to the American public in the most complicated way possible, namely, with imported live animals kept mobile on a schedule. Elephants are not robots or blank canvases, but fragile, often strong-willed individuals. From the beginning, many of them rejected the conditions of their experience in captivity. As the decades passed, circus people responded with increasingly rigid elephant training, containment, and transport routines, which in turn drove elephants to behave in ways that increasingly put people and property in danger, so Americans witnessed a century-long escalation in the numbers of elephants killing people and people killing elephants.

In spite of this difficult history, the presumed naturalness of the circus elephant as living being and marketing sign has effectively gone unchallenged. Amateur and scholarly histories of animal exhibitions, menageries, and circuses are substantial and the product of much diligent research dedicated to preserving knowledge of these companies and the art they produced. Students of circuses have used them to tell the stories of, respectively, a few wily Yankee entrepreneurs, some intriguing American countercultures, or the replication of problematic racist, capitalist, or imperialist ideologies.[9] Most have treated circus animals as fanciful wallpaper, colorful but of no analytical concern as historical beings. Such a methodology is especially inadequate since nonhuman performers feature so prominently in the memoirs and advertising documenting these ventures.

A moment of qualification is prudent here. There is an old American tradition of elephant biography we might examine for clues about the jovial elephant icon and historical beings who supported her. Over the last two hundred years, a long series of journalists, circus people, and fans have compiled informal histories of various historical elephants. Crafted from industry wisdom, local oral tradition, and other elephant biographies found in newspapers, collectively they are rife with inaccuracy. Yet, the primary job of the elephant-biography tradition is not solely to get the dates and other details perfectly correct. Rather, it is to convey nostalgia that reflects the showmanship and marketing messages the circuses themselves offered. They exhibit assumptions, presumably shared by author and reader, that the circus and its customers were inherently benign and that elephants are native to show business. They usually do so by episodic treatment of elephant life highlighting only isolated, if famous, appearances, tricks, or "rampages" rather than the many acts and decisions that led to those events. They do so by perpetuating the myth of captive elephant longevity, marked by the many elephants remembered as Old Bet, Old Columbus, or Old Empress, although these animals died in youth or early adulthood.[10]

Nor have historians of business, labor, and management yet spent much time investigating nineteenth-century entertainment ventures to ask how they pioneered modes of animal management that founded our contemporary practices of interpreting and selling animal celebrity and captivity to consumers. Although an ecocultural history of business and management has appeared that could undertake this work, researchers in this field tend to present a view of the world outside the street shop, factory, or office building that is rather inanimate and seemingly insulated from human politics.[11] Likewise, historians of science and industrialization have tended to minimize animal life and action in the life stories of the people who used nonhumans to invent new practices and build new things.[12] In fact, the questions and insights of scholars in these fields are just what we need to understand the circuses' preeminent role in American history as specialized organizations for the acquisition, management, marketing, and disposition of nonhuman life to mass audiences.

By ignoring how animals have complicated human endeavors while simultaneously avoiding the breadth of elephant life, both amateur and professional historians have actually sold circuses short. This oversight can be addressed only by systematic discussion of some pressing issues that shaped daily life in the menageries and circuses: the global sourcing and ultimate disposition of elephants, debates within the circuses over how or if to keep those animals alive and performing, the constant drumbeat of public complaint about elephant use in entertainment, and—most basic of all—why elephants, in the first place. We may or may not care about historical elephants as individuals of intrinsic value, but without asking about the breadth of captive elephant existence we cannot not know the full experience of the *people* who lived inside these entertainment companies. This is the key lesson of transspecies history and animal studies: human and nonhuman lives exist in symbiosis.[13] What is more, there has never been any purely human space in world history.[14] We must account for nonhumans as living beings—not merely representations—in order to find the fullest possible explanation of history and to avoid simply engaging in self-flattery or self-deception.

Several historians of the United States have been exploring these ideas to seek out glimpses of the interspecific past. They document the hogs who trampled Indian corn fields, horses who died just when their owners needed them most, wolves who were unable to adapt to Anglo-American expansion as the coyotes and gophers did, dogs who followed presidents from room to room in the White House, and other animals whose acts and life cycles drove (or hindered) economies and informed human cultures.[15] From these studies we learn that, on some counts, American use of elephants was akin to that of circus horses, creatures produced by the broader contemporary system of equine energy production, distribution, and management. Especially after the late-antebellum mobility revolution, equine labor drove various kinds of industrial, transport, and urban innovation. With limited veterinary care (even such as it was in those days), equines were employed until exhausted at the age of four or five years or killed by disease. Thereafter owners disposed of them, if alive, either by sale or by euthanasia, or, if dead, wherever one could find markets for horse meat, leather, or hair.[16]

Like equines, the elephants owned by circuses were living commodities sometimes worth more dead than alive, and they carried volatile meanings about how animal use reflected human identities and politics.[17] Striving to make the most of the animals at the least expense, circus elephant management strategies prospered primarily through the use of forcibly weaned juvenile and adolescent animals employed for short periods and maximum productivity. Like equines, as the century progressed, elephants experienced increasing confinement and the restriction of species-typical behaviors that did not produce value. This was a system in which the adult animal was often dysfunctional and in which elephants existed as beings removed from their own means of self-preservation and self-direction, so they were both a tool and a burden to a circus.[18] American consumers mostly knew elephants, unlike horses and mules, by way of their genial icon surrogates, having only minimal personal exposure to their reality as living beings.

The transspecies histories that have emerged in the last ten years, like environmental histories, have certainly questioned the idea of human self-sufficiency that plagues our interpretations of the past. Still, these authors are perhaps too cautious in the questions they ask. They tend to restrict analysis to unconscious animal acts (such as succumbing to disease), to animals taken as anonymous members of broad populations (necessary when this is all that the primary sources permit, to be sure), or as a side note to activities of *people* deemed famous or important (biographies of pets of noted Americans). Can we instead choose a group of animals that can be examined as unique individuals when possible and whose lives are the primary focus of the inquiry, so that humans are players in a story that is equally nonhuman and human? With respect to American show business, that goal requires us to ask some provocative questions: How did elephants shape the showman's work of developing and selling animal celebrity as entertainment product and abstract cultural concept? More broadly, are elephants crucial factors of historical causation? How might we document this? That is, does animal experience matter in history?

Even if we can only glimpse aspects of what elephant life might have been, the fact that captive elephants in the United States had experiences and acted on their environments as they perceived them must shape our explanation of how and why the circuses came to be. Indeed, in this book I maintain, as do some environmental historians, that at a moment of mass extinctions—which includes the grave endangerment of wild elephants in Africa and Asia—it is unethical to argue any longer that historical animals are best considered solely as figments of the human imagination or as raw materials of human agency. Doing so would be to replicate in our scholarship the human attitudes that have come to imperil the planet outside the academy.[19]

Instead, this book creates an elephant-centered history of the show trade by turning the history of the circus inside out. Such an approach endeavors to examine the inner workings and debates of the circuses as much as the finished products they offered and to find ways to look beyond the known stories of how circuses influenced the development of broadside advertising or railway contracting or social values. It means going beyond simply celebrating the people who worked in menageries and circuses, the men who owned these ventures, or the audiences that patronized them. It means doing more than simply documenting the human ideas about and uses of elephants in the past. The task of finding real elephants in American history requires that their lives and movements determine the beginnings and endpoints of the stories we investigate and direct many of the questions we ask. It means seeing elephants as sentient beings who responded to a context or some internal motivation to produce events and effects that mattered.

This approach leads some scholars to speak of animal "agency," or nonhumans' role as "historical actors" empowered by "consciousness" or "subjectivity." Others have pushed back, arguing that these terms should be reserved solely for humans and the writing of "history" conceived solely as a record of human agency.[20] Others ask if nonhuman animals can be taken as "subalterns" that, by acting, can "speak" back to human power and thus become more knowable subjects of historical analysis. All these attempts ask a well-meaning but self-interested question by implying the omnipresence of human supremacy in global history and by imposing human politics of "resistance" on nonhuman life.[21] Elephants do not comprehend human constructs like "capitalism"; they do not perceive "race," "class," or "gender" (although they certainly have preferences about individual people and settings), so it is problematic to argue that they resist these human things. In fact, the confusion about the issues of animal agency and nonhumans' place in American history (whether these issues are best analyzed within social or environmental history, for example) may result from a lack of clarity about the difference between *individual agency* and *human social and political power*.[22] As sentient beings acting on their environments as they perceived them, elephants had agency. Yet significantly, because they had no understanding of the human cultures that created their captivity in the United States, they could not possess any (human) social and political power. This was the crucial fact that made their captivity possible.

This book refuses any notion that elephants understood, endorsed, or resisted the world of show business as a human cultural or business practice. Rather than speaking of elephants rejecting the circus or capitalism per se, it speaks of them as rejecting the conditions of their experience. Circus elephants experienced a kind of parallel reality that was defined by their interactions with people, other elephants, horses, bull hooks, hay, the weather. Further, as historians, we can take elephants *as elephants* without needing to know definitively what a given elephant's intentions or internal experience was at every moment. We can employ recent research on contemporary elephants to determine what we can infer about the acts and experiences of their historical counterparts (more on this below). More importantly, in doing so we will finally get to the work of documenting what historical animals have done, that is, what they have consumed, produced, destroyed, and chosen when presented with options, and how, when, and (possibly) why they did so. Armed with such information, we will be able to argue about the possibility of animal causation in the past, not just as a theoretical exercise, but from a body of evidence.

Moreover, we need not burden elephants and other nonhumans with the responsibility of having created primary source materials, as a person might. Although elephants never kept diaries or wrote newspaper columns, they marked the historical record with footprints and dung, the accounts of injured or amazed bystanders, broken bridges and barns, images in photographs, the shape of harnesses and fetters, and other "traces" that we will find if we look for them.[23] At the same time, we can be frank that the sources depicting historical elephants consist in large part of evidence collected by people immersed in cultures of seeing and interpreting animals that are very different from our own. While we may ask about an elephant's welfare or use research on wild populations of elephants to explain the behavior and health of captive elephants, nineteenth-century circus people could make no such studies.

Early American observers of elephant life nonetheless created a valuable body of evidence residing in promotional newspaper pieces and advertisements, photographs, broadsides, contemporary diary accounts, and local news reporting on elephants. For instance, circus press notices were some of the most prolific and explicit sources from which we learn of elephant behavior after the Civil War. Intending them as strategic marketing communications that would drive ticket sales, company agents compiled press notices in a journalistic style and offered them as free "news" content to newspapers and popular magazines. Agents often wrote them as templates and later filled in the blanks (e.g., SHOW-ING IN ____/ Boston / New York / Charleston / St. Louis / Spokane), so these notices may not necessarily be the most accurate accounts of specific individuals at a specific moment or location. Nonetheless, press notices provide firsthand evidence of typical elephant behaviors in those circuses because, as publicity documents, they had to be broadly comprehensible to consumers. And, indeed, they do not contain gratuitously fictional reports of elephant telepathy, verbal expression in English, or other incredible things that children's literature, for instance, displayed with respect to its elephant characters. As a result, if carefully explained as an aggregate of elephant action and human culture, this evidence helps us propose open-ended explanations for what elephants did and why, and why a given elephant matters to American entertainment history.[24]

Several bodies of scholarly research make this doable by opening the extant source base to new kinds of analysis. First, to verify historical statements about elephant behavior, intentions, or character, we can begin with the animal studies literature explaining how humans have perceived and represented animals to express human cultures in specific times and places. American circuses are part of a global history of elephant interpretation and use, especially in the case of Asian elephants. Although captive elephants are usually wild-born, over many centuries humans have taken thousands captive, shaping and being shaped by their behaviors and physicality. Accordingly, this work gratefully draws upon the works of natural history, behavioral ecology, and human cultural history charting those elephants and their representation.[25] Additionally, the elephant's long nineteenth century in America runs concurrently with global histories of elephant conflict with humans, their live capture and exportation, or their being hunted for ivory, phenomena that are important here and will be raised as appropriate.

Second, we can use recent ethological and animal welfare science research (AWS) on free and captive elephants as a theoretical base for the interpretation of historical elephants. Conducted in earnest over the last forty years or so, this research cannot yet provide all the answers we might seek about elephant physiology, behavior, or psychology. Nevertheless, these studies have produced findings that have revolutionized understanding of the complexity of elephant experience, culture, and biology and have led us to the questions we must ask as research continues. In most general terms, ethological and AWS researchers operate from a position that assumes, among other things, that elephants have versatile mental abilities, even if these are not immediately apparent to human perception (unlike twentieth-century behaviorists, who believed that studies seeking complex mental abilities in animals were inappropriate because animals were stimulus-response machines). The assumption of ability is a scholarly premise grounded in the idea that complex mental and social abilities are an evolutionary advantage in many species, not simply humans.

Focusing on captive settings, AWS studies are driven by a respect for the scientific method—being "led by data, not feelings," as such researchers say—and an ethical/practical argument that all sentient animals should have "five freedoms" (from hunger and thirst; from discomfort; from pain, injury, or disease; from fear and distress; and to express normal behaviors), which, if accommodated, can improve human and nonhuman life. This approach assumes that since animals are products of evolution they can be trusted to know what is best for them in responding to immediate contexts. It posits that complex mental lives—including experiences of emotions like anxiety, grief, and joy—should be taken as a given in other species since they promote social cohesion and survival and that researchers must find ways to inquire about them. Indeed, long dead among the vast bulk of ethologists, veterinarians, and other animal scientists is the old notion of animals, especially elephants, as unfeeling automatons or cognitively simplistic. Further, all problems we define as "animal welfare issues" are caused by people (out of ignorance, inattention, lack of resources, etc.).[26] Thus, AWS projects often "ask" animals with preference testing and other means what they prefer in housing, food, access to open spaces, companionship, and more, frequently furnishing results that are indicative of nonhuman priorities but counterintuitive from a human perspective.

Like all scientific work, the ethological and AWS research on elephants is a product of particular political and cultural contexts that yield different kinds of humans and nonhumans. Indeed, there is no "natural" or static ideal elephant, African or Asian, across time and space.[27] Like all species, elephants are product of genetics, environment, and experience. However, they have evolved to survive in certain ways in certain places. Although they may over generations adapt (or not) to environmental change or human pressures, they nonetheless come with certain hardwired needs and abilities that are always non-negotiable, such as the need to constantly move and forage or to abide by modes of social organization developed by elephants over the centuries to promote peaceful group cohesion and reproduction.[28] So, this book does take free populations of elephants as more normative than captive populations simply because they are self-sustaining.

In any event, different environments or communities of captivity will produce different kinds of animals and people, and humans use those processes to produce human cultures and identities defined by particular modes of animal use and interpretation.[29] Therefore, using scientific literature documenting late-twentieth- and early-twenty-first-century elephants to interpret the records of nineteenth-century elephants does require some caution. Spanning an elephant timeline of a dozen generations or so, there will have been no evolutionary change between the two populations, so it is primarily the human and environmental contexts that shape the people and elephants in question. At this point—indeed, we may be better informed in fifty years—the scientific literature helps us avoid imposing human needs or assumptions onto historical elephants as much as is possible.[30] And still, as it turns out, there are many aspects of elephant biology, behavior, and culture that people reported in 1835 or 1905 that scientists have since "proven" over the last few decades, even though nineteenth-century elephant observers approached the species with very different methodologies and motivations.

Last, we can use the work of historians of business and management to understand how elephant and human behaviors were each in turn shaped by life in a for-profit traveling show. The logistical and financial realities of large, mobile business organizations in the nineteenth century imposed many demands on human and nonhuman alike, under which many individuals struggled or flourished as they could. Indeed, what this book shows is the degree to which animal life, human life, and the drive for profits in a geographically expansive nation were symbiotically linked in a triangular tension.

*Entertaining Elephants*, then, presents a cultural and management history of elephant use in American entertainment informed by animal welfare science research, which serves as a theoretical tool for finding living elephants in the historical record. It seeks to approximate the entire past, allowing that even captive elephants probably had moments of pleasure or relief and that some humans did the best they could with their elephants, however inadequate many of those attempts may have been. In this history individuals of all species are understood to be complex characters, neither heroes nor villains, but beings who acted on their environments in their own interests as they understood them.

Materializing on the American entertainment scene as juvenile captives of traveling exhibitors, elephants were at first presented to audiences as a naturalist's curiosity and, within two decades, as exotic Asian visitors. The latter mode of presentation offered the happy circus elephant as genial performer—an animal actor, some even claimed—whose comedic or dramatic narratives could be varied to keep customers coming back year after year. In spite of the premature deaths of the elephants in the exhibition and menagerie trade, impresarios and animal dealers continued importing Asian and African elephant juveniles. By the Gilded Age, although large rail circuses had developed into highly capitalized industrial circuses, like their forerunners they struggled to balance limited budgets of time and manpower with obligations to audiences and the bottom line. As often as they might grow attached to them, the keepers and performers who worked with the dozens of elephants in those companies often lost patience, ran out of ideas, or otherwise came to fear and hate their charges. They struggled daily to modify elephants into uniform and "interchangeable parts" for circus entertainment production systems.[31] Consequently, as early as the 1840s, some circus workers had been clashing violently with the late-adolescent and adult male elephants in shows, and, by the 1880s, with juveniles and adult females as well. This was the paradox of the circuses, enterprises dependent upon elephants, many of whom seemed inevitably destructive to anyone who attempted to contain them.

Although a vocal minority pointed out that the circuses were to blame, for the majority of circus people, elephants appeared to be inherently defective and untrustworthy. Thus, a new cliché of circus elephant as vicious brute emerged in the Gilded Age, to coexist uncomfortably with its older genial circus elephant twin. For fifty years thereafter, in their marketing materials and publicity practices, the circuses actually framed their elephants for the public alternately as jovial entertainer or as elephant "gone mad." Unable to let go of either elephantine icon, circuses allowed captive elephants to exist as at once the most famous and most infamous, the most loved and most feared animals in America.

Meanwhile, since at least the late 1830s, people had been using the phrase "to see the elephant" as a means of describing having seen battle or some other dangerous experience.[32] The colloquialism surfaced during the Texas Revolution and the subsequent war with Mexico, remaining relevant in newspaper talk and oral tradition during the California gold rush, the Civil War, and beyond. More informally, when people sought to speak of a person who had suffered from hubris or naïveté only to get more than he'd bargained for, they said he had "seen the elephant." By the twentieth century, it was circuses that had seen the elephant, so to speak. "It is one of the hard facts of life," one circus man remarked at the turn of the twentieth century, "that schemes made on paper never work out smoothly in actual practice, and nowhere is the truth of this fact better realized than in the business of a touring circus."[33] Many believed that the industry had made a devil's bargain by promoting and cultivating elephantine entertainment with audiences. Yet, circus people knew that to ticket buyers, "American circus" meant "elephant," and there seemed to be no turning back.

# CHAPTER ONE

# Why Elephants in the Early Republic?

In 1796, an elephant came to the United States. The children and grandchildren of those who knew her called her Betsy or Old Bet. In her own time, people simply referred to her as the Elephant. She became so famous that many competing biographies emerged to tell her story. It is from them that we learn that her legend goes something like this: Once upon a time, Captain Jacob Crowninshield brought a young female Asian elephant to New York on a ship. "A true Yankee trader," he had purchased her for $450 in Calcutta the previous year and would sell her to someone upstate for a reputed $10,000.[1] For three years thereafter, the elephant walked up and down the East Coast. Her ownership changed several times, and a series of keepers invited people to pay between ten and twenty-five cents to see her. In Boston, Salem, Hartford, New York, Philadelphia, Savannah, and Charleston, she appeared in empty lots and rooms let by taverns, in newspaper advertisements, and on broadsides found on fences. In the countryside she appeared for the curious in barns, traveling by night so that no villager might see her on the road, free of charge. Or, some said, she traveled overnight to avoid groups of obnoxious, carousing young men who might try to tease her and her keeper. All sorts of people came to see the elephant: incredulous children and their parents, skeptical farmers, drunken urban rowdies, and gentleman philosophers intent on studying her anatomy.[2]

Or perhaps there were "Two Betseys."[3] Another series of stories tells of a female Asian elephant in America in the 1810s. She too became known as Old Bet and worked and lived on the same route after famed menagerie entrepreneur Hackaliah Bailey acquired her for exhibition. This Old Bet became a favorite of tourism boosters in Somers, New York. Bailey and others living there had invested in the nascent animal show business, giving Somers the reputation of "home of the American circus" because of its association with that mysterious early elephant.

With each telling of these stories, Old Bet's personality became more elaborate. By the 1910s, for instance, early lore emanating from 1790s handbills suggesting Betsey could pull corks from bottles had flowered into highly speculative accounts of elephantine drunkenness. One version alleged that Betsy had learned from her keepers where the rum jug was kept. One night in New England during a rainstorm she had guzzled the whole container to keep warm. Soon belligerent, she "lashed about so wildly with her trunk," the story went, that her keepers "barely escaped being killed." They eventually shot her when she injured and killed several people.[4] Another retelling claimed an irate farmer in Maine had shot Betsy because she spooked his horses, or perhaps because she persuaded people to miss church to visit with her. Others said it was a kid on a dare who killed the elephant, tempted by friends to see if he could penetrate Betsy's hide with musket fire. Or perhaps it was in 1821 that "she broke loose in South Carolina and, after being chased several miles by a mob, was shot to death."[5] Although these competing narratives were often untrue in the details, they were honest in their metaphorical comparison of a juvenile elephant with the early American republic, a creature at once compelling, volatile, and fragile.

We also learn from these tales that many Americans sought to describe these first females as founding members of American show business. Circus company press agents looking for a promotional mascot retold and embellished these elephants' biographies many times, spawning the nineteenth-century truism that "an elephant was the backbone of the first American circus."[6] Journalists needing comprehensible content in turn borrowed these stories, similarly confusing the first elephant in America with the second, blending them into one primeval personality. The public listened because, for Gilded Age and modern Americans, the story of Betsey, or "America's first elephant," was a vehicle for patriotic nostalgia about the primitive earnestness of the early republic's citizens: "As money was scarce the farmers dickered with Bailey in all sorts of ways for admittance to the barns. Men pawned their farming implements and boys sold their jackknives to raise the necessary dime," one account reminisced.[7] Such accounts spoke of the public debate over the nature and meaning of a for-profit elephant show, implying that it had been inevitably and wisely resolved in favor of elephant captivity and capitalism.

This chapter documents the brief tenures of the first two elephants to understand how fact and fiction combined to found the phenomenon of elephant celebrity in America. Beginning with the individual known as "The Elephant," it tracks her activities between 1796 and 1799, when she disappears from the historical record. Then, as a sort of conclusion, the chapter accounts for the second elephant, later known as Betsy, who arrived in 1804 and died in Maine in 1816. These young elephants' lives constituted a logistical and marketing experiment that allowed the audience the most direct input into animal displays they would ever have. The elephants shaped this process also, unknowingly persuading showmen of some enduring principles: import juvenile animals; have faith that an elephant will earn more than she costs to purchase, feed, transport, and show; help audiences see elephants as noted individuals; market elephants with fact-rich educational-sounding patter that endorses existing audience assumptions.

The first elephant to arrive in North America was a descendant of a population of the so-called subspecies *Elephas maximus indicus*, which has inhabited Asia for millennia, and she was probably female.[8] As many of her apocryphal biographies asserted, in the fall of 1795 Jacob Crowninshield had indeed purchased her in Calcutta for a reported $450. It was a "scheme," he said, and he hoped to resell her in the United States for $5,000.[9] In a letter to his brothers from abroad that winter, Crowninshield described her as "a fine young elephant two years old . . . almost as large as a very large ox."[10]

Crowninshield's plan for introducing an elephant to his home country was viable in those years because Asian elephants had long lived in captivity as trained laborers. Since at least the fifteenth century the ruling classes throughout South and Southeast Asia had systematically captured, trained and traded thousands of elephants for heavy lifting, the transportation of people and cargo, and as political gifts and ceremonial embellishments. In Ceylon (Sri Lanka) and India, mahouts used highly developed methods of capturing and modifying wild elephants. Driving wild individuals into kraals (enclosures of logs or nets) with noise or captive females and food as bait, men would restrain the new captives with ropes pulled by elephants already trained for the task. While confined, new elephants experienced painful physical discipline, which the mahouts alternated with precisely timed food and water rewards. Thus did these elephants learn that their sustenance came only from humans and that to survive it was necessary to peacefully tolerate a mahout handler and follow some basic commands—experiences the mahouts knew set the stage for further training.[11] Many in Asia idealized a well-trained elephant as a "noble beast" because he or she appeared graciously "subservient to human command," despite the considerable discomfort endured to become so.[12] More crucially, before machines could do any similar work, elephants came with a muscular and dexterous trunk that made them invaluable for construction and logging operations.[13]

As the French, Portuguese, and Romans in Africa and Asia had before them, eighteenth-century Anglo-Indians and British expatriates used their trade networks to import exotic animals to the home country from lands around the Atlantic and Indian Oceans. They produced a robust global trade in animals for royal menageries, exhibitions, elite pet keeping, and livestock development that accelerated during the last three decades of the century. The resulting Anglo-

Indian Asian elephant network saw mahouts share some of their knowledge on captive elephant trainability, health, and feeding with British East India company officials. A few of those men in turn offered elephants for sale with instructions on their care and consequently made great fortunes from the traffic. European traders knew that captive elephants, zebras, birds, and other creatures could be very valuable at home but that they needed special handling before and after purchase.[14] For instance, many wild-caught animals refused to eat in captivity, starving themselves to death when offered unfamiliar food, disoriented, or injured in the capture process. A captive animal trained to take food from humans and presented with advice on its care seemed a less risky investment, and many an investor-sailor purchased animals duly prepared for export.[15]

Crowninshield's elephant had in all probability spent some time in the hands of professional elephant managers before being walked or hoisted with a crane onto a ship in Calcutta. As a two-year-old, she would have been forcibly weaned from her captive mother by a procedure in which the mother was tied to some immovable object, like a tree, and the baby roped and pulled away. This was a job that required great determination on the part of the mahouts, as neither the dam nor her offspring would willingly submit to separation. But buyers tended to prefer juvenile animals. They more easily accepted direction, being less socialized by conspecific adults to fear, avoid, or challenge humans. Moreover, they were easier to control on shipboard without the help of other trained elephants since they were physically weaker and smaller, and their more modest appetites made them economical, as well.

So Crowninshield's two-year-old elephant was more docile but also less durable. Still a calf, she would typically have been suckling and nutritionally reliant upon her mother until age four. In fact, many dealers and keepers knew from experience that young animals were delicate and of unpredictable health. This practical wisdom has since been borne out by late-twentieth-century studies showing that human removal of elephant calves before natural weaning disrupts the development of immune function and social learning. When left to their own devices, female Asian elephants remain for life with close relatives in allomother groups between five and ten in number. There they learn from older elephants "which plants can be eaten; when to migrate . . . ; what to do when threatened by a predator; how to behave when pursued by an ardent bull."[16] That is, they learn how to adapt to novel or stressful situations, find healthful food, and interact safely with other living beings. The gregarious social relationships elephants create correlate with longevity, and for females in particular, solitude in captivity is thought to contribute to chronic illness and early death.[17]

Jacob Crowninshield was one of the first Americans to try his hand as an animal importer and probably had only vague ideas about such elephant needs or their lives at large in Asia. He came from a well-known Salem, Massachusetts, family of sailors and traders, influential citizens of considerable means; Crowninshield would serve in Congress in 1802. Crowninshield's first mate on the ship was Nathaniel Hathorne, father of the writer Nathaniel Hawthorne (who added the *w* to the famous family name). As they made their way around Africa and across the Atlantic between December 1795 and April 1796, Hathorne kept a log of the voyage, although with little direct comment on the elephant.[18] What he does tell us of the elephant's effect is that the crew loaded extra water when they realized she was depleting their stocks at a great pace. And, in mid-February, at the island of St. Helena, Hathorne noted that the crew "took aboard pumpkins and cabbage, fresh fish for ship's use and greens for the Elephant," indicating that they were following instruction on what to offer a young elephant to eat.[19]

Crowninshield's vessel, the *America*, also carried more conventional commodities, such as coffee and textiles. All his cargo was potentially subject to seizure by French or British officials patrolling the Atlantic because the two nations were vying for control of the international sea trade at that time. Meanwhile, American merchant ships took advantage of the situation by skirting tariffs and other trade restrictions to make maximum profits. Crowninshield's ship would suffer no seizures, although after a stop on the Isle of France (Mauritius) it traveled through a series of thunder and lightning storms so strong the winds tore several of the ship's sails.[20] As the crew weathered these seas with their rare elephant cargo they may also have considered whether an elephant would be difficult to butcher at sea. When winds died down and voyages became drawn out, it was common for sailors to eat the animals on board, that is, if they had not unexpectedly died and already been thrown overboard. Desperate crews ate animals no matter how exotic or expensive or anticipated by powerful people, including menagerie or pet stock like monkeys, turtles, parrots, or cockatoos.[21] To explain their actions, the sailors would have noted that the slave and free human death rate on Atlantic crossings was chilling (we now know, around 15 percent). Many would have counted a high animal death toll merely as the price of participating in a dangerous but potentially lucrative business.[22]

However, the passage on Crowninshield's ship was not unexpectedly long, and the young elephant made it to the United States without expiring or ending up as rations for sailors. On April 18, 1796, New York newspaper readers saw the first account of this feat. Among the columns of patriotic banter and announcements of lottery winners, business opportunities, and escaped slaves people had found there, a small story related: "The America has brought home an ELEPHANT, from Bengal, in perfect health. It is the first ever seen in America and a great curiosity. It is a female, two years old, and of the species that grow to an enormous size."[23] That phrase "in perfect health" was one commonly used to describe imported animals and does not tell us much except that the elephant was breathing, physically intact, and conscious. In all probability, almost no one in the country knew in detail what a wild young elephant looked, moved, or sounded like—a fact that not only put her in inexpert hands but also created the curiosity around her. Hence, almost immediately after he landed in port, Crowninshield sold the elephant for a reported $10,000, twice what he had predicted.[24]

The elephant's profile in a local newspaper, even a small announcement, was as important for Crowninshield's speculation as it would prove for later elephants in the trade. When she arrived in New York, the elephant was stepping into one of several post-Revolutionary trade centers on the verge of vast economic transformation powered by technology and information. The entrepreneurialism that inspired elephant importation also produced markets national in scope, boom and bust economic cycles, as well as great geographic mobility, which tended to benefit those free white men who, for a time at least, contrived to master revenue extraction from some part of the developing system—men like Captain Crowninshield. He knew, too, that the American newspaper business was growing by leaps and bounds. New discounted postal rates shipped papers all over the nation very cheaply and encouraged editors to reprint one another's content.[25] The resulting demand for timely news put the elephant and her advertising on a national footing almost immediately; many more people would read or hear of her in the paper than could actually view her in person.

Americans knew little of elephants unless they were affluent enough to have read about them in books of Asian travel or in the natural histories of the period. If more middling, they learned about elephants through excerpts from these publications that occasionally appeared in the newspaper.[26] Those authors often described Indian capture techniques or elephant hunts in Asia or Africa carried out by colonial personnel and affluent European visitors, portraying elephants as fierce but honorable adversaries of the male hunter. If unfamiliar with living elephants, coastal Americans were familiar with the primary elephant commodity: elephant ivory. Combs, decorative cameo sculptures, and cutlery handles (and by the 1840s, umbrella handles) made of "elephant teeth" were often advertised in contemporary newspapers and displayed in shop windows or households.[27] These were products offered to a savvy public known for its predilection for novel and fashionable things at the lowest possible price. In the 1790s, the word *ele-*

*phant* was also in common use as a trade term for a specific grade of heavy paper at the stationers, as well as a marketing icon for taverns, tobacco shops, and other retail spaces marked by the "Sign of the Elephant" as a symbol of something noteworthy or sturdy.[28]

While seemingly ephemeral, combs and the exhibition of a baby elephant were politically rich moments in the early republic, highlighting how early Americans understood patriotism and free trade to be two sides of the same coin. Certainly, Britain and France used tariffs and embargoes to wage war and jockey for political power. Well into the antebellum years, various Anglo-American political thinkers would contend that citizens' access to global products and the requisite sea trade were issues of national security and, from a consumer's point of view, proud economic liberty.[29]

A living elephant further reminded many people of the mammoth and the mastodon (two among dozens of extinct species of *Proboscidea*), whose bones had recently been discovered in various parts of the Northeast. She was consequently embroiled, not only in that tangle of patriotism, national security, and commerce, but also in a desire for mastery over the environment. Of course, the first elephant in America was no mastodon, an extinct being some had imagined as a carnivorous predator who smashed his way through prehistoric forests terrifying animals with his roar. Still, she was descended from similar species and brought to mind Revolutionary-era attempts to refute the Europeans who wrote of the United States as a degenerate place (although without visiting North America to test their theories). Not a few well-read Americans cursed the name of George Louis Leclerc, famously known as Comte de Buffon, who, like many contemporary natural philosophers, had put forward a series of data-free assertions in purported scholarly works. Among them was a claim that the environment, landscape, and weather of the continent cultivated inferior plant, human, and nonhuman animal life, producing specimens that were puny, weak, and even "cowardly" in comparison to their Old World counterparts. Buffon also asserted that Native Americans and Anglo Americans had proven themselves uncivilized and akin to beasts by their inability to reshape their environment to match European expectations. According to those European elites distrustful of American political and social independence, such findings portended badly for the United States as a whole.[30]

Although by 1795 such theories were in the process of being debunked, the perception of such foreign condescension lingered. At the same time, Americans still derived much of their intellectual culture and natural history from Europeans like Buffon, so they attempted to answer him on his own terms. Some argued that perhaps the mastodon and mammoth were not extinct at all, but "yet stalk-ing through the western wilderness," waiting to be discovered.[31] Others reasoned that, even if these creatures were extinct, it was providentially so because God had cleared those dangerous animals away to allow the nation to prosper.[32] Either way, an elephant exhibition in the 1790s offered Americans their first opportunity to recall those debates by examining a being similar to the mastodon or the mammoth, while—crucially—considering how experiences of living animals, not just the "wilderness" of the continent, defined the national identity.

Still, how was an aspiring showman to create and promote an exhibition of a living member of *Proboscidea* that would interest the public enough to earn a profit? Initially, the elephant's keepers took their cue from the growing but still experimental genre of animal shows in their country. After the government had lifted the Revolutionary-era ban on popular entertainments in 1789, Americans encountered many exhibitions of unusual animals, whether deformed, imported, or captured from the fringes of Anglo-American settlement.[33] Exhibitors commonly stoked audience patriotism by inviting people to come as consumers "at their leisure" to see the "first" of a given species kind ever exhibited in their city or nation.[34] In the late 1790s, a Salem minister, William Bentley, noted in his diary the various shows passing through his town, including a bison, an alligator, a juvenile "Catamount" (a North American cougar, castrated ostensibly to make him more docile), a leopard, a tiger, and a moose presented as "a Natural Curiosity for /9d. Brought in from the province of Maine."[35] The entrepreneurs who supplied such nonhuman amusements invested in a potentially lucrative business opportunity that required only a moderate investment and no specific training. Since mobile, they could show in one community then move on before local critics became too loud. Such had been the lesson in the case of Gardiner Baker, forced in 1796 to shut down the stationary menagerie at his famous museum in lower Manhattan when neighbors complained vocally about the smell and noise of the place.[36]

For several years, the elephant's keepers introduced her so astutely into this entrepreneurial context that they earned enough to keep her fed and growing while exposing her to the largest possible audience. They used an improvised business plan to maximize their impact: locate a venue, advertise, show until ticket sales decline, and then relocate to a fresh market, returning to each place within a year to show again. This scheduling tended to put the elephant in the south during winter and the north during spring and summer. After the elephant's first appearances in New York in April, May, and June of 1796, her keeper drove her on foot down the coast to appear in Philadelphia, Baltimore, Charleston, and many towns in between.

In this way, an elephant seemed advantageous to the entrepreneur, who proceeded hoping she would pay more than she cost to keep, feed, and show. The United States of the 1790s was a predominantly agricultural nation and would remain so for more than a century thereafter. The bulk of the population, and thus a showman's audience, lived in the rural and frontier parts, scattered over an impossibly large land area. Focused on the substantial markets in coastal cities, the showman might nonetheless keep the elephant earning by showing her in small towns and villages while traveling to the next engagement. Officials in those out-of-the way places often suspected that traveling exhibitions unfairly drained away local income, in spite of the earnings they might provide to local innkeepers or the farmers who supplied hay.[37] Accordingly, in York, Pennsylvania, for instance, municipal records reveal that the elephant's keeper was obliged to hire a ferryman to transfer her across the Susquehanna River, then sacrificed one pound and ten shillings to town officials for a permit to exhibit her.[38]

Yet, an elephant could be driven and carry herself along the pothole-infested, muddy roads that made wagon travel so miserable. Resembling a horse, a mule, or an ox in this respect, in 1798 the elephant's keepers would describe her in utilitarian terms by explaining on her Charleston bills that the elephant could "either lead, drive, or follow, at the rate of 3, 4 or 5 miles an hour, and is now from a journey of 1500 miles, without any sign of fatigue."[39] Although they may not have known it at the time, walking through the countryside was not an inappropriate activity for an elephant. Elephant family groups at large in Asia travel between three and nine kilometers per day, depending on the season and the situation, while elephants made sedentary by captivity in zoos routinely become obese or arthritic. Consequently, the elephant's rural route from city to city may have helped her cope with all the cake that people loved to feed her.[40]

By April the following year, the elephant had returned north, traveled through New England, and once again arrived in New York City. It was 1797. After a year of shows, she had become famous enough to be known, not simply as an example of her species, but as a noted individual: "The Elephant." Locals in Manhattan ventured out to see her again and noted that she had grown a foot taller. Now age three, the elephant purportedly weighed 3,500 pounds, was seventeen feet long and twenty-two hands (7.3 ft) high, not unreasonable dimensions for an elephant of that age.[41] Her physical changes upon returning to a given city created a shared history between elephant and viewer that for human ticket buyers intensified the emotional pull of seeing her again and perhaps wondering how large she might grow. Observers in the business noted that if one acquired immature animals to show they would be logistically easier to manage but would mature while earning income that offset increased appetite, all the while cultivating public interest because of that change over time. After all, as wondrous and engaging as the animal was, the novelty of petting a juvenile elephant or seeing her eat coins with her trunk did decline once people had visited two or three times.[42]

Over that first year, the nature of an encounter with "The Elephant" had quickly become colloquial knowledge in the places she showed, in no small part due to the experience promised by her advertising. The elephant's keepers invited New Yorkers, for instance, to visit the "convenient accommodation" at the intersection of Beaver Street and Broadway to see an "astonishing animal." To visitors they promised a being who daily ate "30 pounds of rice besides hay and straw, [also] drinks wine and all sorts of spirituous liquors, and eats every sort of vegetable," endorsing the idea that to see an elephant eat and drink was a key component of her entertainment value.[43] People who made a family outing of the exhibition or left work for a while to stroll over were equally captivated by her lightly haired "skin black, as tho' lately oiled," her massive girth, and her slim tail. In August 1797, William Bentley would see her at the Market House in Salem, where she appeared before a "crowd of spectators [that] forbade . . . me any but a general & superficial view of him." Bentley clarified later, "We say *his* because the common language. It is a female & teats appeared just behind the fore-legs."[44] Bentley's attention to the elephant's physical form was typical of many observers who found her body fascinating. A Harvard-educated preacher familiar with the contemporary works on animal taxonomy and natural philosophy, Bentley knew the show was a rare opportunity to verify the accuracy of the literature he knew about elephants.[45]

The pièce de résistance of seeing "The Elephant," however, was the experience of her trunk. "Bread & Hay were given him and he took bread out of the pockets of the Spectators," William Bentley said of the elephant's investigations of visitors. "He also drank porter & drew the cork, conveying the liquor from his trunk into his throat."[46] Elephants have been known to consume alcohol when they find it, so this is not an improbable report. Nonetheless, many people took this act as an incredible "satire on wine bibbers," likening the guzzling little elephant to enthusiastic American tipplers.[47] More importantly, later wisdom based on these early shows was that visitors found the elephant and her trunk fascinating but strangely unnerving. One story from the 1840s recounted a nervous boyhood visit to the elephant, who, "although he looked just like the [stuffed] one that was in the Museum . . . kept constantly moving his feet, head, or trunk, [and] produced an entirely different impression." Urged on by his father, the boy offered an apple to the elephant, who "immediately stretched out his long trunk and took it from Albert as easily as though he had a hand. Albert repeated this several times, because it gave him a good opportunity to see how conveniently he could use the fleshy finger which terminated his proboscis."[48]

What Americans did not know was that at large in Asia and Africa elephants use the trunk to learn and remember specific ways of altering and consuming an incredible variety of plant materials, more than any other species of land animal. The combination of memory and dexterity allows elephants to use their uncommonly powerful trunks also to perform delicate tasks in preparing food. They pull out plants from the ground, thereafter washing the roots off in water or beating the dirt out against an available surface before consuming. They tear away rough outer leaves of plants to eat the less fibrous parts inside, while crushing fruits under foot to produce smaller pieces transferred by the trunk into the mouth.[49] To identify different plant materials, elephants collect information through the "finger," or "dorsal extension of the tip of the trunk," which has dexterous and sensory abilities produced by a concentration of nerve endings, blood corpuscles, and whisker-like hairs within the skin.[50] In New York, the elephant's trunk was an amazing evolutionary gift employed simply to sniff out gingerbread and rum.

Since onlookers had no understanding of elephant trunk use in the wild and could not imagine what it was capable of, the elephant mostly demonstrated how her proboscis functioned much like a human hand. Humans likewise collect information about the world by manipulating or feeling objects and surfaces with dexterous appendages, so people who paid to see the elephant must have been tempted to imagine the elephant's tactile experiences and, by extension, her sentiment. Showmen must quickly have realized that the elephant act prospered specifically because it provided consumers with such emotionally rich experiences of elephant sentience. Indeed, many people would explain elephant dexterity as confirmation of their reputed "sagacity," namely, logical intelligence and sympathetic interest in human activities.[51] Thus, to feed an elephant was a thrilling elephantine anatomy lesson as well as a fable about interspecies cooperation. In each city the elephant visited for any length of time, word of her trunk habits spread, so before long people began arriving with food hidden in their pockets, hoping she would look for it there. In doing so, the elephant unknowingly created an "authentic animal encounter," during which many customers must have perceived a feeling of intimacy with her that implied that she (and maybe all elephants, some may have assumed) wanted to befriend humans.[52]

In truth, the elephant and her visitors were co-producers of these exhibitions and at times defied the keeper's plans for managing what went on to his own advantage. "Nobody is allowed to give anything to the Elephant but its keeper."[53]

So would the elephant's advertising warn by the fall of 1797, undoubtedly because spectators had been feeding her stones and other dangerous objects—although the warning also inadvertently promised just the elephant-initiated action that was central to her entertainment value. A year later another disclaimer became a fixture in her broadsides and newspaper ads: "The Elephant having destroyed many papers of consequence, it is recommended to visitors not to come near him with such papers."[54] Rumor had it that one distracted man had lost "a comfortable variety of bank notes, confidential letters, and undrawn tickets in a lottery" when the elephant turned trickster, slipped the billfold from his pocket and swallowed it whole.[55] In fact, it was not just the patrons but also the elephant who defied the keeper and complicated his work.

Nonetheless, such stories about "The Elephant" were fantastic publicity, as customers often grumbled that competing curiosities offered visibly ill or sullen animals. William Bentley was disappointed by a spring 1798 display at Boston's Bowen's Museum, where he found "a bear sleeping & slumbering with an insolent contempt of every visitor. A Baboon, more fond of entertaining his guests, an affronted porcupine, & two owls who gave us no share of their notice."[56] Exhibitors typically marketed their creatures by commissioning handbills and advertisements that depicted animals posed at alertness, free of chains or cages. Yet, in practice they provided whatever kinds of animal they could, once healthy perhaps, but after several days, weeks, or months in captivity, lethargic, obviously traumatized or ill, and clearly unaware that ticket buyers craved to look them in the eye, feed, or pet them.[57]

The men who ran this still-experimental trade had an uncanny sense of what customers wanted from exhibit animals, but they were far better at proposing such animal encounters in print advertising than they were at executing those promises with the animals and resources at hand. They usually resorted to exhibiting creatures in "shifting den" cages in order to protect public safety and prevent escape. Wooden boxes only marginally larger than the animal inside, shifting dens came with one side constructed of metal bars through which animals might be viewed, fed, and watered. "Carried in wagons, unloaded by hand and placed on sawhorses for public viewing," exhibitors used these containers to transport and display animals until the development of rolling stock containers around 1840.[58] Because seldom if ever removed from these boxes, the animals inside suffered from imposed solitude (if a social species), prodding by onlookers and unsympathetic keepers, inescapably close proximity to (also caged but still threatening) predators, an inability to find one's own food and water (resulting in chronic malnutrition and dehydration), and a frustrating inability to practice species-typical behaviors. Antebellum observers commonly reported finding animals presented in shifting dens to be emaciated, bald, physically deformed, frantically pacing, or comatose. Indeed, scholars in Europe had for years been debating the utility of observing confined animals, at least for research purposes, since they so seldom exhibited the habits the naturalist wished to study.[59]

American showmen were no doubt aware of these debates and their customers' negative reactions. Some may have wished to provide better animals but were simply unable to do so; others may not have cared as long as a steady stream of paying customers continued to attend. And even as the century progressed, menagerie and circus owners continued to display imported creatures like cut flowers, namely, as long as possible but with no expectation of permanence.[60] For those species, exhibitors could expect a further life span of mere weeks or months, so an Asian elephant's survival for several years and her cage-free presentation could make her unusually profitable in the business. As one menagerie advised in typical form some years later, "Of all animals, the Elephant is the most curious observed in nature. They therefore have very strong claims on our attention, whether we contemplate them in the wild luxuriance of their native plains, *or in the confines of a Menagerie*" (emphasis added).[61] Elephant showmen insisted that a captive elephant in America provided a satisfying experience of wild elephant life, although conveniently located for an afternoon visit.

Theoretically, primate acts presented the trade's monkeys, orangutans, and baboons in similar ways; showmen put them on leashes in taverns and on street corners, where they could voluntarily accept food from onlookers with their hands. However, primate shows were scientifically and religiously volatile in the early republic. Several exhibitors noticed the similarities between humans and other primates as a controversial temptation to satirize cultural authorities by dressing monkeys and apes in human-style clothes and training them to mimic human habits before a crowd. Such shows asserted the inferiority of nonhuman animals while lampooning targeted members of the public. For instance, the moose that William Bentley encountered in Salem in the late 1790s had appeared in a show with "An Ape . . . exhibited at the same place in the full dress of a Sailor," he said.[62] Entrepreneurs used trial and error to measure whether to frame animals, like the moose, as a naturalist's curiosity (although he was for sale onsite) or, like the ostensible-sailor ape, as subversive entertainment.[63] Elephants evaded such infamy and straddled both genres, as they were so radically removed from the human form that their trunk usage never took on the provocative (and later Darwinian) overtones of the primate shows.[64] The elephant's experience and physicality mattered in this as well. A small elephant might cause amusing mischief with the audience by grabbing hats or other objects from spectators just as a dexterous raccoon or monkey would, and yet she came across as endearing because she was unable to bite, scratch, or claw people who became too assertive or pestering—even if at times she may have wished to.

Still, the propriety of elephant shows would never be stable, and some ventured out to them only to realize, once there, that this apparently innocent commercial trifle was actually somewhat troubling. In July 1796, the Philadelphia Quaker Elizabeth Drinker heard about "The Elephant" from people who had seen the show. When the exhibit returned the following November, Drinker admitted, "I immediately concluded to see it." Gathering her grandchildren, she ventured through the city where they found the elephant down an alley in "a small and ordinary room, where was tag. rag &c." Certainly, the elephant's fifty-cent ticket price for adults and twenty-five for children made it pricey, but accessible to people of all means, including unaccompanied youngsters, laborers, and recent immigrants.[65] The "rag. tag &c." assembly she jostled with in the barn that day was also younger than she (Drinker was in her early sixties) and was creating a raucous scene. Drinker was relieved to find a friend of similar age there with her grandchildren, a group with whom to stand and, hopefully, ward off unwanted male attention. In fact, later exhibitions would promise "Good accommodations for Ladies" in acknowledgment of the salty atmosphere at the exhibitions.[66]

"The innocent, good natured ugly Beast was there," she remembered. "I could not help pitying the poor Creature, whom they keep in constant agitation, and often give it rum or brandy to drink—I think they will finish it 'eer long."[67] The elephant's first keeper had periodically extended times to see, feed, and touch her from seven o'clock in the morning until eight in the evening.[68] Today most petting zoos do not expose animals to visitors for such lengths of time, periods of rest and privacy having been shown to improve animal health. Of course, although she sensed this to be true, in 1796 Elizabeth Drinker was operating from contemporary common wisdom that one should protect young livestock from harassment by barking dogs and teasing children, for instance, since it seemed basically cruel and taxing for the animals. William Bentley had similarly noted that when he went to see the elephant he witnessed the keeper unsuccessfully attempting to force her to lie down. When the elephant refused, the keeper climbed on her back but she shook him off, drawing a big laugh from the crowd.[69] Bentley, Drinker, and other middling citizens in the early republic were already critical of the assertive working-class attendees at animal shows. Virtuous citizens, as the logic went, should not attempt to tease or provoke show animals but only visually inspect, pet, or feed them appropriate foods as a kind of polite and instructional amusement that emphasized permitting an animal to show his or her habits voluntarily.[70]

At the root of Drinker's concern over the "constant agitation" of the elephant was a fear that customers with little stake in the elephant's long-term health were directing the show's action and that, in the name of keeping afloat financially, the keeper would allow it. It is difficult to image that it could have been any different, though. Americans were then accustomed to democratic amusements in which the audience interacted closely with entertainers, human and otherwise. A quiet, nonintrusive audience would not become a mark of the exclusive entertainments and arts of the elite until after the Civil War. In the meantime, ticket buyers wealthy or poor, slave or free spent money at horse races and cockfights, participated in wolf hunts or bull- and bearbaitings with dogs, where one could find men of all classes together, calling out to the animals and their handlers. Spectators also participated in theater performances, yelling stage direction and opinions from their seats or hurling rotten fruit at the actors to make it clear what they expected to see.[71] And so it was with "The Elephant," an entertainment phenomenon that emerged organically from the nexus of the entrepreneur's opportunity and the elephant's action to produce experiences of nonhuman sentience that catered to existing consumer habits, as distasteful as they might have been to some observers.

In Drinker's account there is nonetheless some ambivalence, which reveals the degree to which people experienced the elephant with a confusing mixture of disapproval and delight. Over several of the elephant's visits to town, Drinker bought entrance to the show and discussed her experiences there with a number of her friends and family, who also took their children there and said they had been "highly entertained."[72] The elephant show was not morally inappropriate for someone like Drinker, a vigilantly modest woman who worried that her reading of "romances" and other light literature might be sinful. Her concern was how to rationalize one's sympathy for animals in the abstract with one's desires as a consumer to have access to those animals. Drinker later explained this disconnect and how it was exacerbated by urban living. Having moved from Philadelphia to the countryside during a yellow fever outbreak, she said she had seen "many humbling occurrences, that are hid from us in Town—the many droves of Cattle that pass this door, sheep & Hogs also, a drove of upwards of 200 sheep past by the day before Yesterday going to slaughter . . . the sufferings of the inferior animals, as they are called, is here more obvious—as we hear and see more

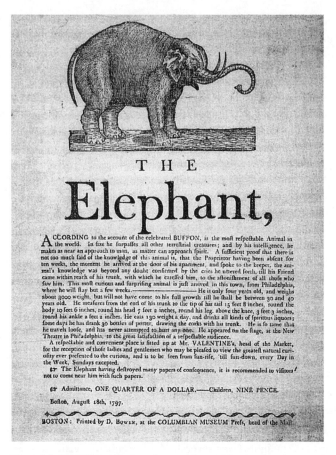

The celebrity of "The Elephant" was enhanced by gratis newspaper
publicity and paid advertising.  "The Elephant," 1797. Early
American Imprint Series, I, 32075.

of them."[73] Thus, "The Elephant" as a curiosity exposed Americans' fascination
with her sentience to be sure, yet equally put on display the many visitors who
nonetheless held no sympathy for a juvenile elephant's experience.

The confusion over whether the elephant and other curiosity shows would
serve to educate or pander to ticket buyers only intensified as the elephant made
repeated appearances up and down the coast. To capitalize on the situation,
boosters began describing the elephant, not simply as a passing fad, but as an
American entertainment staple, who was as instructive and morally uplifting as
she was famous. "Wonderful Work of CREATION," "A Curiosity which needs no
puffing," and "WONDER OF NATURE," the elephant's publicity and advertising

would announce of her.[74] It was in these moments that the elephant's exhibitors tapped into the trade's strategy of using educational-sounding language to legitimize for-profit ventures.[75] The tactic employed fact-rich, natural history–style patter to placate critics while simultaneously creating an emotional experience for viewers that actually confirmed their anthropocentric consumer privileges. While the details of the show might seem new to the audience, the underlying message was always a familiar one. Many exhibitors suspected they could best prosper by thus challenging cultural authorities like newspaper editors and clergymen, who argued there was a basic incompatibility between moral education and secular consumption. Instead, those entrepreneurs would assure Americans that for a small fee there was no need to choose between self-improvement and amusement or between moral animal use and self-interest.

"The Elephant, according to the account of the celebrated Buffon, is the most respectable animal in the world." So began the juvenile's revised biography in the summer of 1797, casting her presence as a democratic opportunity for experiential learning that could give anyone access to the natural histories read by the elite.[76] Despite Buffon's earlier condemnation of the United States, his massive chronicle, *L'Histoire naturelle* (1749–1804), was still the state-of-the-art compendium of knowledge about nonhuman life. Buffon's American readers found his work in expensive bound editions at the bookseller's, or they could read his ruminations on the ancestral interrelation of nonhumans and his tentative evolutionary theories in bootlegged cheap chapbooks and newspaper excerpts.[77] Others were more intrigued by the common wisdom Buffon had obtained from European animal keepers and woven into his otherwise clear-eyed work. Buffon wrote of elephants as potential tricksters, explaining that in 1771 he had seen an elephant rifle through onlookers bags and grab their hats with his trunk and that he had heard they would exact revenge by dousing offenders with water sprayed from the trunk.[78] Such information was a potential gold mine to a person with little experience with elephants but needing credible, interesting copy for his advertising.

The author of the broadside further persuaded the reader of the emotional truth of the show with biographical elements that confirmed the audience's anthropocentric consumer outlook. Tugging at the heartstrings, it recounted that the elephant would recognize and celebrate her owner after his absence, thereby showing her near-human level of "intelligence." "The Proprietor, having been absent for ten weeks, the moment he arrived at the door of his apartment, and spoke to the keeper, the animal's knowledge was beyond any doubt." Confirming the elephant's gladness at seeing a long-lost "Friend," the text assured readers that the elephant's mental state was "confirmed by the cries he uttered forth, till his friend came within reach of his trunk, with which [the elephant] caressed him, to the astonishment of all those who saw him." Such stories suggested to readers that they too might become the elephant's friend by visiting the show multiple times. It promised the viewer that the juvenile elephant being driven from barn to tavern along the coast was an amiable visitor who sought out human scrutiny. By this argument, citizens were indulging the elephant's own wishes when they paid to see her—now for the reduced price of "ONE QUARTER OF A DOLLAR." Although "The Elephant" broadside appeared on tavern walls and in the papers in Providence, Hartford, Boston, and other cities, it is possible the elephant's promoter additionally supplied the document as a handbill or written out as an ad hoc script to guide the keeper who spoke about her to customers on-site.[79]

Detractors who thought animal curiosities to be a fraudulent novelty might have pointed out that the broadside exposed the true motives of the exhibitor by its utter misrepresentation of the actual elephant displayed.[80] The printer D. Bowen, of Boston's Columbia Museum, employed a stock elephant woodcut that gave the elephant the body of a pig, the hind legs of a dog (with rear knees bent in the wrong direction), and muscular forward-leaning shoulders. The image additionally masqueraded her as a male, bearing the tusks that female Asian elephants do not grow. While a matter of printer's convenience, the advertisement's woodcut essentially offered a generic elephant as an icon of the elephantine brand: cage-free and apparently grinning, trunk raised in salute to the viewer. "He is so tame that he travels loose, and has never attempted to hurt anyone," it pledged. And here was the ultimate genius of the broadside: its impressionistic woodcut drew in illiterate people while the text advocating for Buffon and the dog-like loyalty of the elephant spoke to more middling folks who sought morally beneficial information about natural philosophy or animal sentience. Four times the piece referred to the "respectability" of the elephant, her venue, and her audiences, seemingly determined to overcome the unseemly impressions Elizabeth Drinker and others had taken away from the shows.

Americans may have been initially amenable to such an appeal since it aped the culture of the nation's nascent urban museums, for-profit institutions sincerely marketed as educational *and* marvelous. Museum proprietors like New York's Gardiner Baker or Philadelphia's Charles Willson Peale were well known in their communities but struggled along a "quest for solvency and public favor . . . [in the] oppressive conditions of the marketplace," just like the mobile animal showman.[81] Gardiner and Peale often paid for advertisements telling potential visitors of the items they displayed all in one location—antiquities, stuffed and mounted animal skins, rock specimens, public-access book libraries, artwork—linking natural philosophy, aesthetics, and business in a tangle of patriotic and pious self-improvement ideology.[82]

Many showmen gently poked fun at such projects with their own transient productions. Those people who trooped out to see "The Elephant" also knew of the several traveling "learned pig" curiosities performed by a man and a trained domestic hog. Elizabeth Drinker's young male relatives visited one of these shows in Philadelphia, at which "the pig spelled one of their names, by taking out every letter one by one in his mouth, told the day of the month in the same manner."[83] The same day he visited the elephant, William Bentley also paid to see a "Pig of Learning" and took away a similar impression. "The exhibition greatly exceeded my expectations," he said. "It was taught to discover the cards, to assort the letters of words, & to bring numbers for any purpose."[84] That decade at least one dog and a number of monkeys would similarly work the curiosity circuit as sagacious animals. Like the hogs, their performance required them to respond to subtle commands from a handler, which the audience would not detect, so as to give the impression of an animal answering complex human questions (rather than simply fetching a particular object or nodding when prompted).[85] Urban Americans walked streets inhabited by packs of pigs and dogs that roamed at will, fearlessly eating refuse, knocking over children, and biting people who got too close.[86] Thus did the various pigs of learning and others invite sincere spectators to consider their self-improvement in a society seemingly obsessed with industry, moral virtue, and (for whites) democratic access to education.[87] To the cynical, the learned pig genre came across primarily as a proto-Barnumesque opportunity to snicker at the credulity of one's fellow consumers.

Worries that the animal shows were indulging the worst propensities in the American public had also begun to materialize in the press, where journalists similarly labored under the contradiction that animal exhibits seemed both wonderful and dangerous. One Boston wag writing under the pen name "The Economist" complained of fellow citizens' "irrational, and irregular propensity to novelty [by] attendance upon theatrical entertainments, visiting museums, seeing raree shews, &c." American citizens were allowing ephemeral show proprietors to instruct them on what in the world was valuable, yet those entrepreneurs had an ulterior, pecuniary motive that drove them to flatter, not challenge, the public: "While . . . people have indulged their curiosity, in seeing the elephant, the bison, the learned pig, and the porcupine, the bear, and the owls, there are in the works of nature, millions and millions of animals, vegetables, and fossils, which they have never seen, and which are no less curious than those they have idly indulged themselves in seeing," he chided, singling out the lethargic animals that had so disappointed William Bentley at Bowen's Museum in Boston, as well as "The Elephant." He insisted that one should not be persuaded by promoters' promises "to strengthen the faculty of reasoning, or to furnish the mind with ideas useful to human life." For, at the shows, audiences had fleeting, emotionally satisfying but objectively unproductive experiences that merely encouraged more spending on similar entertainment.[88] Perhaps "The Economist" had a point, since William Bentley had himself noted that when he saw the elephant in Salem the elephant's keeper refused to supply reliable factual information about her because he was too busy animating the young elephant for a "crowd."[89]

Although such disapproval of the monetized cultural autonomy of the public appeared regularly in the papers and around middling kitchen tables, it was probably ignored or unknown to many of the elephant's visitors, who were too busy being delighted by a sight that was utterly unique in the early republic. Accordingly, when "The Elephant" left the city in the fall of 1796, a noted New York theater critic lamented her departure, tongue in cheek:

> Of Novelty, there's quite a dearth I'm told,
> The newest fashion is now twelve years old
>
> . . . . . . . . . . . .
>
> Tell me the reigning curiosity
> What shall it be? The Elephant is gone
> Egad, I have it—there's the Air Balloon.[90]

The famed elephant was "gone," but the critic hoped that Gardiner Baker's ill-fated subscription plan—inviting people to invest in the building of said hot air balloon—might supply an equal distraction. Baker proposed that from the balloon, weather measurements could be taken and various animals wearing parachutes dropped to the ground, "without their receiving any damage," as a public demonstration (a plan never carried out since sufficient investors could not be enticed into the scheme).[91] Baker's plan shared with the elephantine phenomenon the dilemma of how to engage audiences and stay in business while being perceived as offering innocent public education. Various related questions would shadow the elephant and animal show trade for a century thereafter: Should animal entertainment appeal to the Americans as they are or as moral authorities would like them to be? Should it endorse the consumer's worldview or attempt to radically challenge and uplift it? Or, could some kind of balance be struck whereby animal entertainment praised the audience (and their critics) with high-minded patter while offering experiences of animals that actually confirmed most ticket buyers' desire to impose their needs on entertainment animals?

As this interpretive divide congealed, the elephant and her keeper would stay on the move. In 1798, traveling south on the way to Savannah, the elephant and her keeper stopped for the night in a tiny village called Asheepo, which consisted of only three or four log houses. Itinerant entertainers and salesmen passed through such communities regularly. One contemporary account described such mobile entrepreneurs as "a welcome visitor. He opens doors without knocking, and enters with the familiarity of a friend."[92] Traveling captive animals and their keepers were rare among these men, however, so animal displays were long remembered by locals.[93] When the elephant arrived in Asheepo that winter, English traveler John Davis happened upon her and later remembered that "inhabitants of every sex and age" came to huddle around. They all said that the elephant seemed very large—except one elderly slave who called the elephant a mere "calf" in comparison to those he said he had seen in Africa. The elephant was then only about four years old, and still a juvenile indeed.[94]

By this time, the elephant had already changed hands several times, and in fact the transience of human-elephant relationships would perpetually complicate elephant acts. During the Baltimore exhibitions of September 1796, she had been in the care of someone known as John Carrier, although we know little else about him. By June of the following year, her owner was an unnamed Philadelphian, who employed another man, "William," who was in debt and accused of swindling money when the pair stopped in Providence. The next year, in June 1798, she was again with someone else, a man named "E. Savage."[95] Now in Asheepo, the elephant's driver was Mr. Owen, a Welshman, who "related the wonders of his elephant" and his global travels to listeners, employing the old show business wisdom that every performer—even an elephant—needs a story to engender his or her celebrity.

Davis said the elephant seemed pleasing in that little village because "in solemn majesty [the elephant] received the gifts of the children with his trunk." She did so in obvious contrast to the monkey Owen also had with him, who appeared "inflamed with rage" against the onlookers and similarly dangerous and ill at ease with Owen. With the elephant tethered at the ankle to a tree just outside, Davis and Owen shared a room that night during which Owen related a history of the elephant, which Davis unfortunately did not record. The next morning Owen was determined to get back on the road, but the elephant, "however docile, would not travel without his dinner," Davis noted with amused interest. As he walked on ahead of them, the elephant stayed to browse from the trees, delaying

Owen's trek to Savannah.[96] Elephants at large will eat an incredibly diverse number of plants, but due to their low nutritional value, the animals typically forage
constantly, consuming up to three hundred kilograms daily. Providing even half
that volume of hay for the juvenile elephant would have taxed Owen's patience
and purse. His travels were made easier, if slower, by the elephant's ability to
browse and graze, lending another layer to industry wisdom that an elephant
would earn more than she cost to acquire, feed, and show.[97]

After four years of shows and several trips up and down the coast, the first
elephant in America disappeared from public view and the historical record. She
was last advertised in October 1799 for an exhibition in Charleston, South Carolina. At only five years of age, it is likely she died from one of the long list of ail-
ments fueled by the suppressed immune function common in solitary captive
elephants.[98] Even then the elephant's carcass would have been very valuable, yet
the primary sources are also silent on its whereabouts. Perhaps she was shipped
to Europe or elsewhere, alive or otherwise.

Then, in 1803, another young female Asian elephant arrived in an American
port, announced in a Boston paper as "a rare chance for SPECULATION!" Many
conflated her with the first elephant, and both in due time were called Betsey, or
Old Bet. This young elephant was born in India, probably around 1799, and arrived in the United States accompanied by a South Asian mahout.[99] For thirteen
years under his care, she plied the same trade the first elephant had, appearing
in *"Now or Never"* limited-time-only engagements under the ownership of noted
menagerie impresario and founding father of the American circus trade, Hackaliah Bailey.[100] After her death, he would memorialize her with a statue outside his
Elephant Hotel in Somers, New York.

Old Bet died young, however. In the end it was musket fire that did her in, at
Alfred, Maine, in 1816. Because newspaper editors were always on the lookout
for free content, when an entrepreneurial anonymous author published a pamphlet describing the event, *Murder of the Elephant*, the very week of Betsey's death,
editors from Boston to Maysville, Kentucky, reprinted its contents in their papers.
Many of those stories carried the sensational headline "INFERNAL TRANSAC-
TION."[101] *Murder of the Elephant* and the news coverage derived from it asserted
that the elephant and her keeper, possibly her original mahout—or a native-born
African American man, it was not clear—had been visiting Alfred, when someone shot and killed Betsey in the barn where she was being shown. The killer or
killers had escaped without being identified. A crowd gathered at the site, many
apparently distraught at what they saw: "It would have melted the heart of the
most obdurate, to have beheld the agony of grief and despair which the *poor black*,

the Elephant's conductor, manifested when he saw the majestic animal in the struggle of death," the pamphlet asserted. It was certainly possible that troublesome young men shot the elephant, although some later conjectured that it was an angry local person who saw Betsey as a sign of unsettling economic and social change and self-fashioning through consumption of goods and experiences provided by outsiders, which challenged local production and entertainments. That is, an elephànt represented the world in rural American towns and villages, whether for good or ill.[102]

The anonymous writer praised Betsy's gentle nature as a product, not of her youth and human management, but of her inherent wisdom as a member of a noble species that meant humans no harm. The pamphlet's author said that for a man to have impulsively killed the young elephant was to insult God. It was to abuse the privilege humans carried in having dominion over animals, creatures who "crouch to [mankind] in homage, or fly with fear from that face which once wore the stamp and impression of the Deity!"[103] Humans were not gods, although they believed themselves to be formed in his image. In truth, they did have great power over a solo juvenile elephant and great responsibility to her. Yet, her keepers and customers had operated with minimal knowledge about how to protect her from the raw realities of life in the early republic.

Using the most inflammatory epithets one could in print in those days, *Murder of the Elephant* vented the frustration and regret many people appear to have felt at the time. The pamphlet's author called the shooter a "vile monster," a "shameless villain," "justly compared to the reptiles that crawl upon the earth, and the vermin that infest us on every side," and finally "a scoundrel to his country," who had exposed the dark, crude element in the public. "On such an occasion a whole community is disgraced. The act would disparage a nation of savages," the account in many papers read.[104] *Savages* was a highly pejorative term usually reserved by middling and elite Anglo-Americans for rebellious rural folk and Native Americans who injured or killed white settlers. In this instance, the term conveyed the sense that the killing of the elephant by one individual marked the whole community as essentially uncivilized. Although animal exhibitors would have noted that their urban and rural shows were still frequented by many well-meaning people, they would also have admitted that traveling exhibitions continued to attract rowdy young men and others bent on disrupting the proceedings. Such characters commonly attempted to enter without paying, pick fights with showmen, harass other patrons, or endanger the animals.[105] William Bentley came to the same conclusion, recording some of the gossip New Englanders traded about the killing of the second elephant. As citizens of a young nation eager to be taken seriously in the world, "We believe our manners very correct generally, but we have Savages still," he observed.[106]

Like her predecessor, for many Americans the second elephant raised the question of whether animal amusements would bring out the best in citizens during the Era of Good Feelings or would reveal their rough element to the embarrassment of all. While this quandary remained unresolved, showmen continued to gamble that the public attention to Betsey's killing actually boded well for the business since it enhanced her notoriety and the public's emotional identification with her. Once the flesh was removed from Betsey's bones, her owner arranged for her skeleton to be assembled and exhibited, now without the expense of watering or feeding—what great economy! And indeed, spectators did pay twenty-five cents to see the bones of the "unfortunate Elephant that was shot the 26th of July last, in District of Maine, [and] so well known by the public."[107] These were the foundations of the industry in elephant celebrity, which would boom in the antebellum years. To be sure, even alive, elephants made money in ways and places that other animals and amusements could not. Cage free but friendly, exotic yet endearing, robust-looking and physically unique, these juvenile Asian female elephants encapsulated the species' distinctive show business brand.

# CHAPTER TWO

# Becoming an Elephant "Actor"

One night in 1820, a small traveling menagerie made its way along the road lead-ing to the village of Plymouth, Vermont, where it was scheduled to give a show in the barn beside the Lakin Hotel. Caged animals, keepers, and horses in tow, the small wagon train also included a young male elephant. The next morning, children from nearby farms went out to inspect the road, hoping to find "the tracks of the huge beast." Indeed, his exhibitors claimed he was over nine feet tall, weighing at least three tons, and "one third larger than the female which was shot at Alfred."[1] In fact, this elephant was outmatched by his reputation, one resident remembered:

> There had been a little spatter of rain in the night, . . . and it was easy to see where he had passed along by a line of distinct footprints. But that could only be one side of him, we thought, and we hunted the road over for the tracks that should have been made by the opposite pair of feet. We were finally made to admit that the one line of tracks, showing a breadth not much greater than is made by the tracks of a horse, was all there was of it, and my imagination of the magnitude of the hug-est of created beasts received a decided corrective.[2]

To be sure, he had no way of living up to the expectations of his youngest fans because, although the first bull elephant in America, he was a juvenile like those who had come before him.[3] He had arrived in New York City in late November 1819 on a ship from Bombay called *Horatio*. Abraham Roblin, a speculator like Jacob Crowninshield, had purchased the young male and named him after the ship to remind customers of the elephant's journey to the United States.[4] Roblin quickly sold the elephant to a group of men who had assembled a small traveling menagerie, Mr. Curtis and Mr. Campbell, from Windsor, Connecticut, and the Emerson brothers, of Norwich, Connecticut.[5]

These Connecticut investors, after considering the earlier fame of the young female elephants who had traveled the United States, began marketing Horatio in some novel ways. Although they presented him comprehensibly as a "Natural Curiosity," the group also explained to their customers that Horatio was an exotic visitor trained for service to humans. Shipped from Poona near Bombay "at great expense and risk," they said, Horatio was a spoil of war captured from an Indian monarch, "one of the Maharatta chiefs," who had recently resisted British colonial authorities in India. His main trick was to kneel down so that a person could climb upon his back. Elephants in India have done this for centuries, so the act may have been routine to a South Asian or Anglo-Indian viewer. In the early republic, however, that or *any* directed elephant behavior seemed extraordinary to spectators—at least on first inspection—because it demonstrated the puzzling contrast between Horatio's physicality and his advertised "most perfect subjection" to human command. Thus did Horatio's owners assure the public that, although an elephant of war, Horatio was happy to be their pet. Encouraging a naïvely custodial attitude toward the elephant, they tempted the public: "Any person may approach him and lay their hands upon him without the least danger."[6]

Then, not long after Horatio's sale, the Connecticut group contacted Roblin to ask for help, saying that both they and their hired handlers were "unable to control him." By September 19, the menagerie was at Westmoreland, Vermont, attempting to cross a bridge over the Connecticut River. Finding the gate at the far end of the structure closed, Horatio and the menagerie's various riders and wagons stood on the bridge while someone ran to find a townsman to open it. Once the gate was swung aside, Horatio refused to move. Curtis and an unnamed keeper on horseback "advanced and were in the act of spurring him forward with their whips," yet once again Horatio either did not get their point or was simply refusing it. He began leaning on the bridge's railing and pressed so hard that the railing post snapped off, taking a portion of the bridge deck with it. "The Elephant, the two horses and their riders [plunged] . . . together with the falling timbers and planks a distance of forty-six feet, on to the rocks!" the papers reported with dismay.[7] Witnesses claimed that the horses died instantly and two riders were severely injured.

Horatio lay immobilized on the stony riverbank below for many hours while his keepers tried to decide what to do. The next morning the company improvised a crane with ropes to hoist Horatio onto his feet. Once lifted, the elephant collapsed under his own weight, with observers noting that he appeared to have no power over his hind legs. As with the killing of the second elephant, Betsey, many locals appeared to be sincerely upset about the elephant's injuries. Some suspected that Horatio was suffering great pain.[8] One paper explained with regret, "In the afternoon they got him upon an ox-sled, assisted by men with drag ropes, drew him up the steep bank, and took him to a barn on the hill in the village of Westmoreland, where this noble animal now lies, in much distress."[9] Horatio died later that day, after only ten months in the United States. A later dissection revealed that he had a broken back. Horatio's fall from the bridge certainly exposed the simplicity of the nation's frail infrastructure, designed as it was only to support horses and modest wagons. The incident may also have encouraged the wisdom that elephants will eye bridges carefully before crossing them.[10] To stem their financial losses, Horatio's owners had no choice but to salvage his skin and skeleton, selling or leasing it to Boston's New England Museum.[11] There visitors could contemplate Horatio's remains as notorious publicity for a menagerie trade that would strive to supply the public with bigger, heavier, and stronger elephants year by year.

Nevertheless, some show business people must have surmised that Horatio's case told a story that was ultimately unproductive for the shaky menagerie business. Horatio's disobedience and death showed that his claimed "most perfect subjection" to his owners was puffery. Although the young elephant had engaged gently with menagerie customers, like many "trained" animals, he unpredictably resisted the conditions of his experience—or so it seemed to his keepers. Twenty-five years after the first elephant's arrival in America, animal exhibitors still had not developed any systematic knowledge about elephant management or motivation. As onlookers perceived it, that day on the bridge Horatio appeared to have leaned sideways on the railing because his handlers addressed him as though he were a balky horse. An exasperating villain in American life, the spooked or stubborn horse threw riders, bit and kicked people, upset carts, broke tack, and otherwise endangered human health and industry; and yet, as a known character, the balky horse was a way to frame the young elephant's behavior. None of Horatio's promotional materials indicated that Horatio had been defying his keepers in such seemingly familiar ways or that his owners were at a loss over what to do about it. Only when the elephant died in a public setting did country newspapers in the vicinity finally expose the full range of the elephant's behavior, hinting that perhaps he was not as sagacious as he seemed.

Of course, antebellum menageries boasted that their elephants were extraordinary animals, educational natural curiosities, tractable and safe in all situations (even if this was not always true). Showman also knew that many Americans cared about elephant experience, a fact revealed in the public distress over the dramatic deaths of Horatio and the two Betseys before him. In order to wrest interpretive control over their elephants from the audience and the elephants themselves, menagerie showmen developed the nonhuman persona of the performing animal actor. The performer guise offered an elephant as a consumer-friendly animal with a biography, noted individual habits, human friends, and a desire to travel around America for the audience's enjoyment. This show business ideal also gave menagerie men increased cultural authority as experts on the species, as well as the power to vary their productions from season to season. Showmen and menagerie performers invested immense energy and imagination in finding ways to make imagined animal actors manifest with ad copy, broadside illustrations, and—crucially—actual living elephants. The transition from elephant as exhibit to elephant as performing actor marked a solidification of the show business wisdom that the animal shows' primary product was not simply exotic animals, but consumer experiences of animal celebrity.

When Roblin bought and sold Horatio in 1819, the informal American trade in imported animals had come to include a greater number of exotic species. Beginning in the 1830s, Americans would even find the occasional giraffe or rhinoceros at menagerie shows.[12] The single animal exhibits of the early republican years had become less prominent. The entities people knew as "menageries" had proliferated, offering stationary or mobile collections of various species, some trained for performance with humans, some kept merely for exhibition. By contrast, a "circus" was a show of clowns, acrobats, and trick horse-riders only, a performance genre that had emerged out of the equestrian hippodrome shows pioneered in 1770s Britain by Philip Astley and Alexander Ducrow and in the new republic by John B. Ricketts in the 1790s.[13] To American entertainers of all kinds, the early horse circuses demonstrated that most Americans sought out novel variety in spectacular shows of ambivalent character, allowing some to view the performances as "wholesome" while to others they seemed adult and provocative, if one reads between the lines of the jokes and patter.[14] And when they appeared in a town, equestrian circuses drew audiences away from local theater houses and other for-profit amusements, as did the menageries.

The entire antebellum entertainment industry was thinly capitalized, and a man might be an employer and a show proprietor one season, an employee of a competitor the next. In January 1835, to share the risk, 123 investors paid to organize the Zoological Institute. Merging the nine extant menageries into one joint stock company based in Somers, New York, it had an initial capital in cash, equipment, and animals of almost $330,000. The group took advantage of a growing national market for shows, ranging from South Carolina to New England to Missouri. Zoological Institute animals also often appeared in stationary productions with names like American National Caravan, which appeared at the company's 37 Bowery location in Manhattan for an extended period every year.[15]

Traveling on different schedules to share audiences that might have several shows pass through town in a given year, these men navigated a nation with no common currency and notorious for bank collapses, which periodically made local banknotes worthless. The menagerie cartel made navigating that economic landscape less dangerous for men like Lewis Titus, a Zoological Institute founder, who invented and collapsed shows from season to season as he could afford. For instance, in 1845 he brought to New York on the ship *George Washington* an elephant, a giraffe, and a lion and two cubs to lease to Isaac Van Amburgh's menagerie company.[16] Later, Titus ran a show under his own name and, after that, another show with other investors, billed as "June, Titus and Company Combined Under the Management of Mr. G. C. Quick." They toured in the late 1840s with an elephant, possibly the same one that Titus had imported in 1845, advertised now as "The Performing Elephant Romeo."[17]

The Zoological Institute's approach to the national market for entertainment was typical of many communication enterprises of the era, which similarly sought access to the largest possible customer base. Showmen took note of how canal and road policy determined where cities got built and how state governments in the northwest (now the Midwest) engaged in "a binge of canal construction," much of it fueled with foreign capital investment, which in turn created the economic boom of the 1830s.[18] Like other business people, menagerie proprietors additionally took advantage of new processes that expanded their advertising reach with inexpensive and quickly produced print, matching them up with an entrepreneurial distribution network of shops and peddlers who took books, pamphlets, and other printed materials to every city and village.[19] The development around 1850 of paste with which to stick bills to fences and buildings, rather than tacks, only increased the volume of advertising the animal shows disseminated.[20] Like the printers and newspaper editors who helped reproduce and distribute handbills and advertisements for elephant shows, menagerie owners in the early republic were "speculators on the frontier of economic development."[21] Men imbued with the antebellum "go-ahead spirit," they sought to make a living by finding ingenious ways of extracting surplus value from the nation's growing population—living "on the fat o' the land," as it went at the time.[22] Recall that the elephant known ultimately as Betsey appeared advertised in a Boston paper after she first arrived in the United States as a "rare chance for SPECULATION!" and a tool for extracting disposable income from one's fellow citizens by showing the elephant, building her fame, then flipping her to the next investor.[23]

The unregulated boom-and-bust cycles that plagued the economy made this kind of entrepreneurialism a dangerous proposition. The Zoological Institute would stumble during the panic of 1837, although its operators continually reorganized, muddling through the 1840s along a "feast or famine circuit."[24] In this fluid economy, the institute cartel supplied the majority of the menagerie-style entertainment in the country. Suspicious gossip around the industry held that from the beginning the whole Zoological Institute venture had been (like the incorporated Peale Museum company, some speculated) a stock swindle capitalizing on the perception that the animal show trade had great profit potential.[25] Some believed it had essentially been designed to raise money from American investors who had little understanding of the day-to-day logistics of exotic or wild animal management and transport. A few town councils did pass laws banning traveling shows or enact taxation policies that effectively did so, to protect their communities from being drawn into such perceived wasteful spending, even simply as customers of these businesses.[26] Nonetheless, thanks to the entertainment offered by the menageries, by the 1830s most consumers interpreted the menageries as a fine aspect of city culture made mobile for rural and frontier regions. Even conservative villagers—"the most straitest of sects," as *Knickerbocker* magazine would joke—patronized menageries to inspect caged animals and, in conversation with family members, show off their knowledge of each animal's natural history.[27] At mid-century the clergy, too, had come around to the idea of a menagerie as an instructional amusement that improved a community.[28]

How did the menageries manage to change public opinion after the deaths of the Betseys and Horatio? First, they continued purchasing juvenile elephants to present as the centerpiece of their shows, although not all the sixty or so enterprises that would come and go over the antebellum period could afford an elephant. It was still true that adult elephants were expensive to ship, while many wind-powered vessels had below-deck ceiling heights of only eight to ten feet, precluding adult elephants in the hold. Although later elephants would come across in tents or other temporary shelters pitched upon a ship's deck, the vast majority still came in their youth, even after transatlantic commercial steamer services in the 1840s began to shorten these voyages considerably.[29] In the hunt for more animals, speculators and showmen like Roblin still acquired elephants from individual shipmen or from agencies like the East India Company, which had routinized a growing exotic animal trade to Europe and the Americas.[30] Thereafter, these men would contract the animals to existing menagerie companies or investor-operator's like Roblin's Connecticut buyers, who exhibited the elephant on their own or with assistants.

Some of the elephants imported in this way lived into adulthood, but they did so without reproducing. The overwhelming majority died before the age of twenty-five, only to be replaced with new imports. The nation contained a transient collection of unrelated individuals of both sexes and of various ages and points of origin, mostly consisting of Asian elephants. Menageries exhibited them singly, in pairs, or in trios up and down the East Coast and into the interior as far as the Mississippi.[31] Elephants working for these ventures showed outdoors, in barns or, by the late 1830s, large tents lit with kerosene lamps. In winter, they traveled through southern states or to warehouse farms circus people came to call "winter quarters." These elephants included Columbus, imported in 1818, and Tippo Saib, who arrived in 1821 and was named by his managers after the famed Indian ruler of Mysore today known as Sultan Fateh Ali Tipu. In the 1830s there was the (first of several) "Great India Elephant, Siam," Josephine, and Mogul, and in the 1840s Bolivar (likely named for political leader Simón Bolívar), Columbus, Romeo, and Jenny Lind (named for the famous Swedish singer *even before* she visited the United States), later renamed Juliet, then Lalla Rookh.[32] In the 1850s and 1860s, people also knew another Elephant of Siam, a.k.a. Lalla Rookh, a female called Ann, and a group of males known as Hannibal, Tippo Saib (probably the second to go by that name), Virginius, and Pizarro. Like Horatio before them, these elephants were among the few animals routinely named in advertising and billed alongside famous human performers, mostly trick riders, acrobats, or a famous big cat wrangler.

To their owners, these elephants were not pets or family livestock, who, having been reared from birth on the family farm or in the urban alley way, might be affectionately named Spotty, Billie, or Brownie.[33] Instead, animal showmen appear to have believed that no opportunity for shaping public perception should be missed and so eschewed allowing their handlers to publicly name elephants with fond names. Nor did they fall back on the old practice of naming elephants after the ships that delivered them to North American shores. With more than a few elephants afoot in the United States, they instead used naming as a marketing tool to cultivate the interest of the powerful but fickle American audience. Some named elephants by reference to entertainment characters, like Romeo, Josephine, or Jenny Lind, so as to endorse customers' preexisting popular culture. Alternately, they used an exoticized naming mode, like Mogul or Tippo Saib. Those Eastern-inspired titles suggested to ticket buyers that even wealthy and powerful people—or at least their elephant surrogates—would come to America to perform for a privileged American populace made worldly by the experience.

Because showmen named and promoted individual elephants as noted entertainers, elephants became the most famous animals in nineteenth-century America. The tactic drew from two developing ideologies of the period and combined them into a show business variant that asserted the menageries' role as experts on exotic animals and the audience's role as benevolent consumers of animal personality. First, making an elephant famous was a way to add value to him or her and was loosely related to contemporary American animal value theories. Beginning in the 1820s, many cattle, dog, poultry, and horse breeders began seeing their animals as opportunities for speculation. As the theory went, anyone with some investment capital could potentially make a fortune by creating purebred animals whose health or reproductive viability mattered less than the perception of rareness or a broader fetishization of a given blood line recorded in breed registries and studbooks.[34] Breeders and buyers described fancy livestock as "improved," even when sickened by genetic disease, since they were monetized "animals of enterprise . . . , commodities [whose] value is exchange rather than use."[35] In adding personality to individual elephants, the menageries would likewise improve them in the sense that their value came in communicating the stories the menageries wished to tell, which were a product of the size of the audience market a given elephant might serve.

The menageries' commerce in audience experiences of emotional identification with famous individual elephants also interacted with the concurrently burgeoning pet species trade and the ideals of benevolent human supremacy enacted through it. The antebellum pet market facilitated Americans' cultural instrumentalization of various birds, fish, dogs, and reptiles whereby consumers came to perceive individual personalities in those beings. This was a process that transformed animals from pests, predators, food, or labor into "pets," friends, and family members. Moreover, owners put limitations on those animals, only tolerating them if they served human needs for contact and cleanliness and refrained from injuring people or livestock (at least no people or livestock the owner cared for). For many, such practices seemed out of place in communities where citizens lobbied local governments to cull packs of feral dogs or decimated wild bird populations through intensive hunting.[36] Nonetheless, dealers of pet stock, cages, collars, and other accessories understood that a great cultural shift was occurring. They encouraged the emerging contradictions in Americans' relations to nonhumans by endorsing the idea that some animals were sagacious "pets" with human-style sensibilities and therefore suitable for in-house living.[37]

Like the pet dealers and speculative stock breeders, the menagerie managers portrayed themselves as animal lovers and aficionados, not as the unsentimental traffickers in living property that they often were. They likewise invited Americans to take on a custodial role with nonutilitarian animals, often by assuring them that those animals wanted both "dominance and affection" from their human owners.[38] To monetize such messages, antebellum menagerie men offered some experiences to consumers gratis. For instance, they began announcing their arrival in town with a street parade. A free entertainment that would hopefully bring paying customers to the performance, the parade was part of a marketing program that included bill posters and hawkers distributing handbills featuring images and text describing the wonders of the show. Especially in small towns, menagerie advertising often provided the most elaborate and ubiquitous graphic illustrations people knew, thereby enhancing the cultural authority of the menageries.

Showmen seemed to agree that an elephant's body and origins could be used as an anchor for adding an extravagant and exotic theme to menageries. Literate Americans held very romantic ideas about the opulence and color of life in Asia, especially as it pertained to the pomp and circumstance of the South Asian elite. People found these ideals described in the various editions of the tales of the *Arabian Nights*, as well as travelers' accounts and histories of the classical world in books and the periodical press, often documenting captive elephant use over-seas. For instance, in 1859 the *Historical Magazine* reprinted a 1791 letter from an American representative in Hyderabad that was representative of a longstanding perception in the United States that India was a land of fabulously wealthy rulers, whose elephants served as decorated symbols of their power and cultural sophis-tication. Of a procession for Tipu Sultan involving a claimed "150,000 Infantry, 60000 Horse & 500 Elephants," the American explained in 1791 (and again, two generations later):

> This was the noblest sight I had ever seen, each Elephant supporting a large Castle, containing a Nabob, & four servants. The size of the Animals, the glittering of the Castle, & elegance of the equipage, was the most brilliant sight the eye could behold. . . .The method of traveling in this country is very luxurious, as you must have heard. My servants & escort in all consists of 50 men, 16 bearers to my Palan-quin, domestic servants, runners, &c., are included, & all but 20 are paid by my Prince, who is the greatest Prince now in India.[39]

Elephants had long served as totems of a ruler's power in various parts of Asia and additionally as a sign of British rule in India when the British co-opted the elephant with howdah to carry their own royalty.[40] This was just one of hundreds of similarly breathless accounts that appeared in American newspapers, maga-zines, and books—content anxiously raided by scriptwriters for opulent theater and menagerie productions.[41]

Menageries decked out elephants as "Elephants of War" by way of easternized costuming, large howdah saddles, and exotically dressed human actors in order to bring such representations to life. The elephant of war act simply required elephants to walk when and where directed, stop or turn when directed, kneel to accept a rider, and refrain from introducing new behavior into the routine. With an eastern biography describing him as a war elephant confiscated from one of the "Maharatta chiefs," Horatio was a prototype for the many elephants who would play the same part in menagerie processions and shows containing pastiches of domestic and foreign content. When they arrived in Baltimore in November 1834, the Waring & Tufts menagerie offered the public a "*Grand Entre*" featuring "the Elephant SIAM, dressed with his India Saddle, and an entire band of music mounted on his back playing National airs as the great mammoth of the forest passed along."[42] The menagerie companies were getting bigger by the year, and the Waring & Tufts show that month also boasted a purported eighty grey horses, some pulling a large bandwagon, along with many people and other animals in elaborate costume. They often provided readers with the parade route in advance so that onlookers would know where to stand on the day, while helping newspapers stay relevant to the growing consumer audience.[43]

Glamorous menagerie parades featuring "gilded and painted" elephants calmly pulling ornate bandwagons attempted to flatter spectators by implying they were citizens of a powerful nation.[44] The United States had no royal family or ancient aristocracy, or any occupation government ruling colonies in eastern lands, so showmen entrepreneurially took on this role in the global context of European imperial practices of collecting exotic animals from overseas colonies (there would be no large-scale public zoological parks in the United States until after the Civil War). However, the idea that foreign animals were a sign of state or monarchical power predated the early nineteenth century. Privileged animal collectors from the Aztecs to Louis XIV and the board of the London Zoo had long assembled private menageries.[45] In Europe, curators of zoological gardens reminded visitors of that heritage by presenting an imported animal with a building or a name designed to represent the human cultures in the land of the animal's origin, say, North African sand gazelles (*Gazella leptoceros*) displayed in a "Moorish Algerian"–styled barn.[46]

In those years Anglo-American civilization was pressing out into the North American west, absorbing territory, humans, and other animals into the nation's economic and political systems. Certainly, western expansion put many Americans in an imperial state of mind, yet not with respect to European colonial territories that held elephants, like British India, French Indochina (Vietnam, Cambodia, and Laos), or the Dutch East Indies (Indonesia). Menagerie processions containing easternized elephants probably made sense to audiences in the 1820s or 1840s because some believed that Americans had an equal right to acquire animals from Asia. An "Elephant of War" parade gave spectators access to far-away lands, a privilege supplied, some said, by the most industrious and clever traders in the world, Yankees.[47] The authority of menagerie representations of eastern life were in turn fortified for the bulk of the population, who had little access to expensive books and literary magazines, because the parades came structured like the democratic rituals of militia, club, or political party parades. It was also the era of dioramas and panoramas, shows at which observers stood or sat stationary while viewing a city or landscape scene that scrolled past them on an enormous canvas—just like a street parade.[48]

The nation's hippodrome horse shows of trick riding and acrobats similarly employed exotic parades and historical pageants, and the two genres—tent show, menagerie and circus—competed for the same entertainment spending.[49] As a result, companies boasted of their service to the public in offering "novel and magnificent entertainments, *far surpassing anything ever presented to the American public*" (emphasis added), as a Sands & Lent show hyperbolized in 1848. That season the company appeared in a vacant lot near the city hall in Brooklyn, with elephants boosted as "the leading new features of this extensive company." The company publicized the 11 a.m. parade preceding their shows with varying font sizes to communicate a consumer-flattering sense of regal abundance:

> This mammoth corps will enter town on the morning of Exhibition at 11 o'clock, in procession and cavalcade, proceeded by the Sacred Egyptian DRAGON CHARIOT of ISIS and OSIRIS, drawn by EIGHT Egyptian Camels, containing the splendid full Band attached to the company.—Next in order will be the EAST INDIA CAR, to which will be harnessed the two Elephants, flowed by the magnificent STUD OF HORSES, and all the numerous, costly and highly ornamented vehicles belonging to the company. The beautiful FAIRY CARRIAGE, drawn by TWENTY LILLIPUTIAN PONIES, driven in hand, will bring up the rear of the whole procession, the tout ensemble of which surpasses anything which language can describe."[50]

As the oral tradition goes, the commotion caused by these colorful processions was substantial enough to spook equines unfamiliar with elephants and other strange animals, inspiring menagerie men to alert passersby, "Hold your horses!"[51]

The glamour of menagerie parades contrasted sharply with the plain clothes, workaday livestock, and modest households of most ticket buyers before the Civil War. It also contrasted unfavorably with the mundane backdrop against which the menageries and their elephants appeared. "The streets of the City were in a horrible sloppy condition," a writer for the *New York Times* noted as he watched a parade pass by the newspaper's offices on a rainy day in December 1858. The routing of the parade by the *Times* building was no accident, of course, although the rain would demystify the live experience of the procession to some degree. Still, the *Times* writer obliged the menagerie by assuring readers, "But neither the mud, nor the rain, nor the east wind, were sufficient to prevent a procession of gorgeously-caparison elephants, in scarlet overcoats, marching past our office, accompanied by a band of musicians in a gilt chariot, and by a gay cavalcade. It was a vigorous attempt at Oriental magnificence, but the intention was probably to indicate the resources of some enterprising circus proprietor."[52] Indeed, the newspapers and the menageries and circuses would establish mutually beneficial relationships in which free publicity was traded for expensive advertising purchases in the local papers, complimentary show tickets, dinners, and alcohol as part of a "barter system" of paid content and puffing publicity.[53] Regardless, many saw through claims to temporarily easternize American cities, whose muddy roads and thoroughfares often threw into high relief how rough around the edges both these companies and the nation still were.

The competition between traveling shows would drive showmen to develop a kind of mammoth marketing theory, whereby companies strove to create an authoritative sense of the incredible around their performances. The materiality of elephants, even young ones, with their vast bulk and unique look, made them a crucial sign of any production promoted as a "MAMMOTH SHOW." Thomas Raymond, one of the dominant personalities in the antebellum menagerie trade, pioneered the practice. He put three units on the road working different routes and produced and promoted them all with featured animals—giraffes and elephants—symbolic of what he termed promotion by "grandness" and "giantism."[54] Certainly people knew that elephants were the largest of land animals. So too, most showmen believed these creatures to embody the aesthetic menageries were developing; elephants spoke of the immense capital (or to insiders, the resulting debt) of a menagerie company in ways that few species, other than the rare giraffe or rhinoceros, could match.

Images of elephants, as the most momentous, endearing, and classic of menagerie animals, often appeared at the top of newspaper column advertising. Promoters continued to build personality around noted individuals by discussion of their weight and dimensions, often daring the public to see the show as anything but incredible: "The Proprietors challenge any company in the United States to produce his equal."[55] People had seen the Waring & Tuft Company advertise their famous "Great India Elephant SIAM," purported to be twenty-one years old in 1827, with a claimed height of nine feet and "supposed to weigh

Hannibal as oversized Elephant of War in a mammoth marketing program, ca. 1840.
T. W. Strong, no date, *Van Amburgh: The Mammoth Elephant Hannibal*. Ink on paper, 9½
x 11¾ in., ht2000718. Collection of the John and Mable Ringling Museum of Art Tibbals
Digital Collection.

7000 lbs."[56] Inaccuracy was rife in such advertising, and by 1829 the elephant's
promoters billed the same individual as twenty-nine years old while encouraging
the reader to be amazed by offering the elephant's weight with an incredulous
tone: "7,000 Pounds!"[57]

The technique of persuading by superlatives coexisted seamlessly with the
educational pretensions of the menageries, whose ventures leveraged their cultural authority by linking the financial and logistical feat of acquiring elephants
and other exotic animals to their claimed expertise with such incredible creatures. The 1830s advertising patter for the Asian bull elephant Mogul was emblematic of integrated mammoth-spectacle/public-education communication
strategies. Named for the Mughal Empire of South Asia, the young bull elephant
came to Boston from India in 1831. He would exhibit in the United States for at
least a decade in a variant of the elephant of war mode as "The Great Hunting

Elephant," purportedly trained for chasing game while carrying an armed rider. He also suffered ill health and narrowly survived a fire that burned six cages of Macomber & Company animals aboard a steamer crossing from Saint John, New Brunswick, to Portland, Maine, in October 1836. Tied on deck when a boiler fire overtook the vessel, Mogul was heard to trumpet in terror, and several passengers and the ship's captain frantically worked to cut the ropes around his ankles and neck. Mogul swam to shore to save himself that day, but he died a few months later of unnamed causes (perhaps injuries sustained on the ship or an infection exacerbated by his frigid swim) in Macomber's Baltimore menagerie grounds, at which time the papers claimed he was thirty years old.[58]

In the meantime, promotional materials for Mogul ignored such behind-the-scenes accidents to emphasize his size. Citing the ancient Roman chronicler Appian, Macomber & Company boosted Mogul by erroneously explaining that African elephants were smaller than Asian, and "much less courageous, and are afraid to combat with those of India."[59] In fact, in ancient times, a variety of Eritrean elephant, the proposed subspecies *Loxodonta (cyclotis) pharaoensis* (now extinct or absorbed into modern African forest elephant populations), was smaller than both Asian and African bush elephants (*Loxodonta africana*). It was no doubt these elephants to which ancient writers had referred, since many believed that those elephants had originated in the famous elephant hunting grounds just north of the horn of Africa, where the ancients had captured the elephants that were driven with Hannibal's armies.[60]

In any event, most broadside readers and ticket buyers were probably unaware of current scientific debates over how many different species of elephant existed or had existed in the past, how they had originated (one common source or different sources) or how they might be distinguished, and even how "scientific" the category "species" actually was.[61] Few Americans were in a position to argue, however, especially if they were illiterate and merely going to a show they saw advertised with an elephant's image in the newspaper or on a saloon wall. For literate Americans, the menageries marched in where the schoolmaster or naturalist could not—the plebian world of live entertainment—in order to make the most of the "commercial aspects of natural history."[62] This was a fabulous business opportunity. Citizens were already in the habit of purchasing educational books and pamphlets from peddlers and shops, and they acted accordingly when showmen added their own promotional publications to the market. While providing some promotion for free, such as handbills and the opening day parade through town, the menageries sold other marketing materials as "programs" and "souvenir" books, in effect asking the consumer to pay for the show's advertising. Throughout the century these documents served quasi-educational purposes by supplying detailed naturalists' information about a production's captive animals, while engendering good will with the public.

Both urban and rural Americans were very receptive to the menageries. In 1836 a traveler making his way to Carlinville, Illinois, came round a bend in the muddy trail only to find the little town in an uproar: "I found it absolutely reeling under the excitement of the 'Grand Menagerie.' From all points of the compass, men, women, and children, emerging from the forest, came pouring into the place, some upon horses, some in farm-wagons, and troops of others on foot, slipping and sliding along in a fashion most distressing to behold. The inn was thronged by neighbouring farmers . . . while little unwashen wights did run about and dangerously prophecy the recent disappearance of the big elephant."[63] Even in frontier areas, elephants had come to define the traveling show brand. They evoked the luxury and "mammoth" pretensions of such mobile entertainment and the nervous excitement around menagerie day. The evolution of the elephant from animal curiosity to animal show icon also pointed to the fact many citizens had never known America without elephants. Audiences were growing due to increased immigration to the United States from Europe, Asia, and Mexico, as well as high birthrates and a preponderance of women of childbearing age. Thus did the United States become a global leader in population growth with a youthful society that averaged only nineteen years of age in 1860 (in the late twentieth century the average was thirty years).[64]

This generation of ticket buyers demanded perpetual novelty and was indulged by an entertainment scene growing more crowded decade by decade. It featured a broad variety of theater, from Shakespeare to minstrelsy, melodrama to satire. This was also the age of P. T. Barnum's famed American Museum on Broadway in Manhattan. There and in his controversial traveling curiosities, Barnum and his agents teased citizens into considering the nature of fraud and the meaning of democracy, race, temperance, and natural history with globalized displays of people, animals, and objects. Popular print forms like the "Mammoth Weekly" newssheets measured a dozen square feet and came packed with a cornucopia of illustrations, half-fictional news, and political theater. They coexisted with penny papers offering hoaxes and crime news that spoke of working-class American street life, middling diorama and panorama shows, rural religious revivals, and traveling salesmen offering Bibles and bootlegged novels. Then there was the masculine sport universe of saloons, dogfights and cockfights, hunting parties, bare-knuckle prize fighting, horse racing and boxing, much of it reported in racy illustrated papers like the *New York Clipper* and *National Police Gazette.*[65]

For this more worldly late-antebellum audience, it was still fun to be sniffed by an elephant or to admire his costumed figure in a parade. Nonetheless, menagerie proprietors and performers understood that they needed to innovate their offerings—especially in the cities—in order to keep people coming back season after season. That meant creating new material by increasing the scope (and unavoidably the expenses) of their productions while hopefully keeping ticket prices reasonable.[66] Various companies engaged in aggressive investment in those years. For example, in the 1840s Isaac Van Amburgh began promising New Yorkers that he could entertain one thousand viewers at a time at his menagerie, and urban crowds numbering up to two thousand became common.[67]

Profitability came not only from volume ticket sales but also from mobility. Those ventures that stayed in permanent venues found it very difficult to survive financially, and several well-known hippodrome shows performing in wooden arenas closed in this period. With no real tourism industry (such as would keep Chicago's Columbian Exposition busy in 1893 or amusement parks like Disneyland in the twentieth century), animal shows had to keep moving to find paying customers. Traveling shows also had to provide a comfortable show rain or shine, so they began using open-topped canvas enclosures in the 1820s and covered tents lit with kerosene lamps in the 1830s. Impresario W. C. Coup introduced two performance rings to circus shows in the 1850s.[68] After the Civil War, more complex productions emerged, featuring three rings, an outdoor midway of food and novelty acts, and various out-tents containing fortune-tellers, freak shows, magicians, and the zoological-style animal exhibits people would refer to as "the ménage" in twentieth-century circuses.

The financial and logistical realities of running a profitable mobile show for a fickle audience in a crowded market fueled a snowballing relationship between intermenagerie competition and audience demands. After the demise of the Zoological Institute cartel, that pressure also saw the equestrian and human acrobatic "circus" companies merge with animal shows or absorb individual animal acts. In 1854, one industry insider explained that, although not as highbrow as other national pastimes, such as theater, patriotic toasting, or publishing, the new circus captured the cultural sensibilities of the age and was widely admired as "still the most popular of public amusements, . . . conducted on a magnificent scale as a regular business speculation by enterprising citizens."[69]

Those entrepreneurs believed that sensational consumer experience came not only from better venues and liberal investment in one's productions but also from more complex animal acts. Wisdom inside the trade held that menageries told extraordinary stories with extraordinary animals in size, coloring, origin, rareness, and—increasingly—behavior and that elephants trained for "tricks," that is, displaying modified behavior framed with costuming, music, and a human-made narrative, could be a powerful medium of communication. Each season trainers could create new acts that referred to cultural changes in the country, like the latest humor or a bit developed by a rival animal troupe. Their shows could return to the same ticket buyers year after year, expecting those people to pay to see the same animals again and, not incidentally, using repeat visits to build the celebrity of a famous elephant.

Thus was born the American phenomenon of the "sagacious" elephant and later the "performing elephant actor." This sagacious elephant persona was an age-old concept, now monetized by antebellum menageries (as well as natural history and children's authors). Such creatures had human-style intelligence, although not defined in terms of problem-solving skills, human language adaptation, or tool use (those being twentieth-century fixations). Animal sagacity was defined as "the ability to adapt to human surroundings and to please people," particularly with respect to dogs, horses, and elephants, whose superior intelligence ironically obligated their service to humans.[70] Thus, as elephants became standard fare in menageries and their advertising, Americans were asked to intensively anthropomorphize them, a habit that was crucial to the cultural viability of trained elephants as mass entertainment.

The human cultural practice of anthropomorphizing other species is a complicated and historically contingent one, and the commercial antebellum menagerie variation was specific to its time and context.[71] In the menagerie context, the anthropomorphization in play was more than a perception of shared mental abilities or emotions between humans and elephants. People routinely assumed as much, of course, not least because they identified with the elephant's trunk. That tendency certainly was crucial to the entertainment value of elephant shows since ticket buyers had to be willing to believe that humans and other animals shared some characteristics and sensations if they were to be compelled by demonstrations of nonhuman sentience. Beyond that, the crucial shift was for the audience to humor the idea that elephants understood and endorsed the *motivations* behind human activities—such as the human desire to make money with a trained-animal show. The idea of sagacity made the contrast between an elephant's power and apparent obedience to a trainer a superlative example of the phenomenon indeed, setting the stage for the more abstract elephant persona of performing animal actor.

Showmen knew they could make trained elephant acts profitable because the animal sagacity concept undergirded the late-eighteenth and early-nineteenth-century "zoological pantomime" tradition, which the equestrian circuses in America and Europe had been using with great success. In those shows, human and animal performers played out a human story together. Typically, one saw horses trained to fall down, jump up, or bow on command and thus appear to understand and support the performance's narrative.[72] These equestrian dramas were themselves derived from an even earlier tradition, which existed at least as early as the Renaissance in Europe, in which playwrights and actors moved dogs "from exhibit to performer, and thus from prop to character, or at least to narrative engine," by presenting them on stage with directed behavior that supported the human plotline in theater and carnival shows.[73] Such shows, however, were still a minority, and as a rule in Europe and America animal characters were played by people in animal costume.[74]

Nonetheless, here was a show business tradition that simply required adaptation to exotic species and some cajoling of the customer base for it to pay. Exhibitors had set to work almost immediately experimenting with this idea by cultivating relationships with theater people. In December 1796, just months after the first elephant's arrival, up-and-coming English actor Thomas Apthorp Cooper was starring in a Philadelphia production of the tragedy *Alexander the Great; or, The Rival Queens*. In the face of low ticket sales for opening night, he booked Crowninshield's elephant for an onstage appearance as an "Asiatic auxiliary," revising the program to include "The Grand Triumphal entry of Alexander into Babylon; in which will be introduced for this night only a Real Elephant, caparisoned as for war."[75] Contemporary theater owner William Dunlap recalled, "Those who had declined to take seats to see and support the best tragedian . . . that had yet played in America, filled the house to overflowing to see the stage dishonoured by an elephant," alluding to the fact that the energy an animal brought to the stage was considered sensationally manipulative and thus unbecoming to serious theater.[76] Producers hired the second elephant (the first Betsey) and her trainer to walk onstage for an 1812 Baltimore play, an 1813 Philadelphia performance of *Blue Beard*, and five years later, *Forty Thieves*. Stage personality William Wood noted that these productions were "strengthened" by the presence of a "sagacious elephant," explaining how in Baltimore the elephant helped the theater take in $711 in one evening, instead of the routine $400.[77] These sorts of elephantine presentations showed industry insiders that elephants could temporarily save otherwise doomed ventures. Yet, by the 1830s they did so "at ruinous expense," since they were too costly for theater productions with only one sitting per night.[78]

Indeed, it was the animal exhibitors, circuses, and menageries that made a business of developing sophisticated narrative pieces out of the colloquial "learned animal" genre, which was still current and had come to include goats, dogs, "Trick Ponies," "Comic Mules," and "Educated Monkeys."[79] In fact, monkeys dressed as horsemen had been in show business since at least the 1780s, with "Captain Dick" or "Dandy Jack" being perhaps the most famous stock names for such primates. They were known to ride ponies or large dogs disguised as horses "around the ring [while waving] diminutive American flags in transit, to the delight of the small boys in the audience."[80] In 1827, the Great India Elephant had appeared with a monkey driver on his back to display "feats of horse-monkey-ship" by "the first horse-monkey in America."[81] Showmen knew such gags would appeal to audiences because they drew on industry tradition that presented trick riding, pony races, or musical dressage as horse "sagacity" demonstrations portraying an equine love for music and mankind. These acts mostly required a horse merely to walk, trot, canter, or gallop in a predictable fashion and not try to throw the rider or exit the performance venue.[82]

In the 1820s, "all the tricks usually done by these very sagacious animals," as the showmen promised of elephants, mostly constituted demonstrations of trunk dexterity and tractability: eating food taken from humans, getting petted, moving on command, pulling a cork to drink from a bottle.[83] Beyond this self-taught elephant repertoire, the most popular act was simply derived from the logistical needs of elephant labor management in South Asia: kneeling down to accept a human rider. Some young elephants appear to have arrived in America trained for this act, which was useful for mahouts wishing to ride an elephant or needing regularly to examine his feet, back, head, or other hard-to-reach area. The first elephant in the United States had reputedly knelt in a hippodrome show, "Mr. Larson's CIRCUS," in Charleston shortly before her death or exportation from the country. Larson's promotional notices explained her movements as an instructive example of "the use to which the Orientals put the noble animal, either for pleasure, profit or grandeur."[84] Horatio, too, came trained to bow down to be mounted by a rider, or to lie down, so his advertising claimed, as a sign of his "subjection" (something Horatio disproved when he broke that bridge in Vermont).

One American described an early menagerie—a show of "imprisoned beasts," he wryly called it—that allowed visitors to experience for themselves this articulation of animal sagacity:

Sometimes a small African [sic.] elephant is made to kneel down, and receive a tower on his shoulders. Those of the company who desire to ride, are requested to step forward, "ladies first, gentlemen after-*wards*." After a deal of hesitation, a servant-maid gathers courage, and simpering and dimpling, ambles into the arena. She the showman politely assists to ascend. Another follows, and another, until all the seats are taken up. Then the beast moves once around, with his slow heavy tramp, the ladies descend from their airy height, and are able to go home and say that they have "ridden on the elephant."[85]

When they allowed paying passengers to climb aboard an elephant's back to ride in a howdah, showmen invited Americans to imagine themselves as part of one of those Oriental-tale parades. Here was a kind of participatory zoological panto-mime, whereby living elephants facilitated Americans' adaptation of South Asian ceremony into a domestic entertainment tradition through which customers could physically experience the ideal of sagacious elephant as exotic and willing servant to Americans, as well as the privileged human status implied by it.[86]

The handful of gifted animal trainers who emerged in the antebellum indus-try employed this anthropomorphization of elephant motive as the main premise behind trick development but took it to new show business extremes. They in-novated new ways to convincingly portray elephants as more than simply saga-cious, but as "performing" animal "actors," especially as those animals' move-ments became more abstract and absurdly divorced from the basic elephantine activities of browsing, defecating, sleeping, communicating, and reproducing that were central to their survival. The performing elephant actor was a being who was native to show business, an animal whose essential persona was anthro-pocentric and probusiness. Some children may have sincerely believed the ele-phant they saw in the ring was a knowing entertainer. At the same performance, adults in the audience could take that claim with a grain of salt while nonetheless believing wholeheartedly the underlying political truth that human supremacy over animals was inherently benevolent and economically productive.

The elephant actor performed bits in combination with other species, includ-ing humans, horses, monkeys, and dogs, to support the narratives told by clowns, acrobats, and famed animal trainers that claimed to show how human captivity brought out the best in elephants, even improved and refined them beyond a natural state to realize their true potential as friends of humanity. It was the con-trast between elephantine power and emotional sentience that made these pre-sentations work as instructional amusement, while simultaneously the elephant's exotic origins and physicality helped menageries continued to communicate their "MAMMOTH" brand as spectacles of the incredible.

It was necessary to remind audiences of this dualism over and over and to claim to have the most sagacious and powerful elephants in an opulent show. Of "The Great India Elephant," a typical Zoological Institute notice exclaimed, "This wonderful animal, which, for size, docility and sagacity exceeds any one ever imported into this country, will go through her astonishing performances, which have excited the admiration of every beholder."[87] Even those males the menageries presented as glamorous elephants of war would be expected by their owners to submit to more complicated training in order to learn the sagacious role for future "performances." At mid-century, the better capitalized menageries and circuses often carried pairs of elephants, one of which they presented as a sagacious elephant (male or female), the other as a war elephant (always presented as male). Some companies had elephants play elephant of war in parades, then trained them after hours for a sagacity act in order to maximize usage of all the show's elephants immediately. The bull known as Bolivar first appeared as an elephant of war but in time would do two shows a day for the Van Amburgh & Company Menagerie in the guise of the "highly trained Elephant Bolivar."[88]

The Macomber & Company menagerie used the young bull Mogul to deliver all these interrelated impressions to their customers and make a claim for the uniqueness of their entertainment offerings. Mogul was incredible among elephants in the United States because he "surpasse[d] all others in magnitude and strength" while "more gentle and tractable," as one example of Mogul's newspaper publicity insisted. Warning Americans not to miss out on a show that was a defining cultural event of the moment, Mogul's advertising challenged readers: "His performances would scarcely be credited, were they not daily witnessed by hundreds." To create such an effect in the audience member's mind, Mogul's act had him, in effect, impersonating a dog: "He caresses his master in his best manner, and will not so readily obey another person. He knows his voice and can distinguish between the terms of command, of commendation and of anger. He received orders with attention, and executes them cheerfully, though with great deliberation. All his motions are grave, majestic, regular and cautions, partaking in character somewhat of the gravity of his body." For a dog to perform these movements was not so novel, but Mogul's size gave the show a sensational element and, in effect, asserted that it was possible to domesticate a wild-born elephant.[89]

The Macomber team also used Mogul to perform acts that were legitimately dangerous but equally represented the idea that the elephant chose to live peacefully with humans and somehow appreciated the human concept of showmanship. The commercial gossip framing Mogul explained that his love of show business was evident in his demonstrations of nimble and careful manipulation of his own body, one reporter said:

> He kneels on either side, raises his master to his back with his trunk or tusks as directed; reclines at length in the ring, or walks over the prostrate body of the keeper at the proper bidding. This last scene is one of the most impressive we have ever witnessed. From the situation of his eyes, he cannot see his fore feet, and calculate the distance to the object over which he is to pass without injury; so he carefully measures the space back and forth with this trunk, then divides the distance so accurately that the last step before reaching the body is just near enough to afford him opportunity, with a long stride, to accomplish his feat to the wonder of every beholder. All this is done with so much care and wisdom, that it would seem to proceed from a *higher impulse than that of mere animal instinct* (emphasis added). [90]

Here was sagacity made manifest by a trained elephant who performed the idea symbolically by endangering but not injuring a human performer. Undoubtedly, the sight must also have inspired a plain sense of "how did they get the elephant to do that?!" for children and some adults in the audience, who looked in wonder on animal trainers who could apparently communicate with such a strange and powerful being. And Mogul was in this way an animal celebrity: a consumer-friendly individual with a biography, noted habits, human friends, and desire to entertain audiences.

By contrast, the menageries encouraged Americans to perceive other wild animals as sworn enemies of mankind, no matter how long in captivity. Certainly no patron was permitted to reach into the tiger cage or pet the lions because it was simply too dangerous. Equally, no elephants would ever appear in a "fighting" show, a presentation style native to the menagerie business in the United States. Initially, the form had menageries presenting tigers, lions, leopards, and cougars in large traveling cages on wheels, then, later in the century, in a round cage in the ring. An animal presenter would incite these cats to appear ferocious by inserting whips or other objects through the bars of the cage. Others might actually enter the cage and "deliberately violate [a cat's] personal area—the 'flight distance' or space that it needs for any escape in order to provoke a violent response."[91] The point of a big cat act was to have the huge felines apparently threatening the trainer and, by extension, the audience. Advertising materials and show patter often called leopards, tigers, and lions "ferocious" or "wild," wherein "the primary emphasis was on the man's fearlessness, not the animals' cleverness."[92] Elephants could have been trained to smash furniture, kill small animals, or otherwise perform as destructive beings, but with their cage-free presentation such an approach would have inflamed fears over the public safety risk posed by traveling animal shows.

As the menageries competed for variety and novelty, people would direct elephants in an expanding repertoire of fantastic action that came to include comedic and acrobatic acts. In large part producers and trainers facilitated this by crafting the stories the animal shows told specifically to straddle the fuzzy divide between education and entertainment and also to openly endorse commonly accepted religious or secular patriotic values. Beyond the glamorous elephant of war spectacles, which asserted a citizen's right to be a consumer of animals and the exotic, a second set of stories functioned as children's morality tales in a kind of performed animal fable. A common act in this genre was the apple and fork trick, which capitalized on the dexterity of the trunk but told a story about the need for proper table manners, among other things. The act had a clown and a decorated elephant apparently eating together at a table set for a meal, with dishes, cutlery, and food, including apples and bread. The elephant's part was to sit at the table by "squatting on his haunches" and eat the apples on the table using a large fork bearing two prongs held with the trunk.[93] Trainers in this formulation proposed that they had removed an elephant from his unrefined state much as parents and school teachers sought to do by educating children or industrious citizens sought to do by going to lyceum lectures or Sunday church service.

Many people would thereafter tackle the artistic issue of how to create a beginning, middle, and end to stories built around the acts of elephantine trunk use and structured as comedic but moral narratives about cleverness, pride, or patience. In such formulations, the elephant could be a trickster seen apparently outsmarting the clown while his back was turned by emptying a wine bottle (which contained a molasses and water mixture) brought by another clown dressed as a waiter. Other gags had the elephant making a buffoon of the clown by appearing to put on airs and ringing a bell to summon the clown waiter or stealing the clown's food and then fanning herself with a paper fan held in the trunk so as to appear satirically innocent.[94]

Humor was a powerful means for engaging audiences and had the effect of making more palatable and natural the underlying premise of these bits, namely, that the animal actor affirmed human privilege and the naturalness of the menageries' ownership of exotic animals. For instance, in the late 1830s the Asian elephant bull known as "the Beautiful Elephant, Tippoo Sultan" did various artistic and comedic bits, including apparent "Waltzing" to music, which saw him high step and skip with musical accompaniment (a gait that became central to the gregarious circus elephant iconography), "Walking over his Keeper," and of course the bottle and cork trick.[95] His trainers further employed Tippo in a bit known as "Ringing the Bell," which soon became a circus standard. What was "the plot of this piece," as one magazine would ask of the narrative structures animal shows employed?[96] Circus historian John Kunzog described Ringing the Bell as performed by circus impresario Dan Rice with "Old Put," a probable Indian Rhinoceros[97]—a similarly powerful and often unpredictable animal—that Rice put in his show in the 1850s:

> Rice made his entrance followed at a distance of ten feet by the rhinoceros, heavily shackled and led by an attendant. The chains were removed as the animal entered the ring. A pair of platform stairs, three steps in height, were placed in the ring which "old Put" would ascend and standing at the top would let out a deafening bellow when Rice asked: "Did I train you to obey my commands?' While the animal stood on the platform Rice lighted some red fire inside a small paper house. "Fire," he yelled, "ring the bell." At which command the animal clambered down from the steps, ran to a rod on which hung a swinging bell and would toll it with his horn. This alarm brought out the clown fire department riding in a pig-drawn cart.[98]

There was a distinctly anthropocentric sense of the absurd in such an act in which an animal—elephant or rhinoceros—awkwardly performed such human-oriented tasks only to alert clowns in a vehicle drawn by pigs.

The bit must have had multiple meanings, on the face of it satirizing the allegedly undisciplined fire companies of nineteenth-century urban America, just as likely to loot a burning house as put out the fire within it. It also essentially displayed the anthropocentric judgment that other species would never seriously approach the assumed perfection of the human form and mind, thereby serving "the human need for animals to prove [the viewer's] humanity."[99] The animal who performed human tricks—regardless of all the incredible-sounding menagerie marketing talk about elephant sagacity, grandeur, and power—emulated humans awkwardly and was only extraordinary because (unlike the first few elephants in the United States) he or she was not evaluated primarily for his or her own elephantine nature. Thus did antebellum menagerie men begin to argue that the performing elephant was at once incredible and inherently imperfect.

These ideas about human supremacy were more than just vanity in antebellum America. The nation experienced the prosperity, security, and geographic expansion it did, in part, because citizens used animals as a crucial source of energy (horses, oxen, mules) and a vital component in food production (cattle,

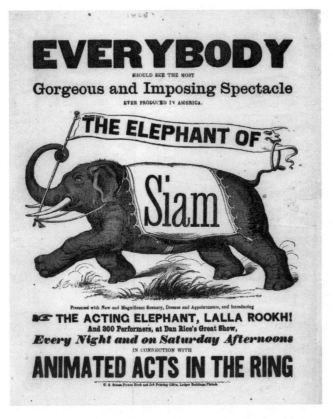

"The Acting Elephant, Lalla Rookh!" portrayed with a flagpole in her trunk and a probusiness persona as industry mascot. American Broadsides and Ephemera Series, 10674.

pigs, chickens and eggs, sheep, fish). Consequently, ideologies of human supremacist animal use lay at the core of a whole package of beliefs and practices defining what it was to be human, Christian, and American. Animal show proprietors understood this ideology very well, making that knowledge apparent in every aspect of their productions, from the script to the venue.

Raymond & Waring showed at Niblo's Garden in 1846 and advertised their show with language that tied those marketing practices and ideas to the older curiosity exhibit mode of presentation:

In the beginning of the world, as we are taught by the Holy Scriptures, God said unto man, "And to every beast of the earth, and to every fowl of the air, and to everything that creepeth upon the earth, wherein there is life, I surrender to thee control; and thou shalt have dominion over the fish of the sea, and over the fowls of the air, and over every thing that moveth upon the earth, shalt thou have dominion." When this power and dominion was given by the Almighty to man, it was intended by his wisdom that he unto whom he had confided so much should become acquainted and familiar with the subjects over which he should exercise his rule. Hence it became necessary that man should study the history of animated nature, make himself master of a science on which his own happiness depended., . . . England, France, Italy, Germany—in fact all modern Europe—have, for centuries back, paid unremitting attention to this subject: and that nation in Europe which has not its Natural Historical Institute is regarded as on the backward course that leads, in the end, to semi-barbaric ignorance and folly. Admittance, 25 cents.[100]

Here was the show business argument for monetized human supremacist amusement in a nutshell. Raymond & Waring suggested to people that one could actually fulfill one's destiny by visiting a menagerie to enact symbolically their "dominion" over other species—some of whom were trained to ring a bell, as it happened. Such marketing appeals insisted that there was much at stake because citizens who improved themselves morally made the nation wealthier and more secure on a global scale by avoiding the "backward course" that led to "semi-barbaric ignorance and folly."

A generation later, famed trainer Isaac Van Amburgh would take this argument to new heights by openly claiming to make his living collecting and taming wild animals primarily as a public service that might remind viewers of their own divinely ordained duty to dominate the natural world.[101] His was no shabby amusement but an experience of civilization, he emphasized. Horace Batcheler, a British visitor to Van Amburgh's shows on Broadway in New York, judged the enterprise "a menagerie of the first order" and was impressed equally by the animals and their venue. "There are two performing elephants, with mules and ponies. I went on the day of the opening. The cages were permanently fixed, cleanly, and airy," he said.[102] The better capitalized shows demonstrated their moral-improvement claims by creating safe and tidy environments for visitors and captive animals; not for them the rough and ready shifting dens and tavern barns utilized by earlier traveling shows. Unfortunately, the Briton did not go on to say if he in fact found Van Amburgh's "GREAT STUDY OF NATURAL HISTORY" spiritually uplifting. Yet, the company certainly asked the audience to interpret their shows as serious moral instruction, not simply frivolous animal tricks, when they explained their use of captive animals as "a study which terminates in the conviction, the knowledge, and the adoration of that Merciful Being, to whose goodness alone we are indebted for every happiness that we enjoy."[103]

Americans may have taken the religio-civilizationist rhetoric of the menageries to heart or chuckled about it as just another come-on designed to fool the credulously pious by pasting a sham morality onto a for-profit business. The emergence of elephant agility acts certainly pushed the piety-claims of the menageries to the background for the many observers who saw plain showmanship on display. Mid-century audiences began finding elephants advertised posing on pedestals and the backs of other elephants, walking on elevated planks or large-circumference tightropes, standing on their heads, and serving as platforms for other performing animals. No longer were they just doing bits that simply endangered people (like Mogul's act of stepping over his trainer), but they now performed acts that endangered elephants, too. The financial viability of these more complex performances was facilitated by a corresponding decrease in the cost of producing representations of the new acts in visual patter in graphic handbills, broadsides, and souvenir pamphlets. Made cheaper by the growth of engraving as a career and technological changes making printing and traveling more efficient and less labor intensive, now circus elephants could be depicted "in a more lifelike way." Their new tricks could be depicted explicitly with elephants frozen in poses that, like the taxidermist's stuffed animal skins, suggested how an elephant looked in live performance.[104]

Victoria and Albert, two twelve-year-old Ceylonese elephants named after Queen Victoria and Prince Albert of Great Britain, were purported to take the elephant act to this new level with high-risk agility balancing acts choreographed by their trainer, one Mssr. Lagrange of France. They and similarly trained elephants of that era marked the emergence of the "spectacular elephant," a being presented as a natural performer evocative of the incredible world of show business.[105] In the 1850s Victoria and Albert were promoted as "Wonderfully Educated Elephants," known to "stand on their head . . . their fore or hind feet at command; march and keep time; dance and play the organ; walk up narrow and inclined planes, and balance themselves on the apex of a pyramid," much of it apparently only by voice command. Victoria and Albert's performances adapted acrobatic and balancing feats normally performed by people, monkeys, and dogs, creating a surreal performance because it seemed to "reverse the order of things," one reporter said.[106] Billed as "World-Renowned and Only Performing Elephants," the two consequently came across to some as beings whose "intelligence is proverbial," [107] such that their "prodigious feats of intelligence and docility elicited continuous and uproarious applause; in fact, each feat seemed to surprise and electrify."[108]

These elephants were fully culturally instrumentalized in their advertising and their performances, barely representative of Asia or the naturalist's curiosity. As full-blown animal actors they had undergone a "cultural marginalization" in which they seemed to have ceased to be fully rounded wild animals with their own needs, behaviors, or experience. Instead, audience members who found their performances entertaining (rather than unnatural, humiliating, or crass, as some observers did in those years) did so because they accepted the menagerie's presentation of Victoria and Albert as entertaining pets, "co-opted into the *family* and into the *spectacle*."[109] They were animals that both endorsed the ostensible inferiority of animals as drawn directly from the natural world while also functioning as "embodiments of Nature's transcendence" by humanity. Acrobatic elephants were interesting to the viewer in large part because they were "symbolically denatured" and so clearly divorced from their original condition in Asia that they presented a surreal vision à la world-turned-upside down.[110]

Sands, Nathan & Company, which owned or had leased the elephants, drew this kind of emotional identification out of viewers by continually reasserting the menagerie's marketing of these elephants as enthusiastic team players. This was especially crucial when an act went wrong and events threatened to overtake the show's presentation of the elephants as gregarious performers. At a stop in Detroit in late April 1856, Victoria took "a bad fall" off the ramp leading to the top of a pedestal twenty feet tall and about five feet across, upon which she was to first rear up on her hind legs then stand upon her head. The next night "she could not be persuaded to walk up the narrow plank" during the show because of "the fright" of the earlier fall, the papers earnestly reported.[111] When the elephants' presenter finally got her to the top and posed, then cued the crowd by turning to face them with his arms in the air and thus "claim their applause," he simultaneously invited them to believe that Victoria and Albert's movements were amazing feats.[112]

Whether Victoria's hesitation to walk the plank was real or part of the show's choreography is not clear. Still, ever since many an animal performer has been scripted to make apparent mistakes or hesitate before a trick in order to inject drama into the show, and the tactic probably goes back to the learned pig genre of the eighteenth century. Such drama increased the viewer's emotional identification with the performing animal through empathy for his or her fear. Yet, the fear at hand was not a fear of humans and the demands they imposed on captive animals—demands in which the captive animals had no intrinsic interest. That kind of empathy had produced the public grief over the deaths of Horatio and the Betseys and was a sentiment that questioned the proposed role of the menageries—and also their customers—as benevolent custodians of exotic animals. For Americans, the fear implied when Victoria fell off her ramp up the twenty-foot-tall pedestal was a presumed fear of injury and failure.

Nonetheless, the elephant's hesitation touted the incredible nature of these new tricks and must have drawn plenty of ticket buyers to the show. When she performed the bit successfully thereafter, viewers were invited to be impressed by her nerve in persevering to try the trick again and her ostensible satisfaction in accomplishing a difficult task. Thus did Lagrange and Sands, Nathan & Company cultivate emotional engagement by offering spectators a story of elephant suffering and transcendence they could understand with little coaching or self-reflection, rather than one grounded in some elephant experience that potentially questioned their captivity—such as how Victoria was coping mentally and physiologically with her separation from her mother while still a suckling baby.

In California, Victoria would perform the walk up the plank again and with Albert appear before audiences of up to two thousand people who paid a dollar a ticket, a price inflated on the distant frontier. In 1859, the elephants were the first to visit California, a state whose circuses mostly consisted of trick riding and trained horse and dog shows. There, Victoria and Albert performed equestrian acrobatic pieces adapted to elephants, for instance, with a human performer named Franklin, who would do somersaults over the pair.[113] And it was in California, on June 28, 1860, that Victoria died while honoring the commands of her keepers. She was then in the hands of the Dan Rice Circus Company, whose handlers had directed her to swim across the Stanislaus River. She struggled in the current, went over falls there, and "died a few days after from her bruises."[114] That venture claimed to have thus lost a $20,000 investment, having just purchased the pair from San Francisco theatrical manager John Wilson, who had toured the elephants in Mexico and possibly further south over the 1859/60 winter. Wilson had acquired Victoria and Albert from Sands & Lent Company, which had in turn paid to ship Victoria and Albert from New York around South America to the West Coast.[115]

The newspapers and the circus people who boosted Victoria and Albert encouraged the public to see these two young elephants' acrobatic demonstrations as "instructional entertainment" and evidence of "what careful training will do."[116] There was no guarantee people would accept such acts as presentations of "elephantine performance" and "the concentration of all that is grand and excellent," as the publicity and advertising suggested.[117] Across the nation some observers did turn critic, arguing that the shows presented neglected animals in dangerous acts that flattered a crude, unthinking crowd. The concern that trained-animal shows were a "humbug" came not from any perception that they misrepresented the nature of the animals in question, since the ostensibly educational menageries sought to portray animals as raw materials for human agency as much as representatives of their wild kin. Instead, critics said that traveling animal shows defrauded ticket buyers because they purported to present family-friendly, instructional amusement while absurdly satirizing the credulity of the customers. One 1844 newspaper comment conveyed this critique in a joke that pitted the wily elephant keeper and his unknowing ward against a sincere, if foolish, member of the public:

> "That's a wery knowing hanimal of yours, is he?" said a cockney gentleman to the keeper of the elephant in question.
>
> "Very," was the cool rejoinder.
>
> "He performs strange tricks and hanties, does he?" inquired the cockney, eyeing the animal through his glass.
>
> "Surprisin'!" retorted the keeper—"we've learnt him to put money in that box you see way up there. Try him with a dollar." (The cockney handed the elephant a dollar, and sure enough he took it in his trunk and placed it in a box high up out of reach.)
>
> "Well, that is wery hextraordinary—hastonishin', truly!" said the green one opening his eyes.
>
> "Now, let's see him take it out and 'and it back."
>
> "*We never learns him that trick,*" retorted the keeper with a roguish leer, and then turned away to stir up the monkeys and punch the hyenas.[118]

Indeed, this joke was funny because many people knew that menageries provocatively addressed people as unworldly spenders, not the frugal producers the shows claimed them to be.[119]

In fact, most citizens and certainly the press, which relied on circus advertising revenue, exhibited a willingness to accept the showman's patter without too much scrutiny. Most chose to perceive managed elephant behavior on stage as an innocent entertainment telling a basically authentic story endorsing human supremacy and the citizen's privileged role as a consumer deserving of an incredible show. The elephant "performers" and "actors" invented in these days were the root of the happy circus elephant icon, which would serve as mascot of the circuses as an emerging mass entertainment for generations thereafter.

When showmen learned they had to manage audience interpretation, they developed a mode of cultural instrumentalization that gave elephants new meanings in the United States as natural curiosities, to be sure, but also as performing actors native to show business whose purported ambitions and exploits would overshadow news of accidents on the road and other events that plagued the circuses into mid-century. It is not clear how fully showmen understood the ramifications of this shift and the ways the product of animal celebrity would facilitate the anthropocentric nature of show business thereafter. Still, they certainly went about their business in ways that showed they understood that profitability came from providing consumers with meaningful experiences of elephant celebrity, which were very persuasive with most antebellum Americans most of the time.

# CHAPTER THREE

# Learning to Take Direction

American animal shows offered consumer-friendly narratives mingled with naturalistic and biographical information about their elephants, all of it contextualized with a glamorous Orientana celebrating the power of the show business entrepreneur. As they could not speak in human language, performing elephants did not explicitly quarrel with these messages. When they took direction by moving calmly in a parade or posing on a pedestal as the band played, they came across to many as comfortable with or even comprehending of the larger story the presentation told—about the affluence of the Asian elite, the immense resources of the menagerie company, the privileged leisure of American audiences, or the elevating nature of homo sapiens's presumed power over beasts.

However, not everyone bought what the showmen were selling. For instance, in 1888 New York journalist and art critic Horace Townsend commented on the phenomenon of animals "advertised extensively as being . . . actors, rather than mere trained animals, and as exhibiting a wonderful amount of intelligence." He confided, "Those who watched them can bear me out when I say that they had merely been taught to obey mechanically a few arbitrary gestures of their teacher. . . . Oftentimes, too, they resolutely refused even to do as much as this."[1] Ninety years after William Bentley had left Boston's Bowen Museum, disappointed by the lethargic creatures he saw displayed there, Townsend and others still sensed that, in spite of the incredible advertising, in the flesh many performing animals displayed body language that revealed no understanding of the human stories their keepers strove to tell.

When he complained that the animals were bad actors because they did not emote convincing human attitudes, Townsend was really working through a crucial and chronic fact of interspecific performance: the lack of actor's trust between human and nonhuman on stage.[2] Animal performers were unlike one's human colleagues, whose identity was invested in a successful performance and an artistic ethic such as "the show must go on."[3] Animals might unpredictably hesitate to follow a cue, perform a move with obvious lack of interest, skip a step, or simply walk away mid-act because they had no stake in audience enjoyment or the take at the ticket wagon. When they rejected the conditions of their experience in these ways, circus elephants exposed their disconnection from events people referred to as "Grand Parade Spectacular," "wondrous show of sagacity," or "circus."

In fact, the most successful animal trainers were those who capitalized on nonhuman performers' incomprehension of the human performance context. "A phase of animal-act training likely to be forgotten by the lay audience is that the animals must be accustomed to them—to the audience," as one vaudevillian later explained this industry truism. "Aftei finally obtaining mastery of the routine—it may take years—the trainer must make his animal unmindful of applause, laughter, and music before the act is presented to the public."[4] Elephants so trained often appeared impassive to many circus goers, although an emotive effect could be produced with particular movements, such as the classic raised-trunk salute to the crowd, which affected an elephantine smile and wave.

This was the conundrum that defined the work of an animal trainer: how to produce elephants who appeared to be both invested in an act's human narrative or humor and unresponsive to audience noise and their own fatigue, boredom, and other needs that might inspire disobedience. In working through this puzzle, nineteenth-century animal trainers developed methods of creating an elephantine parallel reality constituted by props, trainers' commands, and a host of management tools, the experience of which produced desired elephant movements on cue. Elephants operated through a trainer-crafted process of positive rewards (i.e., access to food, exit from the arena, familiar human words and tones of voice) and negative sensations (i.e., pain, memories of earlier discomfort associated with specific human sounds or gestures). That parallel reality drove elephants to focus solely on the trainer (or other human presenters who worked them) while in a ring performance. Unlike many other wild animals, as social beings genetically predisposed to oblige a known individual in allomother groups, preadult elephants tended to be tractable and thus flexible storytelling tools. They exhibited apparent patience and self-restraint when addressed by trainers who asked them to stop eating hay or dozing in the barn and instead take up uncomfortable poses or arrange themselves unconventionally (in elephant terms) on command with other elephants, humans, dogs, monkeys, or horses.[5]

Elephants were not merely stimulus-response machines, however, and trainers knew that compliance was not necessarily an indication of elephant enjoyment. The most productive trainers understood that while learning, rehearsing, or performing a routine, elephants might experience any of a range of emotions, including fear, anticipation, relief, frustration, or some other feeling, but that these feelings could not be allowed to interfere with the desired elephantine movements and should be molded to the advantage of the show. Of course, no trainer spoke publicly in such bald terms about the behavior modification regimes he employed; he spoke of training or, more gently, of "educating" or "improving" his animals through tasks purported to increase their gratification in life. The term *training* would soon come to imply that one used punishment and painful consequences for incorrect behavior, while people inside and outside the circuses would use the concept of *teaching* wild animals to market their work as a more patient, clever, and flexible approach—associations that still mark animal behavior modification language today.[6] That anthropocentric and paternalist rhetoric obscured the risky dependence of the animal shows upon the changeable psyches of their captive elephants.

Performing elephants were powerful icons potentially linking the natural world and for-profit entertainment, so the stakes of a trained elephant act were considerable. Elephants seemed unable to dissemble as a showman might. They could therefore "appear to speak from the universal and disinterested place of nature" and portray the powerful role of the menageries and circuses in consuming rare and fragile nonhuman life as progressive and inevitable.[7] Some audience members were persuaded by this ideal and by the solemn-looking elephants they saw at menagerie and circus shows. Like Horace Townsend, others found trained elephants melancholy and unconvincing in their countenance because their parallel experiences of the performance context peeked out from under the patter, costuming, and showmanship of their presenter. Thus, although seemingly authentic and honest, the trained elephant was a flexible icon that might naturalize or challenge such privatized uses of exotic wild animals, as per the interpretation of a given viewer.

Either way, animal trainers were some of the most influential popular theorists of nonhuman psychology, cognition, and learning in the nineteenth century. Although not as well-studied as thinkers like Charles Darwin, George Romanes, Lewis H. Morgan, or B. F. Skinner, they considered similar comparative psychology questions in using animals as narrative tools. These men (only a few women would work with exotic animals, and then mostly as presenters or decorative "elephant girls") were quasi-celebrities in their time. Details of their physical dimensions, food preferences, and places of birth were not part of their publicity, as it was for the elephants, but the menageries began explicitly naming these men in antebellum advertising not long after they began promoting individual elephants as named celebrities. The most successful trainers knew that essentially they offered the audience an example of a seemingly magical relationship to a member of another species. As performers they asserted that they could shape elephants, big cats, or horses into any performance that humans desired. Animal show proprietors consequently presented animal acts as a sign of the wrangler's bravery, ingenuity, and virtue in the face of an unpolished, even irrational beast. The trainer's persona cued the audience to perceive these characteristics by way of militaristic or, later in the century, safari-style costumes that signified this expertise.[8]

And yet, this show business persona masked the precarious professional status of the animal trainer in nineteenth-century America. Like the hundreds of other men seeking mobile and nonphysical labor as writers, illustrators, publishers, or salesmen, menagerie animal trainers founded their own profession by cobbling together job descriptions and positions at various companies while cultivating bodies of consumers that would fund their services.[9] They were nonetheless a step beyond the ad hoc animal exhibitor, who had simply determined that he made more money if he let visitors pet the elephant or offered her a bottle from which to drink. Later generations of animal men often performed both the creative work of imagining stories to tell with living animals and worked out a technology of behavior that could produce the required animal action. They did so before standardized elephant commands like "Move up!" (walk forward) became widespread in the twentieth century to facilitate the trade in trained elephants from one company or trainer to another. Those who handled elephants in particular contended with extreme logistical complications emanating from the species' sheer size, physical strength, and unpredictable periods of unmanageability, all of which were exacerbated by audience expectations of cage-free presentation and opportunities to feed or ride elephants. In spite of the marketing patter celebrating elephants as sagacious friends of humanity, many trainers knew them to be dangerously clever wild animals of ambivalent character and considerable variation in temperament.

The entertainment experiences trainers provided to Americans were unique and overwhelmingly popular precisely because nonhuman performers were animate and cognizant of their surroundings. Menagerie advertising had emphasized that interactive sentience since James Bailey first produced broadsides naming the second elephant in America as "A LIVING ELEPHANT."[10] Later advertising used graphic arrangements of text in newspaper columns to communicate the vitality of their shows, for instance, by using ad templates packed with "LIVING" animals or by draping the word "ALIVE" down the column to convey a sense of movement without expensive woodcuts or etchings.[11] Visitors reported that a living, breathing animal "produced an entirely different impression upon [one's] feelings from what the stuffed one did," as one children's story related at the time.[12] Live animal participants added a kind of nervous excitement many people probably could not quite define, but knew when they felt it. In typical form, one 1836 production at the Lion Theater in Boston, employing at least two elephants along with horses, one- and two-humped camels, and a military band, promised the public an extraordinary consumer experience in its "brilliancy and effect . . . embracing *the energy of the establishment*" (emphasis added) to communicate this effect.[13]

Mid-century museum proprietors noted how menageries sold out shows and endeavoured to offer the public live animals at their venues in order to cash in on audience fascination with animate nonhuman entertainment. P. T. Barnum famously shipped some beluga whales from the Atlantic via a New England port at great expense. They died within days in their tank in the basement of the American Museum. Nevertheless, that event only prompted Barnum to pragmatically send out another team to wrangle up new whales, whose trip to the museum launched yet another publicity campaign celebrating Barnum for bringing living whales to Manhattan.[14] Barnum and others additionally offered "Happy Family" displays, which enclosed various antagonistic species together. Many people were incredulous at the time, although rumor had it that Barnum's staff had simply "stupefied with morphine" the cats, dogs, mice, and birds used for the curiosity.[15] Nonetheless, these displays were intriguing because they allowed people to inspect a living specimen of a rarely seen animal (as with the whales) or because of the distinct possibility that the animals presented together might actually turn on one another while viewers were visiting. This, of course, was the bread and butter of the trained animal act, a genre of interspecific performance that was compelling because risky and counterintuitive.

The generation of Americans who visited the menageries or Barnum's animal displays had come of age during the wave of antebellum Protestant religious revivals and moral reform activity in the nation. To them, menagerie men continued to explain trained animal acts as lessons in Christian dominionism.[16] Isaac Van Amburgh, the most famous animal man of the century, offered performances that literally presented a young lion lying down with a lamb in depiction of Isaiah 11:6, a passage many interpreted as supporting animal submission to mankind as part of God's plan for the world.[17] British trainer and circus impresario George Sanger would later expose how Van Amburgh's lion and lamb act was done when he re-created the trick by raising a lion cub with a lamb from birth so that they would tolerate each other.[18] Nonetheless, the broad message such presentations offered was—like the Happy Family curiosities—an assurance to Americans that as Christians and consumers they could aid and improve animals by paying to be entertained by them.

Less-well-capitalized traveling shows and urban museums still labored with the costs of transporting, training, and maintaining collections of live animals of any sort. It was even common for menageries to begin a tour fully outfitted but arrive at inland frontier regions with only a handful of animals still alive, with no dependable or inexpensive way to restock. Setting out on a route early in the season with a pricey cache of effusive marketing bills offering dozens of different living animals, weeks later when a company had endured losses, it would roll into town unable to deliver on its promises. "The advertisements were splendid, but the performance was a humbug," one Michigan paper groused of one show stricken in that manner in August 1846.[19]

Many resorted to presentations of taxidermized animal skins or skeletons instead. Designed to suggest some kind of real or imagined live behavior, these animals came posed to represent an animal alert to a predator, screeching or growling, pawing the air as if ready to attack, or at least with legs positioned to suggest how the animal looked when walking.[20] When they offered stuffed animals, even in combination with a few living animals, those proprietors acknowledged both public demand and their inability to manage the expense of showing living animals to meet it. That same Michigan paper had complained of the "miserable hoax" produced by preserved animals, including one particularly pathetic specimen, "a dead, stuffed, common duck; the whole cage being offensive to the nostrils of any person." The only thing worse than the disappointment of the animals, the paper continued, was "the wanton insolence and outrageous impertinence of the attendants [, which] richly merited a tarring and feathering," telling us that menagerie audiences could still be rough and ready, as they had been in Elizabeth Drinker's day.[21]

In the later antebellum years impresarios would begin merging the menageries and circuses into single productions, which became known collectively as "the circus" by the 1850s.[22] Stage actor and show business commentator Olive Logan remarked that, in the antebellum years, the circus had been "a foe to its zoological rival, but like it struggling onward in the race for popularity and importance." Yet after the war the business landscape had changed as another generation of ticket buyers was afoot, "for at this time a traveling expedition is not considered perfect, especially in the rural districts, without the amalgamation of circus and menagerie," Logan concluded in the late 1860s.[23] Born and raised with exposure to both shows' ambitious claims to entertainment, audiences everywhere perceived any venture that could provide a horse circus *and* trained animals as more affluent, glamorous, and up to date. They also permitted people who frowned on equestrian shows to visit a menagerie more often by visiting the animal tent of a combined circus, while refusing a ticket to the more seemingly dangerous and risqué trick riding performance.[24] Smaller "dog and pony" shows aspired to that status, often simply offering one trick rider on a single horse, a few caged animals, and a man with a learned pig as their full lineup of acts.

Many animal trainers realized that their knowledge of how to supply the managed performing animals customers expected gave them considerable power over a company and its reputation. They often knew more than anyone else on their unit what objects, people, or situations made a given elephant compliant as a performer, or not, and how to modify captive animal behavior for new productions. Still, for every dedicated trainer with many years of hard-won experience in the business, there were many more who were no more than hacks. Transient in the business, they moved from show to show over the years as they could find employment, encountering a series of elephants and other animals who might in turn show their reticence by suddenly refusing to perform or learn new routines in the face of a revolving door of trainers. "Keepers often style themselves as [trainers]," one insider noted of the difficulty in identifying truly gifted trainers among the many wranglers living hand to mouth in the industry.[25]

Meanwhile, under pressure from limited budgets and high staff turnover rates, both experienced and hack trainers usually hired men inexpert in large exotic animal management as keepers and barn men for small wages and no fame, often African American men or others with similarly limited employment options. Elephant training and keeping constituted an often-boring yet dangerous existence, because, if one was not injured by an animal, one might suffer at the hands of local roughs, who always seemed to appear on circus days. Furthermore, traveling show workers and managers alike seldom slept more than five hours per night and labored under the possibility that their venture would suddenly go bust and fail to pay back wages.[26] Alcohol abuse, too, was a chronic problem among these crews.

Understanding their power, the most innovative and dependable wild animal trainers guarded their expertise carefully. They developed theories of animal learning and cognition by trial and error or passed them from one man to another during rehearsals so that, for the historian, this professional fraternity can appear inscrutable. Because they were so few in number, nineteenth-century animal trainers had no trade publications or publicized conferences and thus were unlike other professional groups with secrets to keep. For instance, the thousands of antebellum printers and booksellers had early on founded and supported trade journals in which to debate how to develop their writers, customers, and national production, marketing, and distribution networks.[27] Gilded Age stage magicians—made up of some professionals and countless amateurs—packed their trade journals with clearly illustrated explanations of how a given magical illusion could be executed. An army of freelance writers and illusion designers in turn served those men, likewise creating voluminous records of their creative and technical expertise in personal correspondence, countless magazine and newspaper articles, or carefully compiled scrapbooks.

For the menageries and circuses, the famed sporting paper *New York Clipper* served as "a kind of 'organ' for show people," as Olive Logan said of its mid-century columns on entertainment industry news. However, as a trade journal it was problematic for many readers, as "of course [it was] disposed to be very lenient with the shortcomings of the class on which it depends for patronage." Logan and others noted that the *Clipper* contained little that might criticize or even expose the professional modus operandi of any performer, including the animal handlers in the menagerie business.[28] Equally, so as not contradict the marketing messages their shows used to describe Victoria, Mogul, and the others as genial storytellers, antebellum elephant trainers seldom spoke publicly of their training techniques or the parallel animal reality they created for their charges.

Magazine and newspaper editors, however, had for years excerpted in their publications American and British travel narratives of Asia recounting observations of overseas Asian elephant training techniques. And beginning in the 1860s, various former employees and journalists appear to have considered these accounts and, perhaps seeking to trade on what they knew, produced inexpensive exposés of the behind-the-scenes realities in the animal show business. There was Olive Logan's *The Mimic World, and Public Exhibitions: Their History, Their Morals, and Effects* (1871), an account, in part, copied verbatim from newspaper articles. John J. Jennings's *Theatrical and Circus Life; or, Secrets of the Stage, Green-Room, and Sawdust Arena* (1882) in turn plagiarized Logan's account, and both capitalized on the broad consumer interest in circus life. Later, performers John H. Glenroy, George Conklin, and Frank Bostock, as well as circus impresarios W. C. Coup and John Robinson, published original memoirs that similarly proved to be dependable sellers. Although this kind of book would flourish especially after 1900, these Gilded Age editions established the circus life genre. As documents they were carefully crafted to draw upon fans' nostalgia by celebrating the circuses' famous personalities and their work of putting animals through their paces.

Readers knew what to make of insider circus accounts of animal training because they coexisted with a healthy publishing market for horse-training manuals and animal-health almanacs during that century, as well as popular publications like *Frank Leslie's Popular Monthly, Harper's Weekly,* or the city newspaper. Children's literature also regularly offered pieces like "Some Hints on Dog Teaching" and other novelty news with which to amuse or terrorize family pets.[29] Most crucial of all, though, was a popular manual known as *Haney's Art of Training Animals.* Issued in multiple book and pamphlet editions in 1869, 1894, and possibly other years, it was of the same ilk as other behind-the-scenes publications, for example, Tony Denier's *How to Join the Circus and Gymnasium* (1877) or George A. Palmer's *The Secrets and Mysteries of Educating Animals* (1890). Developed with the apparent help of the Van Amburgh and Yankee Robinson circus staffs, *Haney's Art of Training Animals* provided a polite version of the accumulated animal behavior modification techniques used at mid-century by horsemen, small-animal pet dealers, and menagerie trainers.

Haney's volume included chapters on familiar animals like horses, dogs, cats, birds, and sheep, but it also offered advice on working with various wild creatures. Haney admitted, "We do not imagine that many of our readers will have occasion to train an elephant," indicating that his discussions of elephants, lions, and tigers were for the vicarious amusement of an implied child or teen reader. "That money can be made by training animals, is unquestionably true—even a boy can make his pets more valuable by teaching them a few simple tricks," Haney advised. Referring to the supposed equality of opportunity the United States afforded to white males, the manual revealed that many readers probably believed that an animal was of cultural or monetary "value" because he or she served human needs.[30] Likewise, the book spoke to contemporary practices of animal "improvement" through speculative pure-breeding—or in the menageries, through the modification of captive animals for performance—by asserting that "under the tuition of man" animals could "be developed to an extent to which they would never attain in a state of nature."[31] The idea of trained animals as progressive and industrious may have been appealing because an untrained animal, content simply existing without producing anything people needed, challenged American ideologies advocating self-improvement and industriousness.

From this combined literature we learn that all trainers, presenters, keepers, and barn men knew that elephants and other animals structured the training process to a considerable degree and that the successful trainer molded that influence to his own interests. Although thus far this chapter has focused on the persona of the animal trainer, he was only as famous as his animals, and he understood that it was their sentience that made his paycheck possible. Here such men confronted the task of determining how to create narrative art with a living medium like an elephant. Indeed, how would a person—or even a group of people—persuade an elephant to walk around a circus ring before a large noisy crowd, stand on her head, or permit a monkey to pose on her upturned trunk?

The most basic ways of creating narratives around pachyderm activity were the ones established in the 1790s by the first elephant. Audiences loved watching an elephant eat and maneuver her trunk, so handlers found various ways of creating stories around that particular action, putting an elephant to work with minimal human labor simply by taking advantage of his or her dexterity and self-interest. Accordingly, *Haney's Art of Training* advised that the classic corked bottle bit was merely a presentation of elephants who knew how to get liquids out of containers, explaining, "all that is necessary in the soda-water trick is to let him know there is something in the bottle and his ingenuity may be depended upon to get at the contents."[32] Throughout the nineteenth century one might still see the bottle trick or feed an elephant bread from one's pocket in the menagerie tent before the main performance. Seeing the elephant sniff around little sister's face with his trunk—causing her to burst into tears or shriek with delight—was still a good show in itself, and it remained a staple for generations.

However, in the circus ring producers and trainers needed elephants to reproduce the particular movements designed by circus people to support a larger production. Thus animal trainers took away from elephants the work of choreographing their acts. At the same time, the new problem they created was how to tell more sophisticated stories with more complex behind-the-scenes training and rehearsals while funding the extra manpower required to shape elephant behavior to goals that the elephants themselves might not appreciate—although their performances had to convince paying spectators that elephants did appreciate those goals. The show business secrets elephant trainers consequently developed consisted of practical methods for tackling the enormous communication problem inherent in interspecific performance, namely, that "it is impossible to *explain* to any animal what you desire," as the wisdom went.[33]

*Haney's Art of Training* cautioned readers that they would be better at manipulating animals' movements if they came to terms with that fact: "Obedience to his keeper's orders is not in all cases proof of a perception of the object to be attained by compliance." To illustrate, the book referred to a notorious 1826 London case in which the keepers of Chuny, a male Asian elephant in musth (a periodic condition in which sexually mature bull elephants become aggressive), had killed the elephant when he seemed on the verge of breaking out of his enclosure.[34] They brought soldiers into the zoo at Exeter Change and ordered them to shoot the elephant. "When after receiving fully one hundred and twenty balls in various parts of his body, and these proving ineffectual to end his existence, he turned his face to his assailants on hearing the voice of his keeper, and kneeled down at the accustomed word of command, so as to bring his forehead within view of the rifles."[35] That Chuny had submitted to the order to kneel down made for a "touching incident," one that revealed to some the elephant's tragic obliviousness to the motives of the people around him and his immersion in his own parallel experience of the context dominated by the personality of his handler. In show business terms, such an instance would have been framed for a paying audience as a sign of Chuny's perceptive loyalty to his trainer, even at the expense of his own life.[36]

Audiences seemed to appreciate narratives that got across the idea that one could talk to animals who understood both trained commands *and* human motives because they inhabited a shared economy of perception (although it was defined by human interests). From the learned pig shows of the 1780s to the "educated horse" phenomenon of the early twentieth century, Americans revealed a sometimes desperate desire to communicate with other species and share human cultures of self-improvement with them. Animals offered in the "learned" genre were conditioned by trainers to watch for subtle signals from a human handler to perform particular movements, which the human presenter explained with patter asserting that the animal was calculating for himself or herself how to answer questions and endorse human-style reasoning.[37] As they are not distracted by human cultural codes or expectations, members of other species had shown themselves to be keen observers of a person's body language. Practiced in the right way, a showman could in almost undetectable ways cue such creatures to perform without the mechanics of the animal's training overshadowing the formal narrative of the act.

The horse known as Clever Hans would become the most controversial example of the conflation of animal behavior with a presumption of shared perception when he gained international fame at the turn of the century. Clever Hans exposed the fact that many people—even some trainers—had difficulty ascertaining which mechanism caused animal behavior. His owner was Wilhelm von Osten, a former Berlin school teacher, who insisted to the public that he had taught the horse to read, do basic arithmetic, tell time, and answer spoken queries by nodding or shaking his head. Von Osten appears to have sincerely believed that Clever Hans understood in human terms what was being asked of him, as did many observers on both sides of the Atlantic.[38] Scholars in Germany looked into the phenomenon and decided that the horse was actually observing and remembering extremely subtle changes in von Osten's body language revealing when to act—and thus get a lump of sugar. For instance, while tapping out the numbers from one to ten, von Osten would unconsciously relax the muscles in his shoulders and unwittingly show the horse that the correct answer to a math question was on offer. Hans seemed to settle the issue when he proved unable (or unwilling?) to provide correct responses when von Osten asked his questions from behind a curtain, out of the horse's sight.[39] Although to behaviorists of the twentieth century, Clever Hans demonstrated that all attempts to teach human language to nonhumans was scholarly self-deception, to others it demonstrated, not that animals had no language skills, just that attempts to train them to communicate by human language was problematic; they were suited to their own modes of communication and would therefore always come across as inept in American Sign Language or tests of counting or naming objects. Horses like Princess Trixie and Jim Key appeared to display similar talents in the United States. They showed at various world fairs and amusement parks in those years, although their trainers did not believe that the horses were displaying human-style learning—even if spectators did.[40]

Turn of the century educated horses and other "learned" animals demonstrated to animal trainers the importance of the visual or sound command in subtly controlling animal action over a distance. "All performing animals have to be cued for their tricks . . . by a snap of the fingers (unseen by the audience), by the tone of voice, by holding the riding whip in a certain position, tapping the foot, or whispering," one vaudevillian later remembered.[41] Some within the industry even argued that Dan Rice had used finger thimbles to click out commands to his famed rhinoceros Old Put (only after he was conditioned for it by pain-aversion training employing a heated fire poker, circus insiders said) without the audience hearing and thus detecting the workings of the performance.[42]

Menagerie and circus people understood the century-long human hope for insight into how "intelligent" animals like horses, dogs, and elephants thought and felt, yet they employed only command-based animal behavior to develop and execute animal acts. They did not try to communicate to performing elephants why a gag was funny, how to name the color of their costumes, or how to count the elephants in the show's herd. Whether it would be possible to instruct an elephant about such things was explicitly *not* the goal of the trainer, who needed only dependable cued movements from an animal (as opposed to late-twentieth-century animal cognition and language experiments, which engaged in animal training with the primary aim of collecting insights into a given species' brain function or state of mind). Theater people had held this principle since Shakespeare's time, although it probably originated earlier. They saw animals as living props of a sort that could be modified to perform specific acts embellishing or illustrating a narrative. That is, the human performers were required to interpret the animals' movements with patter and music to indicate to the spectators that those movements carried some human-style meaning supposedly intended by the animal actor.[43] Here was a kind of co-production in which each individual, human and nonhuman, had his or her own understanding of how to respond to the other party, while the whole interaction supported a separate story crafted for the audience. In fact, animal trainers have often devised or identified preexisting animal movements first, then created story elements around those options thereafter.[44]

Showmen trained animals by breaking new tricks down into a series of small component actions, demonstrating each one to the animal, then reinforcing that behavior with either positive sensations or relief from negative ones. They developed this practice through trial and error with each individual, but usually with an understanding that animals were products of instinct shaped by experience. As a result of the volume of their practical labors with various species, these men operated as quasi-comparative psychologists who believed that manipulation of an animal's experience grounded in knowledge of traits common to the species could create a predictable behavior in the future. For instance, big cat men who trained tigers to roar and appear menacing on command played on a given tiger's instinct to defend himself when the trainer got too close, but they used a whip, riding crop, or stake to cue the cat to replicate that action by command without being directly confronted during the show.

The trainer's theories of the interrelation of instinct and learning were so powerful because, by contrast, public debates over the nature of "animal intelligence" tended to artificially separate the two, driving people to argue for an interpretation of animals solely as products of reflex or sentiment, respectively. Americans drew from their experiences with horses and house pets or from their armchair ruminations on exotics creatures like elephants, bears, or lions to overwhelmingly measure the apparent abilities of other species against a human ideal of cognition, learning, and communication. Thus, some argued that animals were primarily soulless beings driven by instinct that allowed for minimal memory capabilities and little more. On the other end of the disagreement, especially by the 1880s, were those who veered toward sentimentalization to argue that dogs, house cats, horses, elephants, and other seemingly familiar animals were shaped primarily by culture and experience. By that view, other species held within them an almost infinite potential, an argument to which the "learned" animal shows happily catered.

*Haney's Art of Training Animals* gave a colloquial sense of common show business wisdom about all this. The manual explained that, as far as most trainers were concerned, animals could be manipulated effectively because they were products of both reflex and logic. Haney advised amateur trainers to respect the abilities of their subjects: "An action may be partly instinctive and partly the result of reasoning, but a purely instinctive action never changes except under the influence of reason. Without the possession of these powers [of reasoning] we believe no education of animals would be possible; and we farther [sic] believe that the capacity for learning is in exact proportion to the ability to reason."[45] By "reasoning" Haney and many others would mean the ability of animals to learn and adapt to their surroundings over time through observation, trial, and error. To those who denied animals like horses, dogs, and elephants the power to observe and learn by "reasoning," the manual countered: "We believe the evidence is too strong to be doubted that many animals do perceive the relation between cause and effect," based in their own sentience.[46] It was this middle position connecting instinct and learning that made showmen experts on animal behavior, *Haney's Art of Training* argued. Productive trainers knew how to manipulate the ability to learn without activating an animal's instinctive urge to resist movements that appeared confusing, frightening, or painful. Indeed, the volume cautioned that inexpert training by a frightened or impatient person could awaken just such counterproductive emotions in animals.

Many show business animal men would have agreed in theory. To them, the efficacy of reward training was basic proof of animals' self-interested cognition of their surroundings, through which these men created potentially positive relationships with their subjects. By the later nineteenth century, people referred to such positive reinforcement methods as "training by kindness," whereby a person persuaded an animal with food, "a carrot, or something of the kind," "various little tit-bits," or a predictable end to the training session. Such rewards facilitated learning by giving animal performers a reason to carefully observe the trainer and, believing they could shape the situation, take his direction in order to have power over their environments.[47]

Patient repetition of consistent commands and food rewards for correct performance made possible, for instance, the training of elephants for the table manners bit described in the previous chapter. A particular favorite during the long nineteenth century, as audiences saw it, the piece usually consisted of a clown and a decorated elephant presented apparently eating together at a table set for a meal with dishes and cutlery. In time, food like apples or bread would be brought to the table (perhaps by a monkey or a human dressed as a waiter), which the elephant would eat using a large fork bearing two prongs and held with the trunk. "This appeared quite wonderful, and was hailed with rounds of applause, but it was a trick very easily taught," *Haney's* revealed. "The animal had been first given apples on a fork, and not being allowed to eat them except on taking them off the fork with his mouth[,] he soon learned to do so. Then he was given the fork, and the apples placed before him, his trunk was guided by his trainer's hand to strike the fork into the apple and then he was allowed to carry it to his mouth. If the apples be good ones he will soon learn to do all this without prompting, and will very willingly perform the trick for the sake of the 'perquisites.'"[48] That is, the elephant need not understand the human culture of table manners to perform the act dependably for his or her own reasons. Of course, for the amusement of parents and their children, the patter and props associated with the bit asserted that elephants could know human table manners.

Show people and their customers were in effect asking similar questions about other species' mental experiences as investigators like B. F. Skinner and Martin Seligman would ask in the next century. Later researchers used "stimulus-response" studies employing various positive (food or escape) and negative (pain or captivity) reinforcement to condition animals by drawing upon the combined effect of reflex and experiential learning.[49] Nineteenth-century animal trainers used similar tactics to find quick, practical solutions to the tasks in which various species would learn multiple ways of acting in and on their environment. They held that animal memory and cognition allowed many species to learn "with a reinforcer," as late-twentieth-century researchers would put it, by trial and error, imitation, and observation or insight, and "without a reinforcer," by habituation, imprinting, exploration, and play. (The latter modes of learning allowed elephants to introduce behaviors into circus routines that were unsanctioned by their human managers, such as how to pull out of the ground the stake to which one was chained—see Chap. 5).[50]

Elephants who responded to reward training only enhanced the popular theories of elephant sagacity, conflating an animal's cooperative behavior with the desire for human guidance. For instance, when circus insiders presented elephants walking on bottles or carrying a parasol in their trunks with the explanation that they were "educated," privately they believed it was simply that elephants were highly sentient and able to remember a series of cause-and-effect steps. By the audience's view, which animal shows encouraged, the trained elephant was evidence of the principle that all "intelligent" animals sought to cooperate with people. Following from that, a noncompliant elephant (or horse or dog) was one who simply did not comprehend what the trainer wanted, an argument that rejected the possibility that a comprehending animal might simply wish to refuse for myriad reasons. *Haney's* manual had raised this precise issue in order to warn readers not to become angry or abusive with unresponsive animal trainees, but rather to strategize further on how to communicate more clearly to him or her the desired task and its reward.

Indeed, *Haney's Art of Training* and later Palmer's *Secrets and Mysteries* (plagiarizing *Haney's*) encouraged child trainers that "intelligent" animals could be identified specifically by their willingness to cooperate. Both volumes explained that "a horse or dog can be taught things which with a hog can never learn, and in the lower scales of animal life all attempts at education become failures."[51] Many other animals were unresponsive because of "limited sensory abilities," which prevented them from perceiving a particular direction, for example, in the case of insects or birds. Thus, presumably, many highly sentient and observant animals came across to people as unintelligent because they were simply uninterested in pleasing humans, as were *Haney's* pigs.[52]

Professional animal trainers often coped with similar troubles. Zebras presented a particularly notorious example in the trade of what circus people used to think about the relationship between "intelligence" and cooperativeness. The species came to the United States in the early nineteenth century, and many a showman knew that audiences found them both familiar and strikingly exotic. Not a few would attempt to train zebras to perform, as horses did, fancy stepping, trick riding, and artistic posing presentations. Yet, zebras proved to be impossible, and no manner of food rewards, praise, or whipping seemed to persuade them. Consequently many trainers were not sure whether zebras were just stubborn or actually "stupid." As a result, circus zebras were usually relegated to being led in the parade and displayed in a pen in the menagerie tent, while the few "performing" zebras that did appear often turned out to be white ponies striped with boot polish.

With respect to elephants and the relation between comprehension and willingness, the general consensus among showmen was that initially most juvenile elephants were both "intelligent" and cooperative because they did not yet "know too much" by having been taught by older elephants to avoid or challenge humans.[53] In the industry, however, people agreed that with time many became stubborn, sluggish, or unresponsive. The troubling reality with a being as large and powerful as an elephant was that he or she could discover his or her own physical power and thereafter simply refuse commands or walk away from a trainer. Conceivably, no mere human would be able to intervene or interrupt such elephantine direction and could be injured or killed by attempting to do so. Moreover, there were only a limited number of simple trunk movements elephants would perform solely for rewards. To expand the menageries' storytelling capabilities, trainers devised ways, not only of communicating cause and effect with respect to a particular series of movements, but also of convincing elephants that they had no choice but to comply with direction. In doing so, they gained access to powerful technologies of behavior that allowed a single person to control an elephant over space (by giving a command without touch) and time (by impressing an elephant with the past consequences of following or not following commands, thereby persuading the elephant to assume that the same consequences would follow noncompliance or compliance thereafter).

One veteran trainer pointed out the wonder of the performing circus elephant that showmen thusly produced for spectators: "Has this fact ever struck you—no bridle, no bit, no collar and chain, just the human voice, and in India sharp curved spike or goad?"[54] The "India spike," also known as the elephant hook, was in fact the hook upon which the icon of the genial circus elephant hung. Along with tools like pitchforks, wooden stakes, and other pointed objects, the elephant hook was the preeminent mechanism of motivation and persuasion in elephant management. Many animal trainers held that the elephant's ostensibly insensitive "thick hide" and massive size made these tools necessary for modifying behavior by creating sensations elephants could not ignore. The goad was a tool no trainer, keeper, or circus manager would allow his company to work without, for many believed it was the most effective way to stop an elephant in his or her tracks or motivate him or her with precision. To outsiders, the hook seemed to give trainers a baffling power over these enormous beings and, in truth, masked the fact that minimally trained and inexperienced elephant handlers could accomplish elephant compliance in a hurry no other way.

Adopted from Asian practice in Ceylon (Sri Lanka) and elsewhere, the goad, or elephant hook, was also known (in Anglicized terms) as a "hendoo" or "hawkus," an instrument mahouts had used for centuries. Trainers in America also used the pointed ends of the tool to pull or push on an elephant's trunk, legs, feet, or the especially sensitive spots behind the ears in order to create pressure points of varying intensity. That pressure—or even just the sight of the hook held a particular way—could persuade an elephant to stop or move as appropriate to avoid the pain caused by the hook, much as a human mother might grab a disobedient child by the arm to stop him acting in an undesired way. Early accounts of

**116**        THE ART OF TRAINING ANIMALS.

operation are considerable, and it often requires the sagacious
interference of the tame elephants to control the refractory wild
ones. It soon, however, becomes
practicable to leave the latter
alone, only taking them to and
from the stall by the aid of a

Modern hendoo.

decoy. This step lasts, under ordinary treatment, for about
three weeks, when an elephant may be taken alone with his
legs hobbled, and a man walking backward in front with the
point of the hendoo always presented to the elephant's head,
and a keeper with an iron crook at each ear. On getting into
the water, the fear of being pricked on his tender back induces
him to lie down immediately on the crook being held over him

Elephant hook. *Haney's Art of Training Animals* (1869).

mahout work in Asia appeared in American periodicals and described the tool as a
"sharp-pointed hook, with which [an elephant driver] turns the creature by the head
the way he would have him go," steering the elephant like a ship's helm.[55]

*Haney's Art of Training* also explained how mahouts used goads to force recently captured elephants into a given position by convincing them that comply-
ing with human direction prevented intense hook pressure. "The dread of man's
power thus established . . . by threatened attacks he may be induced to move
in any desired direction." Thus, for example, a hook permitted mahouts to persuade an elephant to bathe in a body of water. Using two previously trained elephants as "muscle," a mahout began

> by lengthening the neck rope [on an untrained elephant], and drawing the feet to-
> gether as close as possible, [until] the process of laying him down in the water is
> finally accomplished by the keepers pressing the sharp points of their hendoos over
> the backbone. . . . This step lasts . . . for about three weeks, when an elephant may
> be taken alone with his legs hobbled, and a man walking backward in front with the
> point of the hendoo always presented to the elephant's head, and a keeper with an
> iron crook at each ear. On getting into the water, the fear of being pricked on his
> tender back induces him to lie down immediately on the crook being held over him
> in terrorem.[56]

There are many delicate points on an elephant's body, and elephants learned
quickly to avoid potentially painful pressure in those places, making the hook a
versatile and powerful tool.

The goad largely solved the problem of motivating elephants, and if used with some ingenuity it could control and modify elephant behavior in several ways. For instance, many elephants learned to stand on pedestals just larger than the area of their feet bottoms when a trainer repeatedly hooked the bottom edge of their feet until they got the impression that it was painful *not* to step onto the pedestal once the command had been given and that more hook pressure and discomfort would come from stepping off or moving one leg off the pedestal before all the others were on it. Elephant performers learned that the exception to those conditions came when the human director gave the signal to move all feet off the pedestal.[57] All this was made easier because, at large in Asia, elephants had proven themselves "nimble climbers," with far more balance than many people suspected.[58] So elephants learned to sit on a pedestal or oversized chair—say, for the table manners routine—by getting hook pressure on the bottoms of the feet and the top of the head at once, learning that if they bent up and back-ward, that sharp sensation would end.

George Conklin was one of dozens of men who used the hook to modify ele-phant behavior. Born in Cincinnati in 1845 to German and French immigrants, Conklin worked his way into the circus business, first as an 1860s advance man. After some years he learned to work elephants from Stuart Craven, a "very strong and very determined" trainer who had worked on contract with circuses owned by famed impresarios Pogey O'Brien, Cooper and Bailey, and Adam Forepaugh.[59] Later he was the director of New York's Central Park Menagerie and a representa-tive for the Hagenbeck Company.[60] Conklin was one of many who learned to use a hook as a way to motivate elephants into both understanding a direction and obeying the relatively tiny human who gave it so as to produce animal acts that kind words, food rewards, or a water bucket were not enough to produce. He knew that elephants could be taught a complex task by being taught a series of simple steps that would in time be integrated and that animals could be trained through both positive and negative reinforcement to remember the steps. There-after, one could reduce the amount of direction, that is, the number of signals a presenter or trainer gave to an animal to produce the same series of acts.

Conklin described how he had learned from Craven that elephant response and human culture could be combined to produce aesthetically pleasing perfor-mances for audiences. Working with an elephant known as Queen Anne while on the staff of O'Brien's Circus, Conklin modified and presented her so that she appeared to march to music in an act called the "Spanish trot." Conklin played the part of parade marshal while costumed in "a black-velvet suit, with gold stripes running down the sides of the pants." In rehearsal, Conklin and two as-

ELEPHANT TRICKS.

Here are some tricks to which the bulky but intelligent elephant has been trained. No. 1 is shamming death ; Nos. 2, 3, 4 are busy with creditable feats of equilibrium. No. 5 goes through one of the most difficult performances attempted, namely, to walk on a rolling cylinder. No. 6 plays on a barrel organ, to the tune of which No. 7 dances on a tub. The two Nos. 8 indulge in a quiet see-saw, whilst the two Nos. 9 are busy replenishing the "inner elephant." No. 10 is no lover of the bottle, for he walks on ten of these placed in a row, a somewhat fragile pathway for an elephant.

"Elephant Tricks," made possible by training with an elephant hook. E. A. Tilley, engraver, after Camille Gilbert, for *Picture Magazine*, 1894. Picture Collection, The New York Public Library, Astor, Lenox, and Tilden Foundations.

sistants used multiple hooks to condition Queen Anne to move her feet to the tapping of a stick:

> I placed a man on either side of her with an elephant hook, while I stood in front with my whip. As I moved the whip they would hook into first one and then the other of her legs, and lift them up. In a very short time she learned what we wanted and the moving of the whip or a stick was all that was necessary to make her lift her feet. After that had been accomplished it was simply a matter of practice to train her

so that when I walked beside her and beat time with my stick she would follow its motions with her feet, lifting them high in the air.

The process demonstrated a kind of systems theory within the technologies of behavior the circuses used by which multiple training tools were more powerful in concert than in series, and the combination of multiple handlers, hooks, whip, music, and costumed trainer produced an apparently dancing elephant. Conklin confided later that the trainer and elephant, not the circus band, really drove these types of presentations because the elephant was probably indifferent to the music: "As we came into the ring this way the band would strike up 'Coming Through the Rye,' and to the spectators it seemed as if the elephant was really dancing to the music, but as a matter of fact the music was being very carefully played to her dancing."[61]

The hook facilitated such advance preparation of elephants that showmen could present audiences with the mystery of the gregarious circus elephant made manifest by a huge animal, not visibly motivated with straps, collars, bits, or other tools familiar to those who knew horses, oxen, or dogs, apparently performing of her own free will.[62] From a management point of view, elephant learning and memory skills introduced this efficiency into show production as the elephant performed the mental labor of remembering the steps in a routine. In this manner, trainers in effect passed back to elephants the crucial supporting labor of remembering training and rehearsal encounters with people, food, and props, so that, once in the ring, no hook, pitchfork, or whip was needed to elicit the desired actions from an elephant. This efficiency often inspired elephants to confound the trainer or presenter by hurrying through a routine or skipping steps altogether. Turning an evolutionary talent for efficiency to their own perceived interests, elephants frequently cut corners to get more quickly to the food, rest, fellow elephants, or quiet surroundings that habitually followed a given routine of movements—a habit that a trainer could attempt to halt by assertive interventions with a hook.

In the United States, many showmen made their own goads out of a wooden handle and any sharp piece of metal they could craft. As the nation's elephants aged and grew larger over the century, elephant trainers would transform the small hand-held elephant goad into the Gilded Age hook, which had dimensions and heft greater than that of a baseball bat. Many extant illustrations or photographs of circus elephants staged outside a circus ring show a man posed nearby, hook in hand. The tool was a constant presence anytime an elephant was unchained, or if a number of people needed to get within trunk's reach of a chained elephant. Behind the scenes a pitchfork or wooden stake could be used in the same way as a hook, and was reputedly used with frequency by keepers and barn men, who always had them at hand for managing the hay supply or staking down animals and tents.

Elephant men further struggled against the sheer physicality of elephants in pushing narrative innovation toward more acrobatic stunts. Some movements and poses were impossible to communicate to an elephant with rewards or even ample use of the hook. Dog trainers realized the luxury they had in this respect. In training for a routine, if they found a paw in the wrong spot or found a dog to be slow about getting up on her hind legs, those trainers could physically pick up and adjust the paw, or lift the whole dog onto the pedestal and give the reinforcing command or signal. One could not so easily or subtly adjust an elephant, and confusing or random (to the elephant) use of the hook could create panic or an elephant who shut down entirely if the desired movement was too disagreeable or bizarre to the trainee. "It will be readily seen . . . that it is a problem not entirely free from perplexities to discover ways to make an elephant understand what you are talking about when, for instance, you ask him to stand on his head," George Conklin observed.[63]

Consequently, special equipment had to be improvised for the tricky work of positioning and bracing elephants for abstract posing and acrobatic acts. Thus Conklin explained the Gilded Age practices he had used to train elephants for the headstand:

> Stand him facing a high, strong brick wall, with his front feet securely fastened to a couple of stakes driven in the ground. A heavy rope sling was put around his hind quarters and from this rope was run up to and over a pulley high above him on the wall, then down through a snatch block near the ground, and the end fastened to a harness on another elephant. When all was ready I would take my place by him, strike him on the flank, and say, "Stand on your head." At the same time an assistant would start up the other elephant and draw the pupil's hind quarters up until he stood squarely on his head. The wall kept him from going over forward. After a moment or two I would slack off on the rope and let him settle back on to his feet. Then I would give him a carrot, or something of the kind. I did this two or three times every morning and afternoon. It was not long before it was possible to do away with the rigging, and at the word of command he would put his head down and throw his hind quarters into the air.

This mode of elephant behavior modification became easier when one had adult elephants to use to pull the ropes, and elephants often became laborers in such

BARNUM'S SHOW IN WINTER-QUARTERS.—From Sketches by James C. Beard.—[See Page 106.]

Elephant training with ropes, pulleys, hook, and other elephants. *Harper's Weekly* (1882).

training-by-restraint systems. These techniques facilitated preparing elephants for incredible tasks that trainers would otherwise be unable to produce—tasks that were later shown to produce arthritis, hernias, and other strain injuries in elephants.[64]

Mogul's old act of stepping carefully over his prostrate handler still communicated to later audiences the idea that elephants understood and savored show business, trusting or even loving their trainers. Other elephants performed the act for post–Civil War audiences, which might or might not have read about animal training in the popular press. The general effect produced by those elephants and all the others presented cage free among circus crowds was as compelling as ever because they enacted a spectacle of sagacious sensibility over brute power.

In 1882 one reviewer for a paper in Minnesota marveled at the mystery of the genial circus elephant thusly produced:

> The more I think about elephants, the more wonderful they seem to be. The great, clumsy creatures are so very knowing, so very loving, and so like human beings in many of their qualities. They know their power well, and they also know just when they must not use it. . . . Keepers and trainers of elephants often lie down on the ground and let the huge fellows step right over them; and that they feel perfectly safe in doing so, because they know the elephants will pick their way carefully over the prostrate forms, never so much as touching them, still less treading on them. Yet the mighty creatures can brush a man out of existence as easily as a man can brush a fly away.[65]

What such circus fans did not know was that in order to create this illusion of elephant consideration for humans, elephants were first conditioned with rewards and a hook to step over a human dummy. Only when a given elephant proved dependable at this task would a trainer lay himself down, and even then he did so in rehearsal with keepers bearing pitchforks standing immediately nearby. George Conklin claimed to eschew the pitchfork himself but revealed that he never lay down for this bit without his goad. Working with Queen Anne, he said she "would walk over me both ways, straddle me lengthwise, and end by kneeling down over me crosswise until she almost or quite touched me. . . . I taught her to do the act very slowly, to increase the impression that it was very difficult. It was not as dangerous as it seemed, for I had my hook in my hands all the time, and if Queen Anne had settled down a little too heavy a touch from that would have raised her very quickly."[66]

Trainers spoke publicly and in ambivalent ways about one additional chronic problem with performing elephants. One circus man remembered that the mythology that elephants "never forget" was fine for the audience, but that in fact many elephants often jumbled or edited out parts of routines they had long known: "They need constant rehearsal lest they forget their routines, but substitute *fanciful ones of their own*"[67](emphasis added). Elephants studied in the last two decades have been shown to have excellent "long-term, extensive spatial-temporal and social memory." Those skills are useful to elephants at large in Africa and Asia for finding other elephants over long distances and for remembering where good food and water can be found many dozens of miles away in times of drought or other environmental change. In captivity, those abilities permitted elephants to remember show routines as a series of movements in space, while remembering the cause and effect of command, pain, and reward.[68] Those skills did not necessarily endear an elephant to humans in all situations or rid him of the need to practice species-typical behaviors. "Even a well-trained elephant will sometimes stop his tricks abruptly and calmly walk out of the arena," trainer Frank Bostock would later explain, as "he simply wants to go back to his house and eat peanuts and biscuits, as he was doing when interrupted for the performance."[69] That is, training and conditioning could teach an elephant to learn a series of movements or consequences but not necessarily motivate him or her to perform indefinitely.

Many circus people seem to have perceived early juvenile elephants as essentially domesticated because of their obliging use of their mental and physical abilities under the direction of a trainer. Yet, by age four or five they would begin to behave like the wild animals they were. With little inherent respect for people and no inborn desire to make use of humans (such as dogs and even horses might possess), any early obedience they exhibited was the product of (arguably difficult) elephantine mental and emotional labor. In all probability, the elephant who appeared to forget his mark, position, or other detail of a performance routine thereafter was discouraged by a new element in the situation, such as a new trainer, or was simply transitioning into a more mature and independent stage of life. Although the pressman's idea circulated that apparent elephantine forgetfulness, reticence, or innovation of movements in a routine demonstrated that the performance "was no fraud," elephant men knew that they needed to limit those inconsistencies in order to keep their jobs.[70]

As early as the 1860s, trainers began letting on to selected outsiders about this aspect of circus training. Even circus press agents, the in-house publicists who wrote complete "news" pieces to be distributed as free content to the newspapers, admitted that the living elephants that produced the circus ideal of genial performer could become lethargic or difficult after some years of training. However, in an industry in which so much could go wrong, showmen demanded that *at least* the animals be "made to respond instantly to the word or signal of command," a *New York Times* reviewer noted. "When the tricks are first taught the elephants like to do them, and will often go through a performance of their own volition. But after the novelty wears off they regard them as work, and are inclined to shirk. When they rebel they must be soundly flogged, and after that they will not refuse to perform their parts."[71] Turning this bit of bad news into a marketing opportunity celebrating the power of the animal trainer as manly authority figure, that reviewer went on to explain (in deadpan tone of voice) that with unresponsive elephants "the keen prod touches up the laggards" and "a sharp jab of a metal point calls him to his duty."[72]

With respect to early republican and antebellum juvenile imports, elephants who ate food from forks and carried howdahs bearing paying customers may largely have been conditioned with positive rewards of praise and food. In those stages, elephants still choreographed their interactions with visitors to a considerable degree. Some may have considered that elephant action as a style of cooperation—as *Haney's* manual implied—since elephants structured the training process by following their own desires. Indeed, over the last few decades scholars and some trainers (of dogs and horses, primarily) have begun to describe such situations as a kind of "social space created by the language shared by two or more creatures" in a training context, as "human-nonhuman, and human-animal cultures and assemblages," or a "contact zone" of "joint, cross-species invention that is simultaneously work and play" in which animals may develop "positive emotional relationships" with particular people.[73] Some observers sympathetic to the circus trade have even gone so far as to call elephant training a kind of "learning and occupational therapy" that captives enjoy (over their ostensibly impoverished wild conspecifics) as "colleagues" of human trainers and performers.[74]

However, in no way do we diminish the agency of nineteenth-century American elephants by understanding that, once menageries added hooks, ropes, and teams of men and other trained elephants to their training systems, the power balance shifted decisively in favor of the humans in the equation. In fact, this was the elephant trainers' very intention, for in that era they could produce the performing elephants their employers and audiences expected in no other way. Whether extension of the patient and limited training from the bottle and cork trick to complex acrobatic acts would have been possible or not, when we consider the limited budgets of time, patience, and manpower the circus trade imposed on mobile animal shows, it is difficult to imagine these companies being able systematically to develop truly egalitarian collaborative modes of elephant training that adult elephants would not eventually reject as they aged. More crucially, the manly culture of mid-nineteenth-century wild-animal trainers shows no evidence of any widespread desire to produce such collaborations between man and elephant. Such ideas appear in the primary sources documenting elephant training only in cases where egalitarian tactics were more efficient in producing a given elephant movement.

In fact, the bedrock of elephant obedience and training, especially with older subjects who began to see performing as "work," was the practice of teaching each one to "mind" his or her handlers. Americans understood the concept of animals who "mind" a human mostly with respect to horses and dogs. The process was often known as "breaking" a horse for the saddle, and for dogs, it involved impressing on canines the authority of their human director. In both cases "minding" symbolized the moment in which an animal accepted human direction by deciding not to challenge with a resisting bite, struggle to escape, or inactivity. A classic account explained how dog acts were grounded on this practice: "[The trainer] takes a dog of, say, a year old, and first teaches him to 'mind'—that is, to come and go at the word of command. This takes from six to ten lessons, but is the necessary foundation, without which all subsequent education is valueless. After he is taught to understand who his master is, and that he must obey him, the work begins."[75]

The idea of humans asserting themselves as "boss" in relationship with working animals was not simply a function of human vanity but a cultural and material necessity for those who survived off the labor or bodies of animals like horses and mules. Thus, both European and native-born elephant trainers in the United States began using the ancient South Asian practice of breaking elephants sometime in the antebellum years, perhaps as early as Horatio's time. Breaking required tying an elephant down and beating him or her with heavy poles or stakes until he or she ceased struggling and vocalized a sound of "resignation," a process that was exhausting for all involved and could last up to thirty minutes. Thereafter, elephant control came from enforcing a balance between reward (to communicate to an elephant, say, "finishing the commands leads to eating hay outside") and pain aversion (such as using an elephant hook to put pressure on an elephant's skin so that he or she understood something like "it hurts to step off the pedestal without a command to do so"), with occasional rebreaking, needed when elephants were noncompliant or assaulted circus property or people.

Elephants still structured the training by their ability to remember orders of steps and the consequences of obeying or not obeying and their ability to suppress their own physical power out of respect for a man bearing a sharp object in one hand and a carrot in the other. Their handlers knew, however, that they limited the range of behaviors open to the elephants and that the elephants had to understand this, whether they resented it or not, in order to survive as long as they did. To be sure, the act of breaking an elephant was a radical human intervention, one that asserted the anthropocentric imbalance of power and attempted to shape elephant experience to the circuses' business advantage.

Accordingly, trainers had begun conceding that in many cases animals performed in the ring regardless of their preference to do so because they had been taught "to mind" particular humans. Influential British animal trainer Frank C.

Bostock trained various species, including elephants, for shows on both sides of the Atlantic. He put bluntly his understanding of why outright breaking and goads and similar tools (like whips and clubs, often employed with big cats) worked to modify an animal's behavior: "[A wild animal] only bows to man's will because man through the exercise of his intelligence, he takes advantage of the animal's ignorance," he said. So would *Haney's* and Palmer's manuals, George Conklin, and many trainers explain that the hook facilitated the trainer's work of taking advantage of elephants' simultaneous understanding of cause and effect *and* their apparent inability to perceive human limitations, thereby shaping that elephantine parallel reality that made performance possible. Although the appearance of unwavering human control was central to the presentations of big cats, in the show ring trainers and presenters seldom made this aspect of elephant management obvious.

Here elephants had agency but no real power, as they were quite uncomprehending of the human business and performance cultures that required their captivity and of the limits to the technical means by which their acquiescence was engendered. Thus, they had no effective way to challenge circus life broadly, and many circus people certainly believed they would if they had the means. A person could in effect fool these powerful animals into believing they had no choice but to obey people they could crush underfoot in an instant. Indeed, in the making of a circus elephant's "tricks," it was the elephant who was tricked into participating. Bostock continued, "Every animal trainer thoroughly understands what the public does not know—that the trained animal is a product of science [systematic training]; but the tamed animal is a chimera of the optimistic imagination, a forecast of the millennium."[76]

At the same time, seeking to continue on with the circus icon of the genial performing elephant actor, some showmen began responding to growing public concern that elephant training was "cruel" by allowing that the goad and elephant-breaking procedure were painful and intimidating but necessary for training elephants. One Barnum & Bailey press release, for instance, qualified that "elephants will not submit to abuse. It is necessary to treat them well, but at the same time it is imperative that they should understand that they must mind their master."[77] That is, industry arguments began to assert that actions deemed "abuse" if applied to a dog or horse were not so when applied to an elephant. Consider how Palmer's *Secrets and Mysteries of Educating Animals* explained the most honorable way "To Teach a Horse to Kiss You 'Good Night' While Lying Down." The manual endorsed a prevalent parental ideal of teaching children kindness to animals, explaining:

Here is an act that any child can teach to a horse in a short time, and without cruelty or torture, ropes, sharpened prods, or any other device, at command, and the equine does it readily and willingly. Some trainers say put a rope around the lower jaw, have a horse lie down, stand over him, and pull or jerk his head up. Others say run a strap around his head below the eyes, with a tack in the lower side; pull; it hurts, a horse lifts up his head, and, in time, will raise it up as soon as he lies down, remembering the tack. This is a shame; it is barbarous and inhuman. Let me tell you how I work him, and if you cannot make him do it, in less time, without pain, come and get the pick from my stables.[78]

The methods Palmer's manual explicitly discouraged for horses and dogs were precisely the methods described for use with elephants in the same volume and myriad press releases, puffing interviews, and industry memoirs.

Turn-of-the-century animal trainer Lucia Zora, one of the first women to achieve fame working with exotic animals, clarified this show business contradiction. She explained the industry knowledge about getting animals to mind as a process that was crucial to the long-term safety of circus workers *and* elephants, brutal as it might seem to some. As every undesirable behavior the trainer or keepers allowed would convince elephants that humans were ultimately powerless, trainers who were lenient with elephants saddled their charges with "the terrific handicap of being 'spoiled.'" That leniency produced elephants who believed they could "do exactly as they pleased, and that is the worst thing in the world for a performing elephant," she said, because he or she would act out in future only to be punished or killed when found unmanageable.[79]

Still, there was no single unified public message about elephant training, and some trainers might tell the press or leak out to book writers that they hooked elephants forcefully, and sometimes at random, while still designing ring performances to make the hook and the training experiences invisible to the audience. An elephant's memory made that behavior-modification system of trainer, hook, and elephant productive. The goad, like the whip, could function with the trainer's body language to direct animal action in the ring without actually being employed to cause pain during a public performance. Therefore, many showmen did not see the hook as a cruel device—at least not in my hands! they usually insisted. Rather, in public settings especially, people sought to use the goad as a kind of "pointer" for an elephant, a way to "get his attention," as the euphemism goes.[80] Zora would talk about this process as producing animal performers who were "trained to the whip" and need only be disciplined occasionally with a training tool and over a distance in the ring, "as reprimand" during a performance with an audience of children looking on.[81] Indeed, in the ring many trainers preferred to present elephants with a fancy horsewhip because the hook, and certainly pitchforks and wooden stakes, did not look glamorous or graceful to spectators.

The training context for people who worked with acrobatic dogs or show horses was more complicated because the audience was more literate about canine and equine behaviors and the probable mental experiences they indicated. Early proponents of state societies for the prevention of cruelty to animals had spoken out about how the entertainment value of an animal performance was destroyed by an apparently "cruel" presenter. For instance, some critiqued the "very poor Circus" of Forepaugh's Menagerie Company, among whose ring stock one informant saw a horse that, "with its high spirit all broken down by the whip, and shivering and trembling over the difficult feats required of it, so far from giving pleasure, almost makes a sympathetic observer sick." Reiterating the main argument made in the recently available works of Charles Darwin regarding the common evolutionary heritage of all living beings, the piece continued, "Cruelty to an animal touches every humane man and woman precisely as cruelty to a human being does—the only difference being one of degree and not of kind."[82]

Consequently, many horse and dog trainers believed that they produced the most pleasing performances for the spectator, in which their nonhuman performers appeared bright and animated, by avoiding the use of force or pain aversion whenever possible. Reflecting that public perception, George Palmer warned his readers, "First, never lose your temper. No matter how hard your subject may be, how discouraging it may be to you, if the animal should step on you, throw you down or push you over, don't fly into a passion and use the whip."[83] *Haney's Art of Training Animals* similarly advised, "If the pupil is in constant fear of blows his attention will be diverted from the lesson, he will dread making any attempt to obey for fear of failure, and he will have a sneaking look which will detract materially from the appearance of his performance."[84] Later vaudevillians agreed about the aesthetic importance of tempering the mechanics of animal training so that it could be concealed from the audience. Renowned for their work with dogs and other small animals, whose body language the audience could examine closely in small theaters, one performer said that trainees should be "for the most part treated kindly, if only as a matter of business. A beaten dog cringes; can't perform, or if he does shows his lickings in perfunctory stunts."[85]

Elephants, on the other hand, did not show their attitude as a dog or horse might. They did not cry or cringe, shiver or yelp, flatten their ears against the top of the head, or put the tail between the legs, nor did they rear up and whinny in a tone that any person would have heard as a sound of distress. In fact, the obvious physical displays of horse discomfort contributed to a mid-century opinion that circuses of trick horse riding were manifestly cruel, whereas menageries of caged exotic and wild animals were educational (even if the animals there often appeared lethargic or withdrawn) because audiences were relatively unfamiliar with the behavior of exotic species.[86] This opinion grew from the same roots as the era's humane societies, which sought to draw public attention to the visible suffering of workhorses in American cities—mere background noise to many busy people trained by experience to see right through the nation's ubiquitous living energy source.

Some worried about human circus performers for the same reasons, a concern that similarly helped divert attention from seemingly impassive elephant performers. Well-known mid-century stage actor Olive Logan more than once critiqued "how children are driven to their tasks in circuses." She and others wrote exposés of how show business parents forced their children to complete dangerous trick-riding and acrobatic acts and beat them when they refused or made errors. Logan said one Cincinnati audience in the late 1860s had interjected during such a performance shouts of "Shame! Shame!" and "That'll Do!" when a young girl acrobat fell several times from a horse, became visibly frightened, yet was being pushed by her director to continue.[87]

In another notorious example, Leo Lawrence, a teenaged boy indentured to a trick horse rider named Samuel Watson employed on a Forepaugh Circus unit, charged Adam Forepaugh Jr., son of the famed circus impresario, with having assaulted him when he fell from an unbroken horse a number of times while training for a bareback routine. The press noted the boy was seriously injured when he appeared in a Philadelphia court to "make a complaint against his tormentors." Lawrence reported that Forepaugh had beaten him with tools normally reserved for the company's animals: a long whip and "the elephant prod [which] was plunged into his body." The "society to protect children from cruelty" was at that moment attempting to gain custody of the young man and complained, "Our officers have frequently endeavored to gain admittance to this training school [run by the Forepaugh family], but they have always been refused."[88] Certainly, as they had learned with Horatio, circus people were cognizant of the need to control which back-of-house realities (whether sanctioned or not) came into public view in order to protect the belief that human and nonhuman circus performers were happy to serve American audiences.

When elephants appeared in the circus ring their faces were inexpressive and their body language seemed the same as ever. The memory cues and actual mechanism of persuasion the trainers had used, whether reward, repetition, or pain aversion, were invisible as far as many circus fans could tell. The elephant's parallel reality supported elephant shows and a belief that elephants had an innate sense of theatricality, an old notion that had been circulating since at least the early modern era in Europe.[89] The menageries of antebellum America helped the idea persist (as it does even today) by constructing scripts and routines featuring compliant elephants by which showmen could more persuasively tell audiences elephants enjoyed show business. When signaled, for example, elephants curled the trunk at the forehead to expose the mouth in apparent salute or smile to the crowd. (More recent wisdom holds that this pose can cause mucus to collect in the trunk if it is held up too long, impairing an elephant's ability to breathe; nevertheless, the pose is still used.)[90]

Notwithstanding the professional necessity to make animals indifferent to the audience, animal performers' body language could be evocative of animals energized because happy to work in the ring. One turn-of-the-century account warned audiences against this impression, insisting that it was in fact "an agony of excitement and terror, which you, my good Sir or kind Madam, mistake for joy and friskiness; look carefully and you will see [the trainer] hit or kick the nearest animal ever so slyly, you will see him raise his whip to indicate what is to happen when the performance is over."[91] Frank Bostock tried tempering this perception by insisting that his subjects showed "a sort of intoxication" because "affected by the attitude of an audience, that they are stimulated by the applause of an enthusiastic house, and perform indifferently before a cold audience."[92] Yet, in the next breath Bostock and other trainers would also say that they suspected animal performers responded to the context with a heightened performance out of nervousness and a desire to get the act over with. Many knew this because they could see an animal read the trainer's body language and the clapping noise (called "applause" by humans) to mean they would soon get a swift exit from the ring, rather than a snap with the whip.[93] The genius in this for the circuses was that members of the public need not witness an elephant being hooked or hoisted with ropes, deprived of or rewarded with food, water, or sleep during the actual performance; this had been done earlier in preparation for shows. Elephants did the labor of remembering these experiences under training and rehearsal, then acting accordingly in performance. The agitation they often inadvertently showed in the ring was explained to viewers by some circus people as a sign of an elephant's enjoyment of performing and of his innate theatricality.

Animal training was a matter of human management and entertainment cultures confronting elephant reality. When productive, such training created a par-

Elephants in rehearsal raising their trunks to give a circus elephant "salute" in preparation for the crowd that will later fill the seats. "Elephant Pyramid, Olympia." Princeton University Library.

allel reality for an elephant over which a narrative could be laid. The mechanics of that training were thereafter obscured by an elephant's ability to remember previous experience, positive and negative, while in the ring surrounded by a noisy crowd, a blaring band, and other animals. By these means, circus audiences could depend upon seeing elephants who "saluted" in greeting to the crowd, stood on small raised platforms, performed headstands, leg lifts, rhythmic marching or "waltzing"; who walked a plank, a line of oversized bottles or large-diameter hemp "tight rope"; who performed "see-sawing"; who untied scarves and ropes, grabbed and manipulated bells, hats, fans, dishes, musical instruments, or horse bridles; who held down the taut cable supporting a spinning human acrobat, served as a performance platform for dogs, ponies, goats, humans, and other elephants, and walked in parades, ran in races, and pulled carts.

The menagerie's goal of producing a performing elephant actor by way of living elephants was not an inevitable part of elephant history but a human choice, which demanded much of elephants. Still, to many outsiders circus captivity seemed to be a very kind home for elephants, which were not visibly whipped or driven like the cart horse that might have brought one to see the circus. Even though exceedingly strong and fast, most elephants most of the time performed more or less on cue, and if disinclined to perform, once unchained did not often run out of the barn or circus tent. Most spectators detected no sign of the elephant's parallel reality because of a common inability to accurately read elephant body language or other modes of communication. Thus could circus men present an elephant performer crafted seemingly without artifice and declare to the public, "Come to our show to see for yourself how well we treat our animals and how happy they are!"

# CHAPTER FOUR

# Punishing Bull Elephants

Before elephants arrived in the United States, Americans were in a position to envisage them as mysterious beings able to be or do just about anything. Elephants were essentially a "rhetorical animal," an imaginary creature that served as a carrier of cultural ideals and moral values.[1] Once elephants began touring the United States, their physicality and abilities opened the door to some new ideas, and people came to interpret menagerie elephants variously as sagacious wonder, exotic visitor, and genial performer. Living elephants also tested some previously obscure formulations that had quietly resided in the background of American culture. An older rhetorical animal, one that even predated the birth of the United States, became newly relevant over the century as a way to think about circus elephant as business innovation. This being was the subdued elephant. Some interpreted him as a fooled elephant, a tragic creature who knows not his own strength because tricked by people who claim benevolent motives and limitless power but who really possess neither.[2]

Early on, this rhetorical animal—the subdued and fooled elephant—had appeared in a *Niles Weekly Register* editorial entitled "How to Tame an Elephant!" which carried a critique of the American market system. The *Register* was the antebellum paper of record, many believed, a balanced publication that offered news of politics and trade while pioneering human-interest stories documenting American culture and values.[3] During the financial panic of 1819 *Register* publisher Hezekiah Niles lamented the growing national crisis of inflation and the many people holding worthless notes issued by insolvent banks. With federal support, that "paper system" had greased investment and the availability of credit but had also facilitated speculation by men who produced no concrete product but merely sought to extract value from the intricacies of capitalism in the nation. At the time, Niles and others shared a sense that common men had sought personal fortune in a system that was designed to capitalize on their labor and leave them penniless in the end. Worse yet, Americans had gone into this scheme willingly and became trapped in an economy that, we now know, has suffered volatile boom-and-bust cycles ever since.[4]

To comment on this mess Niles employed the metaphor of the fooled elephant by telling a tale about a wild bull elephant tempted into a kraal by tamed cow elephants. He is charmed by these females at first, but then finds that humans appear and bind him with ropes, beating him when he resists. Later, the people reappear, but this time they give him food and loosen the ropes so that he comes to believe they are trusted helpers, although all along they are the architects of his captivity. "The people of the United States (but especially those of the west, in present circumstances), may be considered as the elephant," Niles explained of his metaphor. "The paper system as the females who deceive him; their obligations to the banks as the cords that fasten him; speculators as the person who beats him; and the bank of the United States, through the agency of the government deposits, &c. as the master spirit, demanding 'unconditional submission.' " The public did not know its own strength, Niles asserted. Americans labored for the benefit of exploitative others whose commercial interests infected the common sense of the nation, limited the terms of debate, and misled the virtuous man into "kiss[ing] the hand that robbed him of his freedom."[5]

A generation later, the metaphor of the subdued elephant was still current in elite parlance. It would become far more contentious and broadly recognized as actual living elephants intervened into American culture with their bodies and movements. With a constitutional crisis developing and the economy booming from the California gold finds, circus people of the 1840s and 1850s would certainly operate in a very different national context than Hezekiah Niles, although they similarly contended with a changeable economy. To them, the subdued elephant was more than a metaphor; he was an everyday source of income and a safety hazard. And, to them, he was certainly no fool. He was an unruly brute, they said, who required frequent punishment, without which he would become completely uncontrollable and destroy what showmen built. Like the so-called rogue elephants of Asia and Africa, who would be produced through human-elephant conflict over resources and harassment of elephant herds by colonial-era hunters later that century, the American unruly bull was a product of the symbiosis between human-elephant behavior, both an imagined and a very real animal.[6]

The animal shows began down the difficult path that saw the metaphor of the fooled elephant elaborated as they cultivated audience demand for more and bigger elephants to support the menageries' "Mammoth" marketing aesthetic. Companies presented those early bulls as novel because male, interpreting them alternately as elephants of war or sagacious actors. Both personae could be embellished by a set of tusks. By the 1840s, the first of those males was nearing sexual maturity, adulthood, and periodic experiences of musth, wherein bull elephants became aggressive and can be destructive. It was at this moment that the menageries and circuses first came to see the elephant, so to speak. Having invested more than fifty years developing audience demand for elephants, the reality of keeping mature males dangerously taxed the staff enlisted to execute the fantasy of the elephantine entertainer.

Live animal shows and their paid advertising still cultivated citizens' sentimentalization of elephants as sagacious performers, of course. In print, however, the circuses also began to describe their elephants as difficult and stubborn workers, advocating strict discipline for them. Elephant trainers had begun routinely inviting journalists and circus publicists to see how their work got done, asking them to portray it for a broad circus audience of readers and ticket buyers. By 1881, the raw, behind-the-scenes accounts that some trainers had for several decades contributed to the diverse entertainment package the circuses offered was the subject of a *Harper's Weekly* commentary that exemplified how the character of the subdued elephant had turned from simpleton to impudent villain:

> The training of elephants is not always an easy task, and when the animal is really refractory, the keepers have "a heavy hand" with their charge. The most usual method of persuasion employed, when coaxing and feeding have failed, is, we believe, to "job them with pitchfork till blood is freely drawn"; at least this was the explanation given by a trainer of repute of his own practice in his gentle art. . . . Under this or similar treatment an elephant can be made to exhibit the greatest docility in the arena, and will show a touching devotion to his keeper, which cannot fail to render the circus at which he is employed as moral exhibition.[7]

Here *Harper's* pointed out how the elephant's parallel reality and memory helped circus people while in the ring to obscure the means of elephant act production. The commentary also revealed that, although the menageries were increasing in size, technological complexity, and rational management, many trainers within these organizations were still steeped in highly subjective theories of manhood and interspecific dominance.

Circus owners, upper-level managers, advance men, and accountants were, by contrast, notably obsessed with efficiency. They labored to create management systems to monitor the speed of tent take-downs, the length of shows, the most profitable touring routes, and ways of avoiding taxes on depreciating assets like rolling stock. Yet, when it came to controlling the *messages* circus people made public, as a group circus people seemed less disciplined. They appear to have worked out as they went along how to cope with that old problem first presented by Horatio, namely, how to suppress, manage, or even monetize reports of elephant disobedience without sabotaging the industry's happy circus elephant mascot.

Many Americans appear to have been fascinated or appalled by the strange juxtapositions inherent in many circus companies' aggregate marketing profiles. America's most ubiquitous mass entertainment claimed to offer whimsical family-friendly amusement while seeking publicity for elephant men, many of whom boasted of their violent lack of patience with the animals in their charge. To be sure, many people noted that the postwar circus was not the naïve menagerie on wagons with which they had grown up. Imports of males continued nonetheless, and segments of the public reached a turning point of sorts, wherein, to the animal trainer, the subdued elephant was a vicious brute, while to the circuses' critics, he was, like Hezekiah Niles's tamed elephant, a tragic figure who spoke of animal suffering for the crude greed of men. Middle- and upper-class whites with private concerns over the experiences of captive elephants would begin publicly to note the two faces of the industry, their concerns manifested in dryly critical news items and commentaries like the *Harper's* piece. Indeed, as the century progressed, both elephant trainers and their public critics would invest their identities in rival interpretations of the subdued elephant in ambivalent and disparate ways.

Here was another situation in which elephant experience mattered. Bull elephants responded to the conditions of their captivity in the United States and radically altered the work cultures of show business. Accounts of trouble began turning up in the press and oral tradition just as the first male elephants in the nation were reaching early adulthood. The antebellum record of this conflict shows the menageries either in denial or working mightily to downplay a growing problem. For instance, the show patter and press accounts around Mogul, the Macomber & Company performer, inspired various tales about his personality. Most were quaint stories of his sagacity, such as the one that told how he woke his keeper one midnight after the man had forgotten to bring him his evening rations, politely indicating with his trunk that he was hungry.[8]

However, other stories depicted Mogul as an emotional and jealous being. One claimed that the elephant's same keeper brought him a container of oats, but to save time scooped out a portion to take to one of the company's ponies at the other side of the building. When he returned Mogul gave the man "a smart blow with his trunk on the left cheek, just below the eye, wounding the flesh and starting the blood immediately."[9] Colloquial accounts of this episode asserted that Mogul was angry about sharing food with the horse and had struck his keeper in retaliation. Unfortunately, we cannot be certain that Mogul experienced the event that way, or that he struck the keeper out of jealousy or anger over his feed. He may have acted for some other reason that witnesses at the scene did not consider. Nonetheless, these stories did exhibit human reporters' reliance on explanatory anthropomorphism, specifically, a tendency to infer the motivation of animal behavior from its apparent function or results, as a way of coping with the limited means of communication between humans and elephants.[10]

Antebellum menagerie keepers and animal handlers experienced increasing unpredictability in the male elephants working the show circuit, and they began theorizing how to prevent such behavior from recurring so as not to muddle the menageries' claims of offering powerful animals made docile by captivity. In Mogul's case, even after he had reportedly struck a menagerie worker, his owners cautioned the public not to be afraid because, they argued, when elephants struck out at people it was because they were driven by immature if comprehensible emotions that further training could temper. Such actions were anomalies for such sagacious beings, they assured ticket buyers. "Mogul . . . is usually gentle, obedient, tractable, patient of management, and submits to every kind of exercise for the gratification of visitors," the elephant's keepers advised the public.[11]

The press, on the other hand, could not resist a sensationally dramatic story. As they had in the days of the first Betsey and of Horatio, the papers provided Americans with regular and uncensored reports of menagerie elephant activities that the menageries themselves were in no hurry to publicize. For instance, in 1842 they reported on an incident at the Amphitheatre at 37 Bowery in Manhattan in which an elephant on display there, probably the teenaged Asian bull known as Siam, attacked Charles Howes, one of the show's trick riders and member of the famous Howes equestrian circus family.[12] The papers said that a keeper had been whitewashing the elephant's stall and "wishing to move him, . . . pricked him with the prong of a pitch-fork." The elephant "became infuriated," and seeing Howes at hand (or trunk), grabbed and threw him against a stable wall. At that moment, Howes's dog was reputed to have begun biting the elephant, providing an opportunity for show staff to quickly drag the bleeding Howes away.[13] (After only a dozen years in the United States, Siam died over the winter of 1845–46 at the company's warehouses in Zanesville, Ohio, "from being chilled while standing out of doors in a heavy storm.")[14]

Three years later, another report told of Pizzaro, a Hopkins & Company Menagerie bull promoted as "gentle, obedient, and docile," who became "so exasperated" at Baton Rouge that he pushed an abusive keeper off his horse, then "lifted him from the ground with its trunk, and tossed him in the air, and then caught him, as he descended, on its tusks, killing the man instantly." Then the elephant similarly grabbed, gored, and stamped upon one of the circus camels, running into a wooded area with the camel's body, "occasionally throwing it down and trampling on it." Thereafter he also attacked a female elephant, known as Ann or Bess (the sources indicate both names). City officials called in a group of soldiers, who surrounded and shot Pizarro without killing him, although they did slow him down enough that his trainer, Clark Saunders, was able to chain him to a tree, where he remained at the time of the report.

One newspaper piece documenting the incident misspelled the city's name "Baton Rogue." Whether just a typographical error or a pun, the language made reference to the fact that unmanageable male elephants were often known as "rogue" elephants, in derivation from the Sinhalese word describing the hazard of a lone Asian bull in musth.[15] Indeed, Pizarro would later kill Saunders in a separate incident. After ten years of menagerie life, Pizarro died, around 1847, still in his teens. He drowned along with another male, Virginius, to whom he was shackled by his Welch & Lent Menagerie handler, George Nutter, while the two were being driven across the Delaware River at Camden, New Jersey. Nutter had intended the shackles to prevent the elephants from ignoring commands and lingering in the water to bathe, but instead the elephants became tangled in them.[16]

However, the famed Columbus was perhaps the first elephant in U.S. history to deliver fatal injuries to a human being, searing the 1840s into public memory as the era of bull elephants in the United States. Columbus arrived in 1818 as a juvenile and worked the nation as far inland as Ohio and Louisiana as part of the Zoological Institute cartel. In Algiers, Louisiana, in 1839, Columbus and the bull Hannibal engaged in a "terrific battle" that became circus legend.[17] Having been separated on different show units for some time, the bulls were reunited that December and were said, upon seeing each another, to have ignored their keepers and begun fighting. Company staff wrangled Hannibal, but Columbus thereafter "ravaged the countryside. . . During his stampede he killed his trainer, the trainer's horse, a negro man and nearly a dozen other horses, mules and cows." The company paid out almost $30,000 in damages, including payment for the death of the "negro man," a slave owned by a nearby family.[18] Two years later, Columbus assaulted his new handler, William Crumb; the man did not recover from his injuries.[19]

That decade Columbus, Hannibal, Virginius, and Siam worked for a conglomerate of shows owned by the Raymond & Company Menagerie, which at times employed the four together in pulling an ornate bandwagon, which many citizens remembered as a sensation. Later, Raymond & Company advertised Hannibal and Columbus as elephants of war along with another male on the unit, Tippo Saib, who was billed as a sagacious elephant performer. Typical patter sang Columbus's praise to the public as "The Monster Elephant COLUMBUS . . . weighing 10,730 lbs."[20] The elephants traveled with various production lineups, including a big cat act directed and trained by the then-famous "Herr Driesbach." In spite of his German persona (possibly designed to compete with Isaac Van Amburgh's exotic trainer persona), Driesbach was born in Sharon, New York. He had apprenticed to a shoemaker after becoming orphaned at eleven years of age, then worked as a policeman for a time before gaining notoriety as a wild animal trainer in the employ of circus impresario James Raymond.[21]

In 1848, Columbus killed a second keeper, a man named Martin Kelley, in the company's Philadelphia warehousing barns. At first, witnesses alleged that Kelley had prodded Columbus with a pitchfork once too often, causing the elephant to lash out at the person on the menagerie crew that he despised most. Other public comment portrayed the man as the victim of the event—"poor Kelley"—and Columbus as an ambivalent character responsible for a "terrible occurrence" because he was suffering from a foot injury. As the menagerie's damage-controlling version of the story went, after the elephant had assaulted Kelley, only the company's "daring" trainer, Herr Driesbach, had the nerve to approach Columbus. When he did, he found an enormous splinter impaled in the bottom of one of the elephant's feet, explaining Columbus's act of violence against Kelley as a side effect of being miserable with pain.[22] It was true that antebellum American elephants received infrequent foot care and commonly developed overgrown pads that might have wood, metal, and other items embedded in them. Yet, the predominant interpretation of the event the company offered had it that, like the fabled lion grateful to the slave Androcles, who removed a thorn from his paw, Columbus was a misunderstood but well-meaning being learning to live peacefully with humans. Columbus would die in 1851, reputedly after falling through the deck of a bridge at North Adams, Massachusetts, breaking his spine as Horatio had done thirty-one years earlier.[23] The death of "old Columbus," as he was remembered, must have revealed to many observers that the national infrastructure was still not growing as quickly as the nation's elephants.

In any event, there were multiple speakers and much ambivalence afoot in the stories about what male elephants really wanted or did not want, and why. In these early accounts of elephantine destruction, one can hear community reporters and the menageries struggling to control the lesson of these events. Was the elephant a wild animal, dramatic in his unpredictability *or*, as many in the industry would argue, an honorable being who struck out only in confusion, thus emphasizing his need for human care and control? That interpretation saw the menageries' troubles with their male elephants, not as evidence of systematic problems with a business model premised on keeping these dangerous but fragile beings captive, but as an unfortunate accident that could be averted by a brave and skilled trainer.

For several decades the accounts of human-elephant discord would exhibit some common ingredients: a male elephant minding his own business in a barn or tent; a mischievous, malevolent, or clueless person who interacts with the elephant, who thereafter strikes or grabs a person or another menagerie animal. One string of stories told of elephants becoming violent after being fed tobacco hidden in hay or handed to an outstretched trunk. In an ironic subversion of the perceived innocence of the first elephants, who gently sniffed around in spectator's coat pockets looking for bread, here the trunk was a tool of elephant fury. In Lockport, New York, in 1846, for instance, the papers said an unidentified elephant performing there had gone into a "rage" when someone hid tobacco in his food. He "walked out of his tent, attacked a couple of horses, killed one—was checked by his keeper, and chained to a tree—tore up the tree again, and injured a man, &c."[24] It is difficult to know how many details of the stories were completely accurate, and people elaborated them in order to speak to various human issues, the controversy over male tobacco use among them.

Lurking under the surface here was the fact that these bulls were reaching sexual maturity and the dangerous periods of physical discomfort, irritability, and aggression known as musth. Imported to the country as juveniles, most survived and performed under training for a dozen years or more, as the females did. Then, as teenagers, they would begin to experience increasingly difficult periods of hormonal change associated with preparation for adulthood, when normally they would need to compete with other males over females in estrus. This was elephant musth, a periodic physiological and behavior change beginning in puberty, at around ten years of age, although there was probably great variation in this due to differences in diet and other factors. The experience drives bull elephants to attain the necessary "sociological maturity" for reproduction but is also suspected to make them feel irritable, certainly physically and probably mentally. While in this state, a gland on either side of a bull's head drips a noxious fluid that drains into the mouth, and he may "dribble urine" and otherwise appear unsettled and aggressive to other elephants, people, or any animal. Since bulls in this condition cannot be worked safely or efficiently, many elephant handlers in Asian lumber camps simply tether musth elephants to trees, restrict food, and wait the process out, shortening musth from several weeks to less than seven days.[25]

Some American keepers must have had an indication that male elephant unmanageability was an inevitable if intermittent phenomenon by way of the story of the famous London elephant, Chuny. Killed at Exeter Change in 1826, he was known on both sides of the Atlantic as a tragic "public pet" and gentle giant, whose death was a shock to many Britons.[26] Before he died, contemporary newspaper reports explained that Chuny had begun exhibiting "strong symptoms of madness [and] irritability during a certain season . . . and these symptoms had been observed to become stronger each succeeding year as it advanced towards maturity." Perhaps tipped off by Indian or Anglo-Indian traders who had supplied the elephant to the men who shipped him to Britain, the staff at the Exeter Change grounds said that Chuny's behavioral changes came with physiological changes and were "always foreseen by the inflammation of the eye, and a purulent discharge from an orifice near the ear."[27] When killed by his keepers in March that year at age twenty-two, having lived through seventeen years of captivity, Chuny was reputed to have been showing "proofs of irritability, refusing the caresses of his keepers, and attempting to strike at them with his trunk on their approaching him, also at times rolling himself about his den, and forcibly battering its sides." Many people blamed the severity of Chuny's musth on the tiny cage in which he had been placed for many years, demonstrating that members of the public disliked seeing elephants so confined, although other animals might be displayed in similar manner.[28] Chuny's death was such a sensation that inexpensive illustrations were offered for the British public to purchase so that they could ponder the event and what it signified about the abilities of elephants and men.

At mid-century, back in the United States, here was the conundrum: menageries needed male elephants in order to support specific show programs and growing audience expectation for large bulls. However, showmen were coming to know how risky this could be since males in musth were proving unpredictable and dangerous even to trusted, veteran trainers. Two management problems resulted, one logistical and one a puzzle of communication strategy.

First, how should a company manage and transport from town to town an elephant who might be distracted by musth for a few weeks or months every year? Representative of this problem was the celebrated Asian bull Romeo (the third of at least four elephants to carry that name), who, along with two females known variously as Jenny Lind and Juliet, worked for Sands, Lent & Company and June, Titus & Company menageries. (Romeo was himself originally known as Canada, and temporarily renamed Abdullah in 1852 for some shows.)[29] Contemporary trainer George Conklin remembered that by the 1850s, Romeo had begun "to be ugly, and all his help got afraid of him. Finally it got so that none of them dared to go near him, not even to clean out his paddock, and all his food and water had to be put down through a hole in the floor over him." More than one trainer claimed to have broken elephants of such belligerence by cowing them. Stuart Craven, told Conklin and others that when Romeo became too dangerous he strode into the elephant's enclosure and shot him multiple times with a rifle until one shot hit Romeo so that the elephant reputedly "gave in" to Craven.[30] Rebreaking may have persuaded an elephant to suppress behaviors associated with musth, but it could not prevent the physiological and psychological event itself. It may have additionally been counterproductive if it created anxiety for particular bulls around certain contexts and people.[31] Indeed, Romeo lost an eye while being "punished" (rebroken) in 1866 and was "so badly mauled that his legs suffered permanent injury" during a later rebreaking procedure in 1869.[32] He died in Chicago in 1872, after a reputedly "brutal operation was performed on his forelegs," an event that made news even in Great Britain.[33]

Romeo's difficult life sketched out a common scenario for male elephants. Bulls would typically respond to physical assaults from humans by temporarily suppressing the drive to particular behaviors, but they would inevitably become dangerous after some months, thus triggering a new episode of rebreaking. Bulls usually sustained physical injuries and were probably strained mentally during each cycle. And many men in the industry were coming to understand that the first period of musth was an indication of the onset of adulthood and bigger problems to come. At large in Asia, a male would live alone or with a loosely knit but cooperative group of other males fearing no predator. During musth, a bull might challenge those other males in competition over elephant cows and as a necessary articulation of his independence as an adult.

As some trainers understood bull elephants in captivity, it was possible, then, that an elephant in musth would temporarily reject human power and thereby learn that, whether experiencing musth or not, he no longer needed to fear his keepers. These men appear to have often had anthropocentric, even masculinist, conceptions of how bull elephants understood hierarchies, but they also seemed to be informed by elements of South Asian mahout knowledge about bull elephant nature. We now know that, while at large in Asia and Africa, juvenile bulls are shown by older elephants how to interact with other elephants, with the group's matriarch serving as group leader and final arbiter of social relations.

Likewise, as adults living in male-only groups, bulls have a fluid pecking order that simultaneously promotes group survival and effective reproduction, so they might be peaceful companions in one context, opponents in another.[34] Some elephant trainers seemed to understand that bulls were by nature independent actors whose innate desire for autonomy simply had to be squelched if they were to function as entertainment tools. Thus, menagerie elephant handlers desired to be continually dominant over a bull, believing that any compromise or laziness on his part or that of his staff in enforcing human will would cause a bull to challenge all human intervention in his activities. That lack of negotiation, inasmuch as it included no periods of equality, may have strained bull elephants mentally. And, as the menagerie and circus people said themselves, it was indeed amazing to many observers that bull elephants managed to heed their keepers as much as they did.[35]

Furthermore, with respect to the issue of communication strategies, how was the circus to adapt their "Mammoth" marketing strategies to situations in which unmanageable bull elephants threatened the central and compelling contrast of a powerful elephant made docile, which defined the phenomenon of the menagerie elephant as jovial actor? How could circus people adapt to press coverage of elephant destructiveness and these "circus-day-gone-wrong" tales that emerged from them without appearing as inexpert or incompetent keepers of elephants? The male known as Hercules highlighted how a combination of musth and limited human resources produced many of the disastrous circus accidents that occurred during the century. Hercules lived with and worked for Reynold's Great Mexican Gymnasium and Menagerie just after the Civil War. On display in the zoological tent before a show in Forrest, Mississippi, one day in March 1869, Hercules's primary trainer, John Alston, told the circus crew that the elephant had "for several days manifested a disposition of insubordination, and [Alston] begged that no one would approach sufficiently near to receive a blow from his trunk." Apparently, Reynold's company management decided it was better to show an elephant in musth than not, but perhaps did not understand or take seriously the risk of doing so. An account from the scene told of the disastrous chain of events that ensued:

> Mr. Mark Kite, from the northern part of this county, coming in after Mr. Alston's admonition, thoughtlessly handed him a piece of tobacco, which so enraged him that he struck at him with such violence as to dislocate his shoulder. . . . He plunged with such force that he broke his chain, and though his keeper used every effort to subdue him, he was entirely uncontrollable, and he would strike and kick at every

Hercules attacks a freight train in Forrest, Massachusetts, 1869. From Olive Logan, *The Mimic World and Public Exhibitions* (Philadelphia, 1871).

object near him. By this time the scene was beyond description. The vast crowd flew for life. He flew at his keeper, and pursued him from under the canvas.

Then, people said, Hercules spotted a freight train passing on the tracks beside the circus grounds and charged, breaking one of his tusks. He derailed the engine, which in turn collided with one of the company's tents and broke open the lion cage. Although the female inside was killed, the male lion escaped and, locals said, chased one man up a tree, then attacked a "fine pacing horse" ridden by a man traveling with his young son, and then tore open the back of a wagon carrying chickens. Soon a group of local men brought their hunting dogs and set out to pursue the lion, although they did not immediately find him.[36] The story of the day Hercules ate tobacco was reprinted in many a paper and lifted word for word into industry accounts like Olive Logan's *The Mimic World and Public Exhibitions* (1871).

While the idea of the furious menagerie "tusker" made for exciting newspaper copy, the underlying reality indicated how the undercapitalized security of the circuses created public safety hazards. Change that might avert such disasters— say, keeping musth elephants like Hercules in a cage or off-site altogether— would be slow or nonexistent because, as some of the circuses became larger, their schedules became more rigid, their running expenses increased, and they became less adaptable to such situations. Now those circus companies' advance men booked shows months in advance with cities, railways, and hay, grocery, petroleum, and other suppliers on routes that stretched out from the Atlantic to the Mississippi and to California. With less on-site control of scheduling and show advertising, the bigger circuses raised expectations among ticket buyers and the local newspapermen who reviewed their work. These businesses now had more opportunities to disappoint by running late if they became stuck in impassibly muddy roads or showed up with key acts missing or under the weather. Disagreeable elephants were but one of many complicating logistical factors that aggravated on-site staff charged with bringing to life the visions of the impresario and producer, arms-length artists and entrepreneurs who often saw in bigger shows more potential than complication.[37]

Company management structures made the problem of a maturing bull elephant far more taxing. Many owners and senior managers had only limited or indifferent input into the training of their elephants. If not traveling with the tour, they might visit only periodically with the show's elephant man and his crew to ask what was happening with a given elephant or herd, or how a new behavior might inspire a novel show script. This kind of communication was usually not worth the cost of sending a telegram. At the same time, as a form of job protection, trainers jealously guarded their knowledge of and their relationships with individual elephants, who, with the wrong handler, might become unpredictably unresponsive or impossible to move or keep fed and watered, never mind obedient in the ring.

Thus, conflicts of interest often arose when trainers manipulated, or were suspected of manipulating, their knowledge of a show's elephants to their own advantage. For instance, in the late 1860s circus owner John Robinson had argued pointedly with his elephant handler, John King, over just this issue with respect to Chief, the company's bull elephant. Robinson asserted that King had been attempting to protect his position at the company by aggravating Chief. King had been tolerating the elephant's resulting violent behavior, "with an idea that by making Chief mean and at the same time handling him without mercy, he could create the impression that no one else could handle the animal. King taught him so many little tricks of deviltry that he was finally listed as a dangerous 'rogue' elephant." This had gone on for some time until Chief struck with his trunk a boy in the audience during a performance at a small town show on the shores of the Ohio River. Thereafter Robinson was determined to punish the elephant to demonstrate his mastery of his company and its staff—although Robinson did not explain how this event would supposedly affect Chief differently from King's previous alleged treatment of him "without mercy."

On the road the next day Robinson said he had halted the circus wagons and announced that Chief would be held to account for his previous day's behavior. He and much of the circus company staff surrounded and beat the tethered Chief. "He did not have a friend amongst the three hundred men working with the show outside of his keeper King," Robinson stressed, after which Robinson claimed to have lit a fire under Chief until the elephant "gave in," presumably by vocalizing in pain. King had pleaded with Robinson to stop the punishment at the time, but the circus owner averred that King "wanted the elephant to be bad so that he could always hold the job of handling a bad elephant" and that the group assault on Chief would reintroduce the fear of humans. In the self-congratulatory version of the episode he wrote into his memoirs, Robinson declared himself vindicated when Chief killed John King at Charlotte, North Carolina, a year later.[38] Sensational reports alleged that Chief had "crushed [King] to a jelly in a freight car," and indeed many people would be injured while enclosed in small spaces with circus elephants over the decades.[39]

The story of Chief was symptomatic of company attempts to force on-the-ground staff to refrain from manipulating their relationships with company elephants when the owner or top-tier managers were not present. Some trainers surely did advocate for problematic (from a circus point of view) elephants out of self-interest. Others did so because they sincerely wished to spare an elephant from a worse fate than the grind of the show schedule, namely, "punishment" and resulting injury, sale to another party, or even death on the orders of a frustrated circus owner or senior manager. In fact, John Robinson would in time kill Chief (some reports said, by electrocution, as part of a research program for executing criminals in New York state). After King's death, Chief's new keeper, William Wilson, soon left the company, and no one could be found to manage the bull, who was becoming known publicly as "vicious" and a "murderous beast." After his death, Chief was stuffed, his hide and skeleton put on display at the Cincinnati Zoo.[40]

However, as a rule, with limited resources, time, or knowledge of the species, animal trainers routinely punished their bull elephants in order to cope with musth and elephant adulthood. Some rebroke bulls in order to placate company owners, others to extinguish the very real fear that the next show might be a great disaster if the crew did not get this or that bull in line. Rebreakings sometimes occurred as premeditated events triggered by an elephant injuring staff or a circus visitor. Circus people conceived of these rituals as "punishment," since "the attitude toward these beasts was the same as their keepers had toward domestic animals and children; if you misbehaved you were punished and were assumed to understand why," as Stuart Thayer has explained.[41] Additionally, rebreaking events could occur as frustrated bursts of violence delivered by terrified or intoxicated circus workers wary of a particular male.

In either case, rebreaking conceived as punishment was typically enacted by a trainer who saw elephant behavior as a reflection of his own power or respectability in the menagerie business. Since it pained elephants in a generalized way—at times distant from the specific elephant actions one wished to halt, such as striking people with the trunk—rather than instructing elephants to avoid specific behaviors, it was thought to engender a broad "respect" and "fear" of specific humans. While not precisely inducing clinical learned helplessness in elephants, those who punished certainly hoped it would persuade elephants that they had no choice but to obey their keepers. Punishment also imposed on a given elephant responsibility for the emotional state and persona of the trainer in ways that bull elephants could not reasonably have been expected to perceive, although some circus folk seemed to have harbored that expectation.[42]

All circus and menagerie people knew that most of the circus-going public demanded elephants, a desire undergirded by the prevailing ethic among Americans that cast animal use a benevolent enactment of human supremacy. So, instead of openly questioning the very use of male elephants in show business—as circus managers would in the twentieth century—many circus people justified acts of violence against bulls as an experience that was actually in the elephant's best interest. Punishment might *seem* brutal, the argument went, yet it kept a menagerie earning and buying hay for the elephant so was in fact merciful. Animal managers had a point in that, before the 1880s, once exported from their native countries, elephants had no alternative to show business. In the United Sates there were almost no zoos (although elephants there were similarly confined and prone to early death), no elephant sanctuaries where they could live on many acres without hands-on human management, no possibility of repatriation to Asia or Africa, and no possibility of living at large in North America, where they would surely have been hunted by locals.

Thus, after the trouble of the 1840s, because they realized that males would require special, heavy-handed care if they were to keep working, elephant handlers were less likely to explain bull elephants as emotional but ultimately sagacious beings. Since musth and adulthood caused male elephants to resist the conditions of their experience, those bulls unknowingly inspired many industry insiders to expand the elephant's reputation to include a new trope of "unruly

elephant," which grew from and justified the practical realities of elephant management. The antebellum ideals of the docile and sagacious wonder of God's
creation or the jovial elephant actor would certainly persist in ring performances
and circus advertising, but after the Civil War they coexisted with the "unruly"
elephant, a duplicitous Gilded Age brute who required regular experiences of
pain and terror at the hands of humans in order to stay compliant. He appeared
in circus publicity doled out as interviews to sympathetic journalists or written
up by company press agents. Thus could the work of rebreaking potentially be
turned to a circus's advantage as a source of promotional stories that might monetize a difficult by-product of male elephant captivity, just like the display of skeletons or skins could mitigate the losses of an elephant death.

Certainly, the period offered more opportunities for trainers to speak directly
to publicists and journalists, rather than simply being represented in show programs, advertising, or the limited and informal communication opportunities of
speaking with visitors in the menagerie tent. Postwar circuses became some of
the first entertainment ventures to use marketing communication framed as "a
look behind the scenes" in order to build relationships with audiences. The circuses were part of a larger business trend in which companies depicted the
means of production in advertising as an endorsement of the drive toward industrial progress, mechanization, or efficiency. A champagne company, for example,
might offer advertisements depicting delicate ladies enjoying the product in a
parlor while, just outside their window, the company's factory is shown belching
clouds of grey smoke.[43]

With respect to animal shows, in the antebellum period, menagerie animal
training and after-hours rehearsal sessions had been no more open to audiences
than the lady performers' dressing tent. A generation later, circuses offered boldly
self-reflexive publicity and performance that seemed current since they helped
people ponder modern managerial and industrial developments.[44] Staged in
open spaces with little security to keep nonemployees away, circuses offered
ticket buyers access to, for instance, the complex circus-day setups that functioned as a free preshow event, exposing how the tents were raised and the wagons and rail cars unloaded. Thus was born the iconic "Circus Day," a joyous
community event that shut down whole towns as children and adults collected
around circus grounds to watch, not just the paid show or downtown parade, but
also the entire encampment of circus folk going about their workaday prepara-
tions with great speed and coordination.

Animal trainers could participate in these trends because the volume of print
publications, primarily newspapers and magazines, expanded considerably after

the Civil War. Helping fill all those column inches, elephant men offered interviews in winter quarters barns and the menagerie tent in which they sought to position themselves as animal experts while potentially leveraging that expertise with circus managers and owners. They might also cast themselves as stars of the circus as much their elephants, defining themselves by their specific ways of interacting with elephants, especially bulls, through the work of breaking and rebreaking.[45] Experience subduing elephants constituted the central node of their private subculture and public personae, displayed for the consumer as the elephant man's awesome, no-nonsense power over famous tuskers. In this way, bull elephants and circus animal trainers together created the celebrity persona of "elephant man" or, in the corporatized Gilded Age, "superintendent of animals." Famous big cat wranglers certainly self-identified by speaking of having being mauled by a tiger or lion, narrowly escaping death only to venture boldly back into the cage with their animals. So it was for the elephant man, who could claim to have physically and mentally defeated the world's largest land animals. Many elephant trainers additionally assumed they were superior in their knowledge of particular elephants to the often transient and inexperienced barn men who mucked out elephant quarters and filled the water barrel, although in many cases this was merely bravado.

To enhance their political position and broad notoriety, elephant trainers publicized a mid-century interpretation of elephants as conniving brutes through the practice, talk, and even public promotion of "subduing" and "conquering" elephants. This practical and cultural phenomenon marked a turning point; audiences would now begin to fully consider how these entertainment companies offered the public both the friendly elephant performer and a vicious elephant brute. That duality contributed to public perceptions of the Gilded Age circus world as at once glamorous and seedy, a reputation that would follow circus people well into the twentieth century. The phenomenon of subduing elephants as standard practice in circus animal management also turned consumer attention to the experiences of elephants who, their trainers freely admitted, resisted circus captivity and human control.

The case of the famed male Tippo Saib, whose front-of-house persona contrasted sharply with the behind-the-scenes stories told by management, was representative of this pattern. Probably imported from India in the 1850s, the young male worked for the Van Amburgh Menagerie company, billed variously as "The Famous Elephant Tippoo Saib" and "The Performing Elephant Tippo Saib," until his death in 1871.[46] He may have been named for the then-controversial Indian ruler of Mysore today known as Sultan Fateh Ali Tipu, who, in the 1780s and

1790s, had futilely fought the British and other Indian rulers with elephants trained for military use.[47] Because some Americans believed the ruler Sultan Tipu had had a particular love for pomp and ceremony celebrating his rule—and this appears to have been true of the historical figure—his association with elephants of war or glittering processions additionally made Tippo the elephant in America a comprehensible vehicle for messages about abundance and power.[48] Tippo appeared in the celebrated trio with "THE MAMMOTH WAR-ELEPHANT HANNIBAL" and "The Monster Elephant COLUMBUS." He appears to have traveled extensively, appearing with a Van Amburg Menagerie unit in the middle states and American South, then again in 1857 traveling with a different show with Hannibal through the Midwest.[49] In the early 1860s, he was still performing around the country and for several months a year at the company's venue in the 500 block of Broadway in New York City.[50]

The company advertised that Tippo would be seen "Waltzing, Balancing, Ringing the Bell, Creeping on his [rear] legs, Walking over his Keeper, Picking him up, Drawing a Cork from a Bottle, &c." The elephant was also reputed to hoist up his keeper on his substantial tusks, then, as the trainer juggled, Tippo walked around the circus ring. Tippo allowed his trainer, Professor Nash, to put his head in the elephant's mouth. Tippo also held lines and poles for human acrobats, as well as gently taking food from the hands of children. Company marketing patter insisted that Tippo was a jewel of the circus: "No language can convey an accurate description of all the beauties and graces displayed by this Miracle of Sagacity."[51] As a male Asian elephant, he was a formidable-looking tusker, sagacious wonder, genial performer, and exotic visitor all rolled into one incredible being.

Then, in 1867, the glamour began peeling away when the periodical press made it known that Tippo had become dangerous to his handlers. The ostensible cause, as far as show manager Hyatt Frost could tell, was that the elephant's immediate keeper and trainer of ten years, Frank Nash, had left the company and been replaced by a new man, Charles Johnson, himself hired away from a Barnum company circus. While that may have been an answer crafted for the newspapers, it was possible that Tippo had become difficult due to the change in personnel; elephants are often wary or hostile in the presence of new people.[52] He may additionally have been experiencing musth without Nash there to manage the event. Frost told the papers that Tippo was in the company's winter quarters, "chained in a small building," and in "particularly bad humor with all mankind." People who tried to feed Tippo or muck out his space said he struck at them with his tusks and trunk, making him entirely unapproachable.[53]

Frost collected a team of men to "subdue" Tippo Saib, a process the elephant had undoubtedly experienced under Frank Nash, and before that, while being removed from his mother as a calf in Asia. The *New York Times* told the dramatic story:

> The new keeper, with nine assistants, had fully equipped himself with chains and cables for tying, and spears and pitchforks for subduing Tippoo. The first thing done was to fasten a brickbat to the end of a rope and throw it over the end of the tusk-chain, which later is fastened to one leg and one tusk. By means of this rope, a twenty-ton cable chain, (formerly used to subdue the famous *Hannibal*,) was slip-noosed around the tusk. Next, an excavation three feet deep was made under the wall of the house, and while the elephant's attention was attracted to the other side of the room by a pail of water poured into his trough, the cable chain was passed through the excavation and fastened to heavy stakes outside. All this time the infuriated monster struck all around him with terrible ferocity, and tugged at his chain with incredible momentum. The next thing accomplished was snaring his hind legs. This was consummated by the slinging of fresh ropes around those two stately pillars of elephant flesh, bone and muscle, and finally, by stealthy strategy of the keeper and another man, these ropes were fastened to stumps outside. The elephant was now sufficiently pinioned to allow the order "charge pitchforks" to be given. Ten men, armed with these ugly implements of offence, plunged them into the rampaging beast, taking care, of course, to avoid penetrating his eyes or joints. The tenderest spot in an elephant is just behind the fore legs, and that locality was prodded unmercifully. By means of a hooked spear sunk in his back Tippoo was brought to his knees, but he surged up again with such awful strength that he swept his tormentors off their feet, and made his chains whistle like fiddle-strings. After an hour's fighting he was brought down on his side, but for two hours longer he tugged at his chains with frenzied obstinacy. He pulled so hard at times that his hind legs were straight out behind him, and three feet off the ground. At the end of the three hours the giant gave in by trumpeting, which is the elephant way of crying enough.[54]

The *Times* depicted a vivid confrontation that enhanced the reputation of Charles Johnson and Hyatt Nash as brave men whose formidable stature was demonstrated by the ferocity of the elephant they conquered, implying that when the elephant vocalized he was in a way saying as much. Hyatt Frost had likewise re-broken the other male elephants in the company, Hannibal and Columbus, when they struck at company staff, destroyed wagons and other property, or experienced musth.[55] In being assaulted and ultimately "giving in" as a mode of self-preservation, Tippo and other male elephants unknowingly engaged in quasi-public performances that constituted the social world and power of the elephant man.[56]

Cultural changes later in the century helped perpetuate the cultural logic behind this confrontational, dominance-submission approach toward male elephants. Many American and European men living in the United States had internalized a masculine goal of success through triumph in competition with other men, all of whom ostensibly operated as individualists in a largely unregulated capitalist state. As beings who challenged white male power and the economic imperatives of circuses, even if in sporadic and unorganized ways, elephants threatened the livelihood but also the very lives of the well-to-do circus owner and his middling trainers and managers. Even the trainer's working-class barn men and ring performers, many nonwhites among them, may have supported white male cultures advocating physical domination of bulls if they, too, were terrified of the bull elephants with which they worked.

Those mid-century attitudes and practices may have seemed "right" to many Americans outside the circus community because elephants in effect challenged white paternalism and male economic power when they complicated the operation of a circus or the plans of a particular trainer. This logic matched the rhetoric of radicalism many people developed in those years to describe their experiences with striking workers, "hysterical" women, African Americans asking for social and political change, or any person who seemed to challenge the triad of white supremacy, capitalism, and male power. Here was a way to put one's lack of interest in or inability to develop some better means of elephant husbandry and management into familiar language, asserting that the circuses were acting in elephants' best interest when they rebroke a male, since he was inherently imperfect and thus in need of human care and control.[57]

In antebellum and Civil War days, although menageries often included "Ethiopian Delineators," that is, minstrel shows, most productions avoided weighing in on political issues like slavery or states' rights, preferring to remain as broadly appealing as possible.[58] Still, their acts could take on metaphorical functions. Because some Americans perceived the Indian Tipu Sultan to have been a tyrant who opposed the British government with no justification and no victory, the story of the elephant Tippo Saib (which might have really meant Tipu Sahib) could function as a trope for the Confederacy, equally perceived by some in the American West and North (and the South, it might be argued) as an illegitimate power averse to democracy and political or economic progress. Indeed, *Frank Leslie's Weekly* made just such a comparison in December 1865, in such a way that the story of the broken elephant Tippo Saib may have helped some readers consider the nature of state authority and political dominance in their own nation.[59] Like *Niles' Register* and Comte de Buffon before him (with respect to France), the elephant beaten and subdued—or tricked—was still a vehicle for talk about political and economic liberty and whether "might made right."[60]

Further, the Darwinian age dawned in the United States in these years and, to many, justified the continued, if not accelerated, human use of other species. While some argued that the interrelation of all species required humans to be merciful toward animals, which were also able to suffer, others focused on the various interpretations of natural selection that saw all species in competition.[61] Meanwhile, those who denied Darwin might act in similar ways. If espousing dominionism, which held that God had given humans all the animals to use however they saw fit—as Van Amburgh Menagerie shows had enacted—one might justify elephant use as a kind of divinely ordained animal improvement.[62] Other pious Americans argued alternatively that, since humans were formed in the image of God and all the living creatures were a message from God to mankind, one should show kindness and frugality when employing animals and their bodies for human consumption, industry, or amusement.[63] Even the authors of children's literature asked young readers to adopt this kind of genteel dominionism. For instance, Gow's *Good Morals and Gentle Manners: For Schools and Families* (1873) instructed that, "without the assistance afforded by the domestic animals, mankind would be very helpless. The horse, the sheep, the ox, the ass, the camel, the elephant, the llama, the dog, and the reindeer . . . each is indispensable to the comfort or safety of man." Such animals had "lost much of their natural instinct of self-defense and preservation" but displayed "a sagacity that is wonderful," which made them both inferior and indispensable to humans. The text reinforced this ideal of "righteous" human-animal interdependence with text comprehension questions on reading passages, such as "Is it lawful for man to use all the animals for his own welfare? Why?"[64]

Weighing the elements of this spectrum of belief about human supremacy, many Americans adopted theories of natural history to speak to these debates by reconceiving nature as a "vicious" wilderness in which a "drama of warring species" played out. People had recently begun to imagine once again the fearsome and predatory mastodon, which had perplexed and excited Americans in the early years after national independence (an opinion temporarily tempered by Americans' experiences of the gentle first elephant in the country), as one of many huge "prehistoric monsters" believed to have stridden around North America terrifying lesser creatures. Indeed, by 1900 the quintessential dinosaur would be the

*Tyrannosaurus rex*, portrayed as a masculine, savage predator (not the opportunistic scavenger many would portray him to be a century later).[65]

Thus the trainer's work of punishing elephants resonated with a broad ideology of "survival of the fittest," which was culturally relevant and "increasingly understood as the basic law of nature, [and] appeared to legitimate the violence of the human animal toward other species."[66] Such attitudes legitimized the near-extermination of the bison, increasing consumption of meat, processing of whales and horses into energy, and other activities that served to define man as master of beasts and thus nonanimal.[67] For Gilded Age male elephants, these trends saw animal trainers presenting bulls as educated performers in the show and on the billboard, in reference to that older tradition of animal interpretation, and as a brute needing to have fear of mankind's power beaten into him, in reference to more contemporary ideologies.

The period's ritualized stories of elephant punishment were consequently a product of the intersection of globalized American entertainment and the integration of American nature, its land, animals, and people, into the larger capitalist systems of the nation, justified by the ideology of competition between species. Individual circus workers probably did not think so broadly minute by minute, many merely struggling to get through the day or a given performance without an elephant ruining the show or getting out of hand. Yet when famed elephant men spoke publicly in such stark terms about how they sought to subdue elephant performers or allowed outsiders to see such things and publicize them, they played upon common assumptions that human domination of the natural world was progressive and inevitable. From the trainer's perspective, the habit also demonstrated to far-flung company owners and managers, who systematically scrutinized the press notices of their and rival operations, that they had hired a trainer who was keeping things firmly under control in the elephant barn. While women, children, and clergymen could celebrate the old ideal of elephants as sagacious friends, the late–Gilded Age elephant man would have said this was a myth circuses used to sell tickets to the sentimental masses.

Industry experience that the older a bull elephant got, the more trouble he would cause, regardless of repeated rebreaking, became part of the popular natural history the circuses were then supplying to ticket buyers. In the 1870s P. T. Barnum's event programs included an explanation of "an elephantine 'must'" in which elephants "become restive and excited, and then commence a regular tantrum, when they will attack anything or anybody that comes in their way."[68] The 1881 *Harper's Weekly* commentary, which in one part deplored the violence involved in "hooking" elephants, also noted that keepers and trainers knew that they would be increasingly unable to contain a bull elephant over time. When fully mature, "there is only one kind of influence which can be advantageously exercised, and that is to 'influence his head off,' or, at least, to adopt the handiest available form of euthanasia."[69] In 1907, an elephant handler would similarly tell a correspondent from the circus trade journal *Billboard*, "The only way to treat a bad elephant is to subdue it with steel pikes until it bellows. After that it will be good for perhaps six months." Yet, when that elephant began "turning into a rogue" once more, he might kill anyone at any time, and his handlers should ready themselves to dispose of him.[70]

The disposition of elephants due to their eventual unmanageability was a serious financial cost to a circus company. In the 1840s an elephant still cost around $5,000, although in the papers people often reported elephant values at double that amount, a number the papers had been citing since the days of the first elephant. At the same time, $5,000 could buy a company thirty horses, including some fancy ring stock, while many Americans knew that that amount could purchase a two-hundred-acre family farm. For only $1,800, a company could purchase a season's supply of wagons, bills, and a canvas tent and poles to shelter several thousand people, so elephants were still a major company expense.[71] In the Gilded Age, the prices for elephants increased as the ivory and zoological trade ate into elephant populations overseas, then dropped in the late 1880s as the massive importations that decade "overstocked" the American trade with at least seventy-five elephants.[72] Prices declined again during the twentieth century, and by the 1950s one could find a trained female elephant for $3,000 in the want ads in the back of the *Billboard*.

Why not simply purchase solely females for a show—as twentieth-century circuses would do? Industry wisdom held that for audiences a pair of tusks on an elephant was dramatic, but all the more arresting because many people knew the value of ivory as a commodity in its own right. Tusks could not be convincingly faked on female Asian elephants (a few circus men claimed to have tried, with unsatisfactory results), and few African females, who did show tusks, were easily obtainable in those years. Some did argue that bulls should remain a show staple because a musth elephant could be saved by a "clever keeper [who] can detect the change in the animal's manner soon enough to make it possible to confine him by himself until he is good-natured again."[73] However, that approach was never widely adopted, as it would have required far more managerial and financial flexibility, as well as human humility, on the part of showmen than most seemed willing or able to exhibit.

For an entertainment premised on fierce competition between companies and

"Mammoth" marketing aesthetics, to ask the audience to expect less made no sense, especially since the circuses often fell short of their own promises due to illness, bad weather, inept management, or insufficient cash flow. And, in fact, the frequent disposition of adult bulls was possible because it was far easier for a company to kill a male and recover part of their losses by selling his carcass for stuffing and display in a museum than to pay compensation when he injured a member of the public or their property. There was a steady supply of new males entering the country to replenish the stock trainers made or found to be unproductive in the circus tent and winter quarters barns. Moreover, circus bankruptcies were frequent enough that one could buy elephants cheaply at auction (especially if reputed to be man killers) if one was flush with capital or credit at the right moment.[74]

The subjugation of Tippoo Saib in 1867 also inaugurated a narrative genre exposing the public to the most dramatic aspects of bull elephant management (and probably some females as well, although it was only males that were so graphically subdued in print). Representations of hooking and breaking elephants in the press thereafter would depict circus staff routinely resorting to these practices when an elephant became "unruly." This was a battle of wills, they said, in which an elephant man could not take no for an answer from a captive beast. The public nature of this information would provoke critics of the circus and expose the long-term rift between circus people and their (mostly working-class) customers and worried middle- and upper-class Americans. This argument over how the circuses treated elephants primarily took place in popular media for adults and children, venues in which people talked about animals while revealing their assumptions about class, childhood, race, and humanity.

For instance, to weigh in on the debate over elephant management the children's magazine *Harper's Young People* included a graphic illustration of the process of subduing an elephant inside an 1885 issue. Notably, the front cover of the magazine that month presented a line drawing of a young girl feeding a bird from her hand. The image conveyed the dominant culture of the magazine, espousing maternal care and gentle interactions with other species so that children would learn to have sympathy for the suffering of others, human and nonhuman. Such animal narratives were a tried and true way to tell human stories, a mode of progressive literature with deep roots in American culture that advocated for "kindness gendered female, yet . . . something boys can learn."[75] Inside that issue, *Harper's Young People* editors, striving to engage young readers as aspiring advocates for animal use reform, no doubt intended their illustration of an elephant subdued as a cautionary piece demonstrating what *not* to do with animals.

Indeed, the villain of the montage is a hook and pointed-stake-wielding male trainer, who scowls in every frame.

The illustration of an unruly elephant subdued appeared in *Harper's Young People* with no real explanation save juxtaposition with a page offering a booster-ing article about the famed African elephant Jumbo, recently arrived in the coun-try, and his relationship with his dedicated handler, William Scott.[76] On its own, the Scott story may have made sense to readers or their parents because of the "domestic ethic of kindness" of the period, a middle-class and children's litera-ture subculture that fostered in young people good manners and respect for liv-ing things by instructing them to gently care for pets and other small animals. It emerged from the pens of many middling and elite Americans critical of the ways the profit motive had shaped human uses of animals and for whom "the American home became defined as the one place where beings are cared for re-gardless of their economic value."[77] Similar developments facilitated people's col-loquial and commercial sentimentalization of animals in novels, magazines, and advertising in ways that made use of them, or their images at least, to tell human stories critical of capitalism, industrialization, and more.[78] Indeed, by the 1890s, many Americans sympathetic to groups like George Angell's American Humane Education Society looked at commercial uses of animals and saw "cruelty to ani-mals as structurally related to capitalists exploitation of workers, and they fre-quently suggested that most instances of the working classes' animal abuse stemmed from capitalists' greed."[79] Scott had left Britain to travel with and care for Jumbo, and there was no talk of breaking Jumbo, casting the line drawing of an elephant conquered as a cautionary tale.

For many readers, however, the depiction of an elephant beaten may have functioned as a kind of boy's adventure story. This was a broad genre that empha-sized personal triumph, rags-to-riches transformations, and acts of bravery through which young readers could imagine themselves pursuing wild animals, visiting foreign lands, or exploring the American West.[80] Within this body of lit-erature and illustration were numerous "hunting stories" providing daring tales of men who felled bull elephants in Asia or Africa.[81] They usually appeared in a familiar narrative structure, in which a white man aided by nonwhite guides tracked and confronted a variety of animals. The authors of such tales added suspense by describing the event as a perilous struggle in which the hunter is triumphant because brave and intelligent—just like the circus tamer of wild ani-mals. After making his shot he witnesses the melodramatic death of his prey, who thrashes, struggles, bleeds, trumpets, roars, and strikes out at his attackers with claws, tusks, or trunk before issuing a dramatic "last breath."[82]

SUBDUING AN UNRULY ELEPHANT.

1. On the Rampage.    2. Floored.    3. Placed on a Pedestal.    4. Hoisted.    5. Conquered.

"Subduing an Unruly Elephant" with manacles, ropes, hooks, sharpened stakes, and men. From *Harper's Young People* (1885).

Although in the *Harper's Young People* illustration the trainer is cast as a villain by his countenance, subversive readings of the story must have occurred, allowing many readers—largely boys, presumably—to vicariously participate in "Subduing an Unruly Elephant" and dominating an angry animal who is ultimately "conquered" by man. In both hunting tales and stories of an elephant subdued, the male protagonist is only as powerful as the virile creature he defeats.[83] The popular commercial representations of the drama of breaking elephants may only have heightened the public idea that circuses were dangerous to boys, who, parents worried, were already tempted to try at home the acrobatic and trick horse riding acts they saw in the ring or on the circuses' advertising broadsides.[84] Those parents may have had a point because circuses published many spin-off magazines, pamphlets, and books that were at once hunting tale, naturalist's account, and circus promotion. P. T. Barnum would release many such books, carrying titles like *Lion Jack: A Story of Perilous Adventures among Wild Men and the Capturing of Wild Beasts; Showing How Menageries Are Made*, which presented the acquisition of circus animals as a youthful male adventure.[85]

The issue of how humans treated animals was a controversy that many deemed too trivial for the front page of the newspaper or for congressional debate, but it poked through to the surface of public life in the juvenile and women's press, which served as a collection point for a segment of public opinion that was growing increasingly critical of circuses, many of whose trainers and customers seemed to relish intimidating animals. These children's publications carried stories and other content that acknowledged the destructive power of wild animals but celebrated the admirable and helpful behavior of the elephant, who, for example, "received kind treatment from his keeper," as *Godey's Ladies Book* told it.[86]

Typical of the critical genre was an 1884 fictionalized short story for children by John Corvell, which appeared in *St. Nicholas*, "An Illustrated Magazine for Young Folks." Pushing back against the vicious brute cliché, the piece combined a moral lesson with the opinion that circus and menagerie trainers were cruel, ignorant men out for profit. In yet a new rendition of the metaphor of the subdued elephant, it told the story of a Barnum & Bailey elephant. "Queen is an elephant in a menagerie," Corvell wrote. "How she did hate the trainer! And how much more fiercely she hated her keeper! If it had not been for the sharp-pointed iron prod, of which she was mortally afraid, she would have soon shown the puny human beings, who made her do such absurd things in the circus ring, that an elephant was above such antics."[87] The narrative allowed some vicarious revenge and attempted to explain why elephants harmed humans when they did by explaining that one day Queen picked the keeper up and threw him against a wall out of frustration, although she knew she would be punished for it. These circus critics held that the answer to an elephant's inability to cope peacefully with the confinement and training regimes of these entertainment companies was not punishment, but sympathy.

For some observers, performing animals (or children) on stage or in the circus ring were being exploited, especially because they were not in a position to fully understand the production they enabled and were thus unable to give informed consent to their participation.[88] So, as with the Niles elephant tamed because fooled, in this story, the subdued elephant was not tricked, but tragically aware of his or her hopeless situation. On that point, the hard-bitten animal trainer and his middling critics would have agreed. For child readers, the concept put the menagerie elephant into a sentimental paradigm used by many youth-oriented narratives, whose characters spoke in the voices of animals in order to articulate proposed feelings and experience that many parents hoped would encourage young people to sympathize with the sentience of other beings.[89] It also pointed readers to the (usually liberal, northern) reform philosophy, which interpreted the circus workers' punishment and painful training of animals as confirmation that animal cruelty, class difference, and capitalism were interdependent.[90] The circus trainer who terrorized his captives to drive the circus toward profitability, or who at least spoke openly to the press about subduing and hooking elephants, seemed to defy people who judged the circuses to be both cruel and unproductive for citizens. He was not a sympathetic voice who portrayed elephants as wild animal friends, but a masculine advocate for manly control of beasts.

Of course, this was a simplification of the diverse circus community. Circus owner William Coup characterized animal managers who randomly assaulted their charges as impatient and stupid, fearful and unmanly. Trainer George Conklin agreed that some trainers were heavy-handed because inclined to "show off."[91] Some inside and outside the circus community argued the circus audience was equally complicit in any abuses that took place, one journalist noting in 1869 that the elephant Romeo had every right to be mortified by his captivity to Americans.

The elephant Romeo, attached to the [Forepaugh Menagerie & Circus], did a funny thing the other day at Germantown. Going along the road in the preparatory parade, he spied a little pool of muddy water, and drawing up a goodly portion in his trunk, he discharged it all over his keeper, who was riding on horseback near him, muddying his clothes shamefully. Romeo is not a well-behaved animal—we believe he killed a keeper not long since—but doubtless he looks upon this whole matter of his captivity, and his being paraded about to make sport for the philistines, as a very unjust affair. From his limited point of view doubtless he is right.[92]

Indeed, over the century, many middling and elite Americans would argue that wild animal exhibits should be employed to educate the working classes, not to flatter their existing assumptions by displays of animals humiliated by tricks.[93]

After a generation of Americans had seen the secrets of animal training and discipline trickling out into the popular press, criticism of circus shows continued to mount. The circuses did not help the situation by immediately developing an adversarial relationship with critics. In the 1860s, 1870s, and 1880s there was a serious lack of imagination at work in these companies about how to adapt company communication strategies to the changing public context. By and large managers and their ad men had not yet become adept at co-opting their detractors by changing their practices just enough to appear to be honorably concerned. As P. T. Barnum told it, the best tactic in situations in which animal management turned deadly or controversial was to order his men to quickly distribute to educational institutions or other buyers the carcasses of elephants and other imported animals killed in his company's possession. Barnum explained of one such being, whose skin was injured by his trainers' hooks, ropes and pitchforks, "The elephant was cut up soon as possible to prevent Bergh's men from looking for wounds," referring to Henry Bergh, head of the American Society for the Prevention of Cruelty to Animals (ASPCA).[94]

In their talk of punishing elephants, some trainers certainly appeared aware of, or even goaded on by, newly visible ASPCA activity that had begun in the 1860s. Barnum first tangled publicly with Bergh over the feeding of live birds and rodents to a boa constrictor housed at the American Museum. Barnum was victorious in the dispute, on a publicity level, since Bergh employed alarmist language to critique "the scene, demoralizing and cruel in the extreme," produced by a museum employee, "semi-barbarian in instinct," tossing live animals into the snake's enclosure so visitors could watch them consumed. Of snakes, who, it was argued, would only eat live animals, giving keepers no choice but to indulge nature, Bergh said, "let them starve."[95] This would not be Bergh's most persuasive public moment, although he would go on to critique a Barnum company rhinoceros enclosure and a circus act in which a horse jumped through a ring of fire. Barnum would brag in his autobiography, *Struggles and Triumphs*, that he had over the years "vanquished" Bergh and his other critics in the press by appealing to the audience with arguments that today are standard talking-points for circus publicists. Namely, Barnum accused Bergh of being motivated

by uninformed emotion rather than knowledge of the species in question. Implying that Bergh's questions were also a criticism of Barnum's audience, the showman noted that various European heads of state and a broad section of "the public" in America could be trusted to decide for themselves what constituted
"cruelty" in animal display.[96] In spite of the publicity such controversies generated, many of the biggest circuses' animal trainers did not take kindly to wealthy
outsiders like Bergh meddling in their business affairs, and they appear to have
opposed them as an individualistic reflex.[97]

Circus people would observe Barnum's example and begin crafting ways of
responding to critics by appealing to the audience to trust these entertainment
companies, both to serve as authorities on wild and exotic animals and to deliver
them safely to Americans at a low price. Bergh and numerous others, on the
other hand, would continue to ruminate on animal management in show business. For instance, Bergh saw Barnum's quick disposal of elephant carcasses and
other secretive acts as admissions of guilt. Such acts were the hallmark of institutions that were disingenuous to ticket buyers, a perspective that would proliferate
in the Gilded Age. Those who held such an interpretation tended to be middle-
and upper-class citizens who sympathized with or were active in a range of interdependent reform movements in which opposition to animal suffering would
"march hand in hand with intense concern for public morality and work discipline." They also supported the idea that the powerful should not abuse the powerless because suffering by some affected the happiness of the whole in society.[98]

Before the Civil War, similarly concerned citizens made demands for reform
with respect to slavery, child labor, the consumption of alcohol, women's legal
rights, and other issues. After the Civil War, the available print space expanded
and, culturally, the expansion of equine labor and the suffering of those familiar
animals drove people to argue that it was appropriate to advocate for animals as
well as humans as an issue of public concern. Many were outspoken against the
urban cab drivers and other working-class men who seemed prone to beating and
driving horses to the point of collapse in plain view of passers-by on the street.[99]
In doing so, the more affluent and determined Americans defined for themselves
a kind of gentile status by their luxury in eschewing the exigencies of time and
money to refrain from cruel treatment of animals, while chastising as ignorant,
greedy, or uncivilized others who directly made their living with instrumentalized species.

Some also spoke out against the fashion for trophy hunting and other elite
uses of animals that seemed wasteful or crude to them, exposing the many contradictions in human-animal interactions globally. Charles Nordhoff, for one,

condemned the famous account of Briton Gordon Cummings, who wrote of hunting elephants in Southern Africa with abandon. In *Harper's New Monthly Magazine*, Nordhoff wrote:

> How needless to a sensible hunter is such butcherwork as Mr. Gordon Cummings delights to tell of in his Southern African hunts; where there is a narrative of his pursuit of a wounded elephant which he had lamed by lodging a ball in its shoulder-blade. It limped slowly toward a tree, against which it leaned itself in helpless agony, while its pursuer seated himself in front of it, in safety, to *boil his coffee*, and observe its sufferings. The story is continued as follows: "Having admired him for a considerable time, I resolved to make experiments on vulnerable points; and approaching very near, I fired several bullets at different parts of his enormous skull." . . . I never read this atrocious story without a feeling of tears for the poor animal, and execrations on the miserable butcher who could both do such a deed and complacently tell of it. I am sure no American hunter would admit such a fellow of his acquaintance.[100]

Nordhoff's representation of Cummings's account was accurate but infused with an assumption that Americans should honorably avoid or minimize causing pain or misery to animals. They should certainly never cause pain simply for the sake of human vanity or for entertainment based on raw human power.

Nordhoff did allow for hunting in the United States, but he insisted that respectable American hunters eat what they killed or put animal bodies to some productive use, unlike Cummings, who killed many elephants and simply walked away from their carcasses in search of his next prey. Of course, most menagerie animals—including the African elephants who would become more common in the United States as the decades passed—were captured as babies whose adult relatives had been killed for hides, ivory, or meat, so a dozen animals might have died in the process of getting one living specimen to New York or London. This was an uncomfortable secret of the exotic animal trade that even respected naturalists such as William Hornaday believed should be hidden from the public, whose members were "so terribly sentimental," he said.[101]

Nonetheless, elephants subdued for a circus seemed to many middling reformers a wasteful and egregious use of life. To be sure, there was a spectrum of opinion over which animals were "inferior," which animals had "intelligence" or could experience suffering, and whether this suffering mattered to human life. Since the days when an ordinary woman like Elizabeth Drinker had worried about the longevity of the first elephant in America, those sympathetic with reform were only part of a broader minority of Americans suspicious of animal use in entertainment. Where others saw whimsical fun provided at a low price by glamorous animal friends, these citizens looked upon the menageries and circuses and saw animal cruelty and exploitation that bamboozled crude audiences.

The popular science writer William Hoseo Ballou was one of many reformers for whom the material and metaphorical elephant beaten and fooled or intimidated had become current as a classic example of "cruelty." Ballou gained some notoriety as a member of the American Authors' Guild working for various magazines, then became implicated in the so-called Bone War of the 1880s over interpretations of the many fossil remains found in the western territories of the United States.[102] Although he was plagued by controversy because he claimed many accomplishments that historians have since found impossible to confirm, he was reputed to have sincerely worked for reform of animal use, specifically in "crusad[ing] to make animals safe in transport at sea."[103] Called a writer of sensational "pseudo-science for the Sunday-supplements" by one contemporary, he appears to have had typically broad interests, if little formal training.[104]

More important, Ballou's resulting writings threw into high relief a broader consensus among Americans critical of the circuses and other forms of commercial animal use that species like dogs, horses, and elephants might possess more innate moral virtue than some people.[105] Ballou said that animal shows were inherently humiliating to animals, who were trained by force and fear to perform to the down-market crowds the menageries and circuses, half-heartedly pretending to educate, appealed to. At one point he published some of his ideas in a *North American Review* piece entitled "Are the Lower Animals Approaching Man?" in which he argued, among other things, that animals were products of experience as well as instinct and that humans and many of the "lower species" like insects coexisted along the same cognitive and thus moral spectrum. "The elephant, with a brain endowed with visible knowledge, kneels in the circus to an ignorant and contemptuous clown; and the horse, with the wonderful and highly sensitive intelligence . . . is beaten before a cart of coal by a man-shaped brute, who can neither write his name nor mention the common decencies of life," he wrote.[106]

Ballou's formulation clearly put the obedient menagerie elephant before his trainer and claimed that animals that could be improved by the care of man should be respectfully protected from humankind's basest interests.[107] Ballou's was a moral or sympathetic taxonomy, which reflected ideas that had been circulating all century—in *Haney's Art of Training Animals*, for instance—that in considering how or why to be kind to other animals, what mattered was their apparent ability for reasoning, patience, and improvement under gentle human guidance. It also reflected a component of public opinion in which people appeared eager to see in their pets or horses aspects of their own experience and abilities so as to argue for mercy toward them. To many in the reform community, the man who was cruel to an animal was, to them, a "coward," who exposed his own weakness in needing to cow an otherwise noble and powerful being.[108] The elephant that survived "the rough treatment of his keeper" was a noble creature humiliated.[109] Or, in more sardonic but equally disgusted iterations, the bull who might fall victim to "overdose of pitchfork" exposed the uncivilized core of American culture and commerce.[110]

The last three decades of the century, in particular, saw a proliferation of sympathetic animal psychology theories. They were at times advocated by natural philosophers and proto-comparative psychologists like George Romanes, author of *Animal Intelligence* (1888), or even Charles Darwin, in his work *The Expression of the Emotions in Man and Animals* (1872), whose books sold very well to an audience already toying with similar ideas. Some observers countered that these thinkers unreasonably sentimentalized animals and discounted their instinctive natures in order to assert that horses and others were primarily products of culture and learning, much like people.[111] Indeed, many people were still interested in blurring the line between man and animal. The "Pig of Learning" and educated horse exhibitors, as well as the menageries and circuses, with their monkey jockeys and elephant clowns, had played to those opinions for decades, sometimes questioning human uniqueness with sentimental art, sometimes reasserting it with satire. The sentimentalization of other species was contextualized by the rise of a masculinist middle class in which some men described emotion as feminized or childlike and rationality as manly. Thus did many American masculinists verbally animalize minorities and the poor while indulging themselves by sentimentalizing particular animals as noble but doomed challengers to the hunter.[112] In any event, along with the suffering cart horse, the circus elephant conquered became common knowledge among many circus goers and their critics in the 1860s and was, for reformers, a frequent and useful trope for talking about animal abuse and injustice more broadly.

Bull elephants unknowingly helped produce these human cultures and arguments over animal use in a rapidly industrializing nation and, more worrying for animal managers and their assistants, shaped the internal cultures and workings of the circuses. By their nature and behavior, bull elephants made the preparation of shows more time consuming and mentally draining for staff and more dramatic and controversial for the public. Beginning in the 1840s, a generation of animal managers in the industry appears to have been mentally and logistically broadsided by the power of bull musth and adulthood, and they scrambled to cope, sometimes to deadly effect. Many would grow to fear or hate the bull elephants with which they had to work. When they boasted of assaulting disobedient elephant performers, circus people exposed the industry attitude that such males were not sagaciously sympathetic to human motives, as the audience believed. Rather, the post–Civil War bull elephant was a clever and angry brute who disrupted show schedules, injured and killed people, and for which one need have no sympathy. The elephant man or menagerie superintendent who, by extension, defined his persona and authority from the particular work of breaking and training elephants, could therefore become a celebrity rivaling his animals.

Yet, the very abilities of juveniles and female elephants in learning and remembering while accepting food and close contact with humans continued to fascinate ticket buyers and almost guaranteed that elephants would unknowingly perpetuate their role as the quintessential menagerie and circus creatures. Americans would imagine bull elephants—and in time all elephants—as complicated figures: famous, sagacious, and docile wonders of God *and* infamously treacherous beasts striving to compete with man. In engaging in random (to elephants) punishment of increasingly unmanageable bulls in their possession, the circuses were responding to critics with actions that, in effect, argued: "If people want elephants, this is what it takes—so stop complaining." Thus did elephants become some of the most celebrated and most abused animals in American show business. Still, perhaps this approach was not as reckless as it seemed. "Bad elephants, elephants on their travels, and dead elephants are the most interesting by all odds," one newspaper explained. "Except, of course, the intelligent beasts that are paraded in Sunday school books and first readers for the delectation of the young, and which have no existence anywhere else," the author added sardonically.[113] Circus people seemed to have understood that their customers represented the majority opinion in the nation and that the carping of reformers on the outside of that relationship may only have driven more people to more shows to see what the fuss was about.

# CHAPTER FIVE

# Herd Management in the Gilded Age

George Wilbur Peck once recorded an improbable story told by the young son of a circus manager, a man identified only as "pa." As the story went, one year the animals at the circus went on strike. Trouble began at Pittsburgh, where the Teamsters, who loaded and unloaded the show's wagons and equipment from the circus train, walked off the job over small wages and long hours. Pa soon negotiated a settlement with the strikers, and the company moved on to Germantown. But, the boy explained, the creatures in the menagerie had been emboldened by the Teamsters' success and proceeded to ruin the next performance:

> They acted as though they had lost all interest in the success of the show, and wouldn't do any of their stunts worth a cent. The elephants went through their act carelessly, and when they were scolded or prodded with the iron hook, they got mad and wanted to fight, and when they got back from the ring to the animal tent they wouldn't eat the baled hay, but threw it all over the tent and acted riotous. The kangaroos would not do their boxing act, the horses kicked at their hay . . . the camels growled at their food, and scared the people who passed by where they were tied to stakes, . . . the giraffes laid down and curled their necks so they were no attraction to the show, 'cause a giraffe is no curiosity unless he stretches himself away up towards the top of the tent. The zebras rolled in mud and spoiled their stripes.

Staff at the circus were baffled until pa explained the source of the animals' discontent. While allowing that they did not "understand enough of the ways of human beings to be posted on labor unions, and all that," he said that they had "lost respect for the employer." Led by the company's grizzly bear and lion, they had struck.

As the boy told it, pa settled that strike, too. He showed everyone how to manage the animals better simply by giving them decent food. The circus staff lazily cut corners by feeding the lion and bear spoiled horse meat when they preferred fresh, while giving the jackals fresh meat when they preferred it rotten. The elephants, giraffes, camels, and others similarly suffered with "musty baled hay, brought from contractors that may have had it on hand for five years." Pa asked the circus men, "How would you like it if you were served with breakfast food that

**Bolivar Took Half a Watermelon and Put the Red Side on Top of Pa's Head.**

Elephant and African American man equated as subjects of profit-minded circus managers and owners by an elephant that assaults a company manager with a watermelon. C. Frink, illustrator. From George Wilbur Peck, *Peck's Bad Boy with the Circus* (Chicago, 1906s).

had been stored in a warehouse until it was mildewed? A horse or elephant has feelings." As the boy told it, thereafter the animals each got their preferred food, including the elephants, who received "timothy hay, such as mother used to make." The menagerie animals went back to work, and the circus owner was so relieved that he gave pa a promotion.[1]

Of course, this story was fiction. It appeared in Peck's book with an illustration that depicted Bolivar the elephant assaulting pa with a watermelon while a startled African American man and the boy storyteller look on. Bolivar's juxtaposition with a black man, possibly his keeper, in the illustration made his act of resistance comprehensible because it resembled a satirical blackface performance roasting white economic power by likening it to human power over animals. Peck was a newspaper publisher, the Democratic seventeenth governor of Wisconsin, and author of a satirical book series known as Peck's Bad Boy. His tales of the Bad Boy and his pa with the circus were products of his own imagination but also of the broad colloquial culture around the trials of traveling circuses in the nation. Many Americans knew elephants had agency but suspected they had no real power, being unable to understand the human cultures that required their captivity in a traveling, for-profit show. Thus, Peck's tale of circus animals on strike was funny because it built on this knowledge to satirize public anxiety over the labor disruptions, white-on-black violence, and economic instability that came with the push for unionization over the last three decades of the nineteenth century. Americans knew animals could not actually go on strike or shape the nature of their captivity in a unified way. Because human supremacy over animals was so utterly unquestioned and controlling, the comparison of animal captivity to labor or race conflict may have assuaged some readers' fears by minimizing the legitimate concerns of blacks and workers as subhuman and absurd.

To be sure, elephants would never disrupt the circuses by slowdowns, sit-down strikes, or job site picket lines. Yet, that does not mean that the Gilded Age circus elephant was not changed by the advent of enormous rail circuses and the era's broader business and labor trends toward corporate organization. In fact, elephants did serve on their own kind of "picket line." This was the industry term for the arrangements of chained and shackled elephants circuses used to store herds of up to twenty individuals along the walls of barns and tents or in rows on the circus lot. Although elephant picket lines were not a sign of elephantine labor unions, they certainly spoke to many observers of the new industrial rhythms of circus work.

For instance, the year Barnum & Bailey's "Greatest Show on Earth" acquired the famous African elephant Jumbo, one circus publicist wrote in *Harper's Monthly* about what he found at the company's expansive winter quarters: "In one of the rooms, about one hundred feet square, are ranged twenty elephants, nearly all moving about in the restless manner so peculiar to them. . . . Nearly all the animals have a characteristic motion. The elephants move their heads in and out, from side to side, with a kind of figure of 8 movement; the sloth-bear jumps straight up and down; tigers and lions jump over each other in quick succession, as you may have seen the acrobats do in the show; foxes 'weave' in and out with a snake-like movement, and so on."[2] And many circus press agents, circus fans, and some circus workers accepted the swaying of the elephants as a "characteristic," if curious, elephant behavior. "We call it 'weaving,'" said one of the elephant trainers for Barnum & Bailey shows to the press agent that day in 1882, noting how knowledge of the phenomenon had become a basic element of circus subculture. Weaving elephants and other creatures circling endlessly in their cages seem only to have confirmed for some that circus animals were improved by man's care and control because, outside the human-directed performances of the circus ring, their habits seemed to be so pointless and mechanical.

The *Harper's* account was a typical "Winter Quarters" piece, this one created to promote the upcoming Barnum & Bailey spring season, even if it did show elephants and others straining under circus confinement regimes. Of course, at the time, most observers would not have interpreted the scene that way. Circus fans and journalists seem to have perceived elephant weaving as another curiosity of the general unconventionality of the circus menagerie. Circus staff knew that elephant weaving was a byproduct of the voyage across the Atlantic and large-scale management in the United States, although they could certainly accommodate it behind the scenes with little trouble. For us, the weaving of the elephants raises the issue of precisely what kind of elephant the rail circuses created. As human-elephant interaction in captivity had produced the "unruly" bulls of the 1840s and 1860s, elephants and people in Gilded Age circuses, which came to epitomize the classical era of the American circus, created social communities and types of "yard culture" that produced particular kinds of human and elephant action that had not existed before.[3] Large-scale herd management would see elephant men and their crews finding quasi-rationalized ways to make the most of elephant behaviors and abilities that served company goals, like remembering performance steps and lifting, pushing, and pulling heavy equipment, while suppressing elephant activities that did not produce value. In this way, elephants interacted with the mechanics of their management to become a kind of industrialized circus elephant, a being who showed how she labored to suppress species-typical behaviors by swaying and rocking on the picket line and in the rail car. The weaving industrial rail circus elephant would become the normative elephant for many Gilded Age and modern Americans, most of whom had never seen elephants in any other material context.

At mid-century, one of the most obvious signs of the increasing spending power of audiences was the growing size of American circuses, complex entities that made some people nostalgic for antebellum menageries. The older dog and pony show style wagon circus still worked more remote parts of the country, traveling along the trails and roads traveled by peddlers and the medicine shows. Like the prewar wagon show, these small productions moved slowly and could adapt to local humor. The large-scale rail circuses of the 1870s and beyond produced more uniform and broadly accessible content, which did not need to be varied between performances. Many entertainers argued that those bigger businesses were better businesses because, even if saddled with lower profit margins per customer, their higher volume of ticket sales produced greater net earnings than before.[4]

Circus impresarios and upper management gambled that consolidated circuses would indeed have profitable near-monopolies over preferred routes, and they engaged in a rash of mergers and buyouts over the last two decades of the century. A number of groups combined into entities that had ponderously long names, to credit all the previous proprietors, "Barnum & Bailey, Ringling Brothers Circus" being only the most famous early-twentieth-century example.[5] Although such big businesses did much to regulate whole industries and make them more efficient (even if less adaptable to local markets), many people wondered if some were morally bankrupt and deleterious to those employed by them.[6] Circus banter often poked fun at the drive for size in postwar business culture and the precarious ventures were often made to appear more formidable or progressive than they actually were. "Time has been when a 'great menagerie' would consist of a black bear and a blue-nosed baboon, and a juvenile elephant would in itself constitute a 'mammoth establishment.' In those days a bill for a menagerie would be the size of half a sheet of foolscap," one early showman joked. "Now-a-days a capital is required to start a respectable show, which would suffice to set a country bank in operation. And a clown to a circus must travel with more pomp than the President of the United States. . . . Now the announcement of a traveling show will cover the entire front of a moderate sized house."[7]

Indeed, circus ad men and the advance teams that blanketed towns, newspapers, and other public spaces with surreal, hyping circus promotional materials had helped marketers coin the phrase "bill it like a circus" to describe any over-the-top, high-cost advertising scheme exuding a winner-take-all ethos.[8] Such advertising strategies were highly visible to the public and made circuses seem bigger and more modern than they actually were. They were also a reflection of the determination of their proprietors—or their satire on fellow impresarios—in hunting down opportunities in a broad culture industry. Postwar circuses were important innovators in the marshaling of people, animal power, and technology, and their managers learned how to extract maximum revenue from a nation with limited infrastructure and a mobile population spread over a large landmass.

It had taken showmen almost a century to develop the world's preeminent mass entertainment in these ways. In 1815, while the second Betsey walked up and down the East Coast, a few showmen had begun this gradual development toward more technologically sophisticated shows when they introduced the commercial wagon menagerie to the nation.[9] Going inland by the 1830s, early shows traveled on rural trails, and one show brought the first elephant to Missouri and Michigan. To maximize earnings relative to travel costs, shows like the 1845 Great Philadelphia Zoological Garden visited ten advertised stops in the cities, plus, on intervening travel days, little settlements along the way to bring in money whenever and wherever possible.[10]

Thereafter, impresarios Gilbert Spalding and Charles Rogers are credited for having pioneered transport and management practices that could put those bigger shows in more remote places more efficiently. Spalding, as the legend goes, was a drug and paint store proprietor who had a menagerie land in his lap, in a manner of speaking, when it was offered as payment for a bad debt from a customer. With Rogers, a former trick horse rider, he conceived and implemented the first "Floating Palace" circus. Assembled on a barge on the Mississippi and towed by a steamer, the outfit could give three shows a day since the crew did not have to pack up their gear and animals in order to find a new audience. Thereafter various companies used steamboats to efficiently ply rural routes, a habit that predicted the move to rail. In both cases, companies accepted routes limited to the nation's main arteries, fueled at mid-century by a burst of public and private infrastructure development. They found there much of the audience they sought, people who similarly congregated along the paths of trade to find a living. In small towns and in cities, local officials increased the expenses of the business, nonetheless, by taxing shows to recapture community income extracted by menageries, commonly $25 in small towns during the 1850s, for instance.[11]

By this time many rural shows performed for two thousand people at a time, while urban shows might have space for four thousand or more. Two- and three-ring rail circuses appeared not long afterward, the belief being that more than one ring held the whole audience's attention with an unobstructed view of the closest ring. This strategy prevented people from standing up in their seats to get a better look and irritating the patrons seated behind them.[12] It also enhanced industry aesthetics of juxtaposition of abundant spectacle and, some believed, brought the same customers to multiple performances in order to take in everything a show had to offer.[13] These circuses featured "Spectacular" parades, too, and zoological-style menageries and diverse midway acts that became the state-of-the-art in the genre. To facilitate this expansion with efficiency, Spalding and Rogers were reputed to have additionally established for land-based shows the use of portable "knock-down seats," larger tents, ramps for quicker loading of people and gear in wagons, and gas lighting to facilitate night performances.

These were the elements of the modern American circus as it would proliferate after the Civil War, and accordingly, in 1856 Spalding and Rogers' menagerie became the first show known to travel by rail.[14] Rail travel freed show units from the "muddy or dusty country roads and the few macadamized, planked, or corduroy turnpikes along the Atlantic seaboard" that made for slow, uncomfortable, and thus expensive travel. Circus impresario William Coup explained that in the 1860s and 1870s American roads "were in a terrible condition—so bad that slight rains would convert them into seas of mud, and a continued rainstorm would make them impassable."[15] Indeed, often the shows went not simply where proprietors believed there was demand but wherever they could by finding any passable route, hoping to find demand once they got there.[16]

This was a period when a small number of companies was also experimenting with routes in California. Then a land still partly controlled by Native American nations and long-time Hispanic inhabitants, it was increasingly overrun with gold prospectors and others drawn to opportunities there for missionizing, agriculture, ranching businesses, or the white collar professions they supported. The juvenile Asian elephant pair known as Victoria and Albert, famous for their work walking inclined planks, traveled in 1859 to California by the long sea route around South America. For some reason unknown to us today, Richard Sands, the owner of the menagerie company that held them, decided not to show them and sold them to John Wilson, who showed them in California for some years. Many elephants similarly passed between companies and did not normally keep a dedicated trainer or keeper their whole lives. Often renamed, they might be sold repeatedly and shipped great distances.[17] The teams of elephants driven overland by the larger wagon shows grew apace with their ventures until rail circuses, which could contain and move the largest herds of all, appeared in the 1870s. Industry wide, people still believed that elephants epitomized the mammoth show aesthetic, now with an element of patriotism that likened the rising global status of the United States to the American entertainment impresario. The rail circuses agreed that the more elephants the better to get that aesthetic across to the public, with the result that elephants, as purported genial and powerful circus workers, unknowingly replicated their owners' claims to supremacy and modernity.

This was especially so in light of Anglo-American expansion across the continent to the Pacific, symbolized by the "closing of the West," when Native American armed resistance to the U.S. military symbolically collapsed at the Battle of Wounded Knee in 1890. The concurrent proliferation of regional and transcontinental railway lines and industrial conveyor belts employing bison leather facilitated the near extinction of that species and their replacement with commercial animals serving national markets, including cattle, horses, mules, pigs, and sheep. Railways also brought circus elephants into the spaces vacated by the bison. These were similarly living animals marketed and distributed by modern means to consumers, although not a self-supporting population like the feral cattle of the southwest, nor easily bred like urban pigs and horses, for example. Yet, by offering surreal and exotic entertainment, rail circuses may have smoothed over the unsettling power of railway monopolies, volatile and distant commodities markets, or the misery of factory work, even if they used the very same railways, entrepreneurial ideals, and industrial management techniques behind those other phenomena to put the show on the road.[18]

Management practices inside these mobile businesses increasingly reflected the centrally mandated "visible hand" managerial and corporate revolutions of the period, also found in mining, banking, insurance, communication, transportation companies, and other large ventures controlling vast capital and complicated organizational bureaucracies.[19] The rail circuses' contribution to this pattern was to innovate mobile entertainment production through flexible plans by which the circuses might own but also rent or lease acts and equipment to put together all the diverse people, animals, and materials necessary to promote and provide an enormous show identified with a famous company brand for the largest volume of customers. Contemporary wisdom in the business community had it— rightly—that as the scale of a company's operations increased, there were more opportunities for one part of the organization to conflict with another. The coordination of many parties was far more complex than the coordination of two or three, and management needed to grow faster than the company to keep things moving smoothly.[20]

Railway managers had been pioneering these theories and kinds of analysis, although circus managers were well aware of the principle of expanding complexities and accelerating management costs. They had an enormous task in monitoring and managing huge circus staffs, which included tent repair men; accountants to monitor rolling stock, railway fees, telegraph charges, and salary and materials for the advance men; press agents and bill posters who created and distributed advertising and promotional newspaper notices; show producers and choreographers; costume makers; warehousing and winter quarters maintenance staff; and lawyers to create and enforce contracts, payment schedules, and telegraph fees; and more. There were dozens of performers and minders to hire, as well as cooks and servants for the dining tents and hundreds of manual laborers to manage, erect, and take down tents and service wagons.

The industrialized circuses also made other efforts to rationalize their operations by controlling risk, corruption, and operating costs. For instance, with respect to venue choice, the "decision as to exactly *where* to pitch the tents became a matter of intense calculation: planners tried to account for such factors as the weather, the previous year's receipts, rivals' intentions, and the prevailing labor situation. Playing the big cities was less important than maximizing the number of people and minimizing the distance they (and the circus itself) had to travel. The small town of Maryville, Missouri, for example, was considered a more attractive locale than St. Louis," Jim Cullen has explained.[21] That kind of organization went hand in hand with timed logistical strategies that appeared so impressive that even the military in various nations were purported to have studied American circuses to see how to make their own personnel more proficient in setup, take down, and movement of people, animals, and equipment.[22]

We should not confuse, however, the careful structuring and timing of the practical workings of the circuses on the ground with the decision making at every level or in every situation in these companies, many of which went into bankruptcy or were sold off to rivals at a fraction of their initial cost. A sort of elephant arms race had begun in 1851, when the century's entertainment titan, P. T. Barnum, entered the merging circus-menagerie business. In a partnership with circus veteran Seth Howes, he changed the stakes industry-wide by amassing an advertised "TEAM OF TEN ELEPHANTS" in a single production called the Asiatic Caravan, Museum, and Menagerie.[23] Barnum was a famous speculator and in effect posed the troubling question to the nation and the trade of whether a herd of ten elephants on one show was good for his business or even the circus business as a whole. This question would seemingly go unanswered as rail circuses made herds standard and accepted the high costs of purchase, management, and disposition of elephants as unavoidable. And although Barnum and Howes's venture made plenty of money for four years, many in the press criticized it as "a shabby affair" suffering low production values, in spite of the elephantine spectacle it offered.[24]

Elephants were still the quintessential entertainment industry animal, fueling showbusiness wisdom that elephants functioned "as an advertisement" for circuses that distinguished their entertainment product from all others.[25] Elephant-based marketing, by which "every circus manager seems bound to see how extensive a herd of elephants he can collect," as one circus man put it, reflected the business cultures in which proprietors and managers spent liberally on productions, believing that public knowledge of such venturesome spending drove ticket sales.[26] P. T. Barnum had ostensibly advocated this business philosophy when he told his managers, "What is the secret of success? Advertising—advertising—nothing else. That is the sum and substance of the whole thing. I

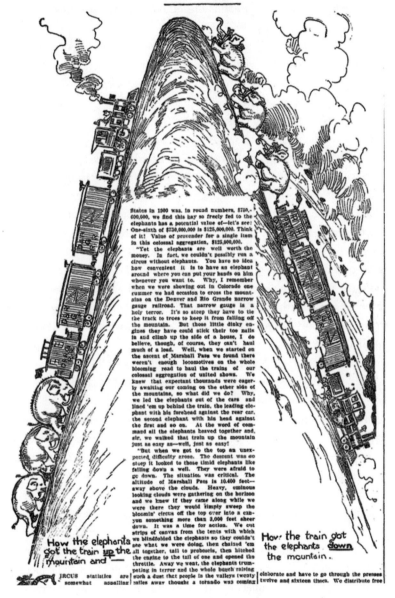

Cartoon metaphor illustrating the reliance of industrial circuses on the imagery and physical power of elephants. *Brooklyn Daily Eagle* (1901).

put advertising ahead of pluck, perseverance and *economy*"[27](emphasis added). Later, Charles Coup agreed that one should not limit the budget for "show and blow" advertising for a circus production, although in any other business such an attitude was surely "reckless."[28]

Whether the largest rail circuses made money because of or despite the massive outlays they made for whole carloads of "paper" (lithographed bills, which were very expensive to produce, distribute, and post along the show route), advance advertising teams, or herds of elephants was never determined, but these expenditures were standard industry culture. Yet, publicly at least, the idea carried that the circus business was about selling not just a ring performance but also an ideal of incredible consumer experience. The broadside advertising, the newspaper press notices and ads, the show programs, the circus day setup, the oral tradition about this elephant or that trainer, the midway food, souvenir, or freak show vendors, and the actual ring performance were all part of the total circus experience package, so the only real question was: When do we ask the public to pay?

Hence, by the 1880s, a few large rail circuses boasted twenty elephants or more as part of a broader plan to draw attention and money away from smaller shows. Adam Forepaugh's outfit even claimed forty elephants for the 1883 season in response to the birth of P. T. Barnum's "baby elephant" at his winter quarters barns in Bridgeport the year before.[29] The "elephant wars" of the 1880s were a result of a generation of gambling management tactics that many people understood as bluster. One joke had a showman bragging to a fellow impresario, "Well, Doc, I had the poorest show on the road last season, but I made stacks of money, and all by advertising." Although "the bill used to say that there was a herd of elephants, including several trained animals and a few wild ones," he admitted to have gotten by with "one little sick elephant, about as big as a horse," a fact ticket buyers discovered *after* they had paid.[30] This kind of humor concerning the industry satirized not only the impresario—always proposing incredible spectacles he often could not provide—but also the credulity of a nation of consumers who seemed to be perpetually fascinated by anything that seemed born of great ambition.

The rail circuses vied with one another for several decades in the "hyperbolic display of elephants."[31] German animal trainer and dealer Carl Hagenbeck institutionalized his elephant collection and distribution company as the most influential in the world by feeding an "exploding" market for the species in the United States during the 1880s.[32] Different grades of animals fetched different prices, depending upon where they were acquired. Cheapest were elephants and other animals acquired directly from hunters in Asia and Africa; an animal for sale in

Asia for $700 could fetch $5,000 in New York. Less costly also were "green," or newly captured animals, because they required a substantial investment of manpower to be ready for a show. Green animals, whether elephants, tigers, hyenas, or unbroken horses, were known to be more dangerous, destructive to company property, and useless as ring stock. Those trading acclimatized and trained animals could demand premiums of several hundred percent. However, when a rival went bankrupt, circuses routinely picked up cast-off animal stock at liquidation auctions for much lower prices, with dangerous "man killer" elephants sold for less still.[33]

In this robust if volatile market, Carl Hagenbeck sought to compete with German dealers Charles and Henry Reiche, who had been importing and selling exotic animals from an office in Hoboken, New Jersey, since 1847. The Reiche brothers and Hagenbeck would have near monopolies on the trade until other men ventured to likewise become dealers of "WILD ANIMALS. Of all kinds," for instance, by placing announcements in the *Billboard* want ads or passing their address and phone number around the circus world more informally.[34] Various specialized traders served growing demand in Europe and the United States. Hagenbeck exported three hundred elephants from Asia and Africa to various nations around the world various nations between the mid-1860s and late 1880s, moving sixty-seven elephants out of Ceylon in 1883 alone.[35] Because elephants from the island were smaller and thus of more uniform size, some circus folk preferred them over elephants from other regions of Asia. They were not exactly interchangeable parts for a circus, but, especially in comparison to a giant like the African elephant Jumbo, they were easier to manage on trains and in the circus lot.

Elephants imported to the United States after the Civil War often came from Ceylon and Africa, where elephant populations had begun to strain under the enormous global ivory trade and hunting motivated by "no object but the excitement of the chase." Various British officials in India and Ceylon became wealthy and politically powerful from their facilitation of the ivory trade, but they also seemed to drive a colonial, manly fashion for hunting elephants who bore no marketable ivory. Some Americans were deeply critical of "the immense numbers annually slain by these gentlemen (butchered seems a better term, when one considers that, while no use at all is made of the huge carcass, to destroy it requires the smallest possible skill of the marksman)," as one *Harper's Monthly* commentary would lament.[36] With up to one hundred thousand elephants estimated at the time to die each year to serve Western consumers, there was a growing awareness that elephants—whether captured for show, hunted for sport, or harvested for ivory—had become components in what today we would call a global commodity chain.

While elephants in captivity in America would create new behaviors and practices to survive, in Africa, elephant cultures were shaped by ivory and sport hunting as well.[37] In the years Jumbo became famous, people in Eastern Africa noted that wild elephant populations there were collapsing. Those elephant communities that persisted contained individuals who adapted to the new reality by retreating to areas in which they knew hunters did not venture or, alternately, by becoming increasingly aggressive toward humans they encountered.[38] Certainly some circuses did not attempt to hide this reality from audiences, self-reflexively contextualizing their own herds with show program accounts of elephant harvesting in Africa and Asia. Around 1869 or 1870, show programs for the P. T. Barnum-backed Great International Caravan, Menagerie, Museum, Aquarium, and Circus, managed by James A. Bailey, included the story of a baby elephant found during an African elephant hunt. Borrowed from a contemporary traveler's account and described as an "affecting incident," it recounted how the tiny calf was the sole survivor of an attack on an elephant matriarch's group. After the guns had gone silent, she emerged from the brush making "mournful piping notes [and] hovering about its mother after she fell. . . . It ran around its mother's corpse with touching demonstrations of grief, piping sorrowfully and vainly attempting to raise her with its tiny trunk."[39] Ethologists have been carefully documenting such elephant "grief" since the 1970s to discuss how group cohesion and survival are aided by elephantine emotions and behaviors, the function of which were sadly apparent to observers even a century earlier.

Barnum's program plainly noted how the hunting crew attempted to keep the calf for export to a zoo or circus, but that "it died . . . in the course of a few days," leaving the moral economy of ivory and animal capture open for the interpretation of the ticket buyer.[40] In fact, such accounts did not necessarily implicate the circuses in the global elephant trade, as they might today, but they introduced to the public the argument that circuses (and captivity to Westerners more generally) were safer places for wild animals than their home territories. This was the nineteenth-century origin of the "zoo ark" concept, which many circus and zoo advocates and their customers have used ever since to argue that circus and zoo captivity is preferable to life in the wild or in nonwhite (today "developing") nations, where animals are endangered by hunting or other human activity.

It is not clear how many circus patrons considered whether their entertainment choices figured into the African elephant crisis or the large-scale harvesting of Asian elephants, since much of the American interest in these issues was a matter for the middle and upper classes. Nonetheless, circus impresarios and their marketing men believed that for most ticket buyers, rationalized elephant labor had become basic to both the meaning of capitalist American circuses and their logistical functioning. Biographical stories detailing where circus elephants originated were apparently less important to modern spectators than they had been in the days of "The Elephant" or Mogul, so the American circus had also become basic to the public meaning of elephants. A press agent for the Barnum & Bailey Company consolidated circus shows wrote a "Winter Quarters" puff piece to appear in *Harper's Monthly Magazine* that captured the sense of how these companies turned the instrumentalization of elephants into promotional patter, claiming that their authority over wild animals made the circuses modern and progressive. "The animals are never ill-treated, but strict discipline is a necessity, in fact a matter of life or death," the *Harper's* account began. The company's staff of six hundred men needed carefully controlled human, horse, and elephant power to move the company's dozens of rail cars, wagons, tents, and nonlaboring animals around the country on schedule. "So perfect is the system of this combination and so skillful the management as a *unit*—[that it is] the result of the experience of a lifetime."[41] The implication here was that in directing animal power to serve company goals while limiting elephant independence, the circuses were serving themselves, their animals, and the nation.

The animal management staff in any given circus company thus included principal animal trainers, keepers, and barn men who worked to keep the animals alive and performing as long as possible, as well as animal presenters who worked pretrained animals in the ring. Animal managers were responsible for the cost of acquiring and managing a wide variety of nonhuman species, from dogs to tigers, horses to elephants. A large rail circus might carry up to five hundred horses (including ring stock), twenty elephants, a menagerie of several dozens of other animals, five hundred human workers, and dozens of various types of wagons and rail cars. And the reality was that, although these companies promoted their elephants and other animal performers as friends or family—that is, as *assets* to their businesses—in the accountant's ledger, they were considered with *variable costs*. Managers took elephants as impermanent inventory and expected to replace animal stock frequently, due to accidental or intentional death, just as they resupplied the hay for the horses and petroleum for the lights. In the business, many insisted that since audiences saw animals as crucial to their enjoyment, while the task of assembling and presenting the most animals possible in a production was unavoidable, the costs of their management should at least be minimized. So did the Barnum & Bailey managers note in one managers' meeting that they were in good financial condition but "determined to keep costs down," a determination that contradicted the company's regular public pronouncements about how "NO EXPENSE HAS BEEN SPARED!!!" in preparing productions.[42]

For the most famous circuses, the costs of business had become enormous. From a rail circus point of view, the simplicity of the days when an elephant would feed herself while carrying her weight to the next exhibition with a single handler was a quaint memory. Properly considered, the principal of accelerating management costs would have required rail circuses to hire, say, two or three new men every time one elephant was added to the herd. That was something these companies rarely did, and consequently the elephant-to-handler ratio did not grow quickly enough to permit keepers to manage elephants simply with a rope and a hook. At the same time, the thinking went, the best way to show herds of elephants and attempt to mitigate the costs of owning, keeping, and containing them was to put them to work in every possible way while restricting any elephant self-direction that cost the company money or frustrated staff. Elephants certainly continued to perform as acrobats, elephants of war, and clowns in the parades and ring shows people paid to see, but, as the circuses and their equipment got larger, trainers put some elephants in their care to work doing the kinds of heavy lifting and pushing jobs these elephants might have performed in an Asian lumber camp or some other work role overseas.

In modern rail circuses the "elephant laborer" worked under the same schedules and efficiency requirements as the equine and human setup teams.[43] This is not to say that elephants perceived their activities as "labor" in the way a human might, although their presence functioned as such. For company owners and staff, elephant work included pulling decorative wagons, performing in the ring, restraining other elephants with ropes under human direction, training for ring performances, lifting tent masts and other heavy objects, pushing rail cars or notoriously heavy decorated circus wagons found stuck in the mud with exhausted horses attached and straining under the whip. In winter quarters, elephants continued their rehearsals and daily "errands of utility, pushing a heavily loaded car here or lifting a heavy weight there . . . [as] a sort of combination switching engine and derrick."[44] This reality exposes the degree to which the circuses were not immune to one of the classic contradictions of industrial history, namely, the degree to which, in order to be logistically or financially viable, technological development was often dependent upon inexpensive human and animal labor.[45] Elephants provided further unpaid work in their advertising and promotional roles. The public appearance of an elephant in a parade helped gen-

Elephants as a power source in modern America. Ronald G. Becker Collection of Charles Eisenmann Photographs, Special Collections Research Center, Syracuse University Library.

erate income for the show, even if that elephant did not appear to be doing anything much but standing by.

Circus managers and animal trainers knew that their charges had to earn their keep as a matter of the basic survival of the company. Some believed that this labor was a fair trade for a life of free food and protection from hunters.[46] In expecting them to do what human workers did but receive in return only hay, the petting and praising words of those keepers they trusted, or physical discipline and restraint when uncooperative, animal managers posited that animals were equals with people, if not in bodily rights or "intelligence," at least in their responsibility to capitalism.

Elephants were oblivious to this human interpretation of their lives in captivity. Yet, the problem of elephants becoming unwilling to cooperate, even in long-practiced routines and management habits, was common in the industry. Elephants periodically altered or rejected movements that were uncomfortable, tiring, seemingly pointless, or otherwise undesirable in some way we cannot know. Animal managers often took noncompliance as evidence of an elephant trying to avoid work and engage in elephant play. For instance, in the 1880s George Arstingstall, the well-known head "Elephant Man" for the Barnum, Bailey & Hutchinson Circus, spoke to the company press agent about one of the company's herd, a young male named Prince, whom Arstingstall called "a great, big school-boy." The press agent looked at Prince and noticed that "the thick skin on his forehead had been badly scraped as though he had been bounding against a stone wall." Arstingstall explained, "I use him sometimes with others when I want to switch off some of the railroad cars on our tracks, and he has scraped his forehead while at work. He is always in for fun, and will quit work at any time to play with anybody."[47] Here Arstingstall portrayed Prince as a naughty child, minimizing the probable discomfort or pain that his work shoving rail cars must have produced. Elephants completed much of this kind of heavy work in years before the use of "heavy leather cushions" for the forehead and special harnesses became more common, in some European circuses at least, to minimize injury to animals.[48]

Either way, the idea of elephant play construed as mischief was a way for circus people to represent to the public elephant behaviors of no logistical value to the circus by giving those acts cultural value as promotional anecdote. The idea of elephant mischief was centuries old, and it attributed to the species a sense of well-meaning humor in foiling human plans. In late-nineteenth-century America it also covered over very real aggravation for elephant men coping with tight schedules and limited manpower.[49] This cultural coping mechanism exposed a human belief that animals like horses, dogs, and elephants somehow understood the human labor cultures they inhabited, interpreting species-typical behaviors that reduced value in the system as intentionally obstructive or impudently lazy.[50]

Elephant dexterity and sentience produced much elephant mischief. While central to the table manners bit or the service of holding a rope restraining another elephant being trained to stand on her head, elephant skills were often counterproductive to a circus when serving elephant interest. For instance, elephants were known to mangle or dirty the expensive head pieces, blankets, or howdahs they required for parades and ring performances by the "habit of amusing themselves by tossing trunkfuls of dirt up over their backs [producing the] sight of them covered with big zebra stripes and leopard spots of yellow mud."[51] Elephant manager George Arstingstall further explained that they would also use their trunks to "spring locks, raise latches, and do almost everything that they see a person do."[52] Anyone who has visited gift shops or museums littered with "Please Do Not Touch" signs in defense against humans driven to put their hands on everything they find visually interesting will recognize the phenomenon of beings innately compelled to explore their surroundings and the effort necessary to suppress that urge. Elephants had abilities to observe and learn that facilitated their imitation of human acts of dexterity in opening locks, untying knots, or otherwise tinkering with their surroundings—sometimes, observers said, with an obvious goal in mind. "They want to know all about everything, and once they do a thing that they ought not to, they will never forget it, but, on the contrary, forever keep it up," Arstingstall would explain. Whether any of these specific incidents of elephant troublemaking, apparent amusement, and "innocent diversion" was true in every detail or not, the basic picture trainers painted was of elephants "mischievous and likewise inquisitive" and desirous of irritating staff.[53] It was also true that elephants were and are born to eat during up to three quarters of their waking hours, since, when at large they must do so to get the nutrition they need. Circus folk accordingly found elephants to be constant fiddlers when chained or contained in a rail car and unable to stroll or browse.

In shows big and small, reports emerged from staff of elephants that seemed interested in knocking over objects and equipment within reach of the trunk or of squashing things with their feet. Some of this was harmless, as with the female who, her keeper claimed, seemed to relish, "when she has been tied up near a row of small electric globes, and left to her own devices *without* a keeper . . . to unscrew these globes one by one and squash them under her foot with the report of a small pistol, her little eyes rolling and twinkling with delight." Sometimes,

however, these trunk habits did come across as mean-spirited. The same informant claimed, "There is a strong sense of humour and cussedness in elephants
and a leaning towards the latter as years roll by."[54] He said that that elephant who
took down the light bulbs had also killed a barn cat with great premeditation:

> With infinite care and trouble [she] attracted the attention of the kitten by slowly
> moving her trunk backwards and forwards along the ground. . . . at last the little
> creature was at the feet of the huge beast. In an instant one large foot was lifted
> some eighteen inches and in a twinkling descended on the unfortunate cat—flat as
> a pancake was the only adequate description of the corpse. Within a second the el-
> ephant had lowered her trunk and thrown the remains twenty yards outside her
> tent. Whether mischief pure and simple or cruelty it is hard to say.[55]

Likewise, George Conklin elaborated that elephant mischief often amounted
to harassment of other circus animals. He said he had trained elephants not to
pester a new horse on the crew by allowing the elephant to get just barely within
trunk's reach of the new horse, who would bite and kick at the elephant as he or
she snuffled around the horse's body. "In this way the elephant found out that it
was better to leave the horse alone, and the horse discovered that he could master
the elephant . . . and it was possible then to put the horse alongside the elephant."[56] Such tactics saved human labor and foisted on the horse the job of
limiting the self-direction of nearby elephants.

Managers and keepers were also keenly aware that much elephant behavior
seemed directed at increasing an elephant's freedom of movement or mental
stimulation and consequently acted against the needs of the circus. "One of Col.
Arstingstall's herd has acquired the habit of pulling up every stake that is driven
to chain it to," one press release claimed. "It watched the manner in which the
men took the stakes up and put the same plan into practice. It kicks the stake
until it is loose and then draws it out of the ground with its trunk."[57] The elephant
who used her trunk to grab at a staff member's hands while being dressed with
a head piece, pulled up her stake to walk off, dragging her ankle chain behind her,
or tampered with circus equipment accidentally left in reach by keepers might be
corrected "by a violent kick upon the trunk from a pair of heavy boots," one critic
would observe. "No doubt this 'little eccentricity' was followed by summary and
severe punishment," in an awkward attempt to deter the elephant from such
behavior in future, he added.[58]

If truth be told, many circus workers actually found elephants' trunks unappealing and annoying, even terrifying. In common form, James Twitchell, an
1880s Buffalo Bill Wild West manager, told of an overnight trip from Chicago to

New York City in a Forepaugh Circus train in which he, a male elephant called Bamboo, and the elephant's keeper were assigned to sleep together in a stock car. As the elephant had been leased by Forepaugh to the Kiralfy Brothers Circus, then about to show in Boston, the keeper was not under surveillance by his employer, and he proceeded to drink the better part of a bottle of whisky and pass out. Twitchell recounted:

> In the still hours of that night, with the train rushing along at the rate of thirty miles an hour, I would at intervals hear a muffled snort from the monster at the other end of the car, and then feel a gigantic foot shoving against me, or the end of his trunk passing inquisitively over my face. Then I would jump up and yell to the keeper, with energetic kicks to emphasize my remarks, "Here, you—wake up! That infernal elephant is going to trample us to death." The drunken keeper would get to his feet, swear, give Bamboo an unmerciful prodding with his fork, the great brute would lie down and cry, and we would have peace. This scene, with variations, is what happened all the way from Chicago to New York. I wasn't trampled to death by the elephant; why I was not I do not know."[59]

Although he surely embellished it for dramatic effect, Twitchell's story reproduced the ambivalence and uncertainty with which many circus people approached captive elephants, interpreting Bamboo's trunk as a tool the elephant would use to carry out nefarious plans with his feet. With limited understandings of their body language or needs, and believing them to be unpredictable, many show people would come to resent the industry's dependence on elephants. To be sure, everyone in the business knew it was negligent, although not uncommon, for an employer to ask men to sleep on the floor of a stock car containing an elephant. Another elephant man would agree (with some bluster), "We 'bull men' live our life with our gigantic pets, sleep near them, ride them, order them around, enter with them into closed cars for long rides. We are not afraid of them. Before many years we become fatalistic. We expect them to kill us some day, but hope it may not be soon."[60]

Like Bamboo's pitchfork-wielding keeper, more than one animal trainer would thusly act to control elephants and their "snaking trunks," which in the cramped quarters of rail cars, barns, and tents seemed to catch people off guard. "I turned and saw the great trunk behind me [—] it had an awful suggestiveness," said one startled visitor to the elephants in a menagerie tent in Utah.[61] For the modern elephant keeper, the interpretation of a youthful elephant's trunk as an endearing and curious appendage that stole gingerbread from children's pockets still remained, as did his enjoyment of a trusted elephant's trunk when it gently took an apple from his hand. Yet, newly relevant was the concomitant idea of some elephant trunks as exasperating weapons, brandished by animals whose self-direction and apparent "intelligence" was a sign of a disingenuous nature. This was an interpretation many Americans had reserved for species taken to be permanently "wild" and unpredictable, like lions, tigers, wolves, or bears, but was now expanded to include elephants in captivity as well—creatures long but mistakenly described as domesticated, although never selectively bred by humans— especially as the domestic population of elephants grew in number, age, and psychological maturity.[62]

Animal managers used their theories about dominance and nonhuman morality to enact punishments against elephants who persisted in asserting their independence by refusing direction or acting to tinker with people or equipment. Yet, on some level, many circus people understood that elephants could not help but move their trunks all the time, snuffling around or interacting with other elephants. "Queen was willfully disobedient and inattentive while Mr. Craven had her in training, and, to punish her he tied her head up with a rope to a great iron rod overhead," one Barnum & Bailey press notice revealed. People in the barn that day may have resented Queen's trunk and ingenuity, but they also seemed to have mixed feelings about how the structures of circus life appeared to make elephants unhappy at times. The notice continued, "Queen's position was an unnatural and uncomfortable one, and excited the sympathy of Hebe, who was chained beside her. She would never have been guilty of the other's fault, but she could not bear to see her punished, and for an hour and a half worked with her trunk patiently but ineffectually to untie the knots that held up Queen's head. When Mr. Craven, more from regard for her evident distress than from sympathy with the culprit, loosed the rope, Hebe was demonstratively glad and exchanged consolatory trunk caresses with Queen."[63]

Of course, late-twentieth-century investigators of elephant kin groups at large in Africa and Asia have documented the same elephant habits of coming to one another's aid when an elephant exhibits signs of frustration or stress, in the wild because threatened by another animal, tangled in wire, stuck in mud, or otherwise in danger.[64] Gilded Age elephant managers had long noted such interelephant "sympathy." Yet, they were witnessing unrelated elephants create cooperative social communities informed by interactions with the humans who kept them captive, sometimes in ambivalent ways. If a trainer was lucky, an older female in a herd would assert herself as "disciplinarian." Since free elephant females and juvenile males live in matrilineal family groups, even in captivity a stand-in matriarch was an authoritative presence. She could calm an uncoopera-tive elephant or persuade him or her to follow human direction, unknowingly functioning as "chief assistant" to the head elephant man, as circus veterinarian Doc Henderson put it.[65]

The bull Hannibal was said to have such great sympathy for the cow known as Queen Anne that his affection for her became a noted moment in circus his-tory. As the story went, the two were owned by one company but kept on different show units until one winter, when they were chained alongside each other in winter quarters. "It was a case of love at first sight," a typical account explained. "The moment Queen Anne was brought into Hannibal's presence, she ran her trunk into his mouth—the elephantine style of kissing. All winter long they were continually caressing each other, and their demonstrations of mutual affection were really extraordinary." Spring came and the unit employing Queen Anne left town with her; Hannibal was inconsolable. "For eleven days he refused to touch a morsel of food, the only nourishment that he received being whisky and water. By dint of a continual swaying or surging against his fastening, he suc-ceeded in breaking loose on the twelfth day, when he took entire possession of the establishment. . . . Hannibal raged around the building, reared on his hind feet, and endeavored to tear down the rafters in the roof with his trunk, but mo-lested some of the animals." It is possible Hannibal was also suffering musth at the time, but, in any event, company staff settled him down the only way they knew how, one magazine explained: "A large force of men was gathered—steel hooks attached to long poles were inserted in his ears and shoulders, and, after great difficulty, he was hobbled and cast, when the customary discipline was ap-plied with the usual satisfactory result."[66]

At other times, the constant close quarters from chaining inspired defiant behaviors and conflict that vied with Hannibal's destructiveness.[67] "Elephantine rivalries" made keepers' work particularly dangerous, especially if a herd con-tained more than one bull elephant. Each male in a herd often had "his special antagonist with whom he would fight if the occasion offered," one press agent explained. The males were "always watching for an opportunity to have a brush, and often in the procession, or when passing into their quarters [they] find an opportunity to strike at each other. More than once it has happened that an ele-phant has had a good 'piece of bark' taken off in such a chance encounter."[68] El-ephants often performed in the circuses showing such elephant-inflicted inju-ries, as well as other sores on legs from chaining or wounds to the body or trunk caused by keepers' heavy-handed use of the hook or pitchfork. This kind of species-typical behavior would add to the growing argument toward the end of the century that herds were best developed with only one male or, better yet, none at all.

Because they suffered various indignities in the business in their own right, circus elephant keepers and barn men often had little patience for elephants who broke equipment, struck people, or got into some other trouble. Often hired en route and given little training, some described going for days without proper food when a circus company employer ran short of cash. They told of having to walk through exhibition grounds turned to giant rutted mud pits by pouring rain, elephants, and wagons, being ordered to work while ill, or being heckled by audiences or the local toughs and hangers-on who picked fights with circus workers. For the many African American men who worked as animal keepers, this was particularly dangerous when working for a circus that traveled the southern routes, where lynching was common in any event.

Like many businesses, circuses often had very high employee turnover rates because they offered seasonal labor to people who went wherever the work was and might not have much loyalty to any given employer. This was a problematic situation, especially for smaller circuses, which were unable to afford relatively experienced elephant men like George Arstingstall or Stuart Craven. Additionally, in the down times between tours, managers of big shows and small ones were either forced or sorely tempted to earn extra income for the company by leasing out their elephants to other companies, as the Forepaugh unit, which owned Bamboo in 1881, had done. While those enterprises saved the expense of paying to store elephants in winter-quarters barns, they also had no guarantee, or perhaps concern, for how that loaned elephant would be treated while away. Men put in charge of leased elephants had good reason to be fearful or uncertain with them, since those elephants might themselves be balky around new people.

The transient nature of human-elephant relationships in the business, combined with elephant mischief and rebreaking practices, produced the cliché of the nefarious elephant keeper. Still a stock character of industry memoirs and circus-themed fiction, the bad elephant keeper was impatient, prone to violence, and usually mentally deranged or a drunkard. The stereotype certainly reflected the rampant alcoholism among men in the nation more broadly, plus the fact that circus work was often lonely and, with respect to large animals, dangerous and stressful. Certainly gossip in the business had it that many of the famed big-cat trainers drank heavily to cope with their work, particularly later in their careers, when they might have experienced several near-death maulings at the paws and jaws of their captives.[69] With respect to keepers responsible for hands-on elephant management, past bad experiences often led to less patience and more hooking as workers cut corners or acted out of fear, resentment, or exasperation with sluggish or balky elephants who in turn might strike them or become even more unpredictable or unresponsive. A fear- and alcohol-driven cycle of human and elephant behavior was in place by this time for many individuals. While some elephants were calm and many keepers wise and patient, the stresses of circus life also produced people and elephants who were less confident or resilient, thus saddling some of the most seemingly modern entertainment businesses in the world with people motivated by the most subjective emotions and habits.

This strained cycle of human-elephant behavior was exacerbated by the inability of circus managers to communicate to nonhuman captives the profit imperatives, marketing message, or scheduling for that show season. Unlike human employees, who accepted a weekly pay packet from the company and (ostensibly) acknowledged that they would abide by the directives of their supervisors, elephants could not enter into such an agreement. Nor could circus people use the threat of firing to enforce their will. Consequently, the heart of the problem animal managers confronted with elephants was one common to those running any company, namely, the question of where to place responsibility for decision making. Circus owners and senior managers overseeing enormous rail circuses understood that they needed to delegate authority to the appropriate employee or else be overrun with work. They hoped that, if they made the correct evaluations of their people and organization, the choice about this problem or that issue would be made by the employee with the best information and the company's overall health in his mind.[70] Disagreements between barn men, trainers, and company owners over the viability or necessity of punishing elephants, over which elephants were too dangerous to be approached except by trusted keepers, which elephants should learn this or that new trick, which elephants should be killed or sold off, exposed exactly this business problem at work. Moreover, elephants who acted out in their own interests complicated staff and management disputes over who should make decisions and who might be acting in the best interest of a given elephant, trainer, or the circus unit taken as a whole.

Joseph T. McCaddon was a senior manager for the Barnum & Bailey Company who differed with some of his staff over the power of elephants in circuses. McCaddon, who had worked with P. T. Barnum, noted in 1891 that "Barnum's favorite story on advertising" was this: "Get yourself talked of and written about, even if you are abused and traduced, if you hope for success in this world."[71] Yet, McCaddon does not appear to have agreed with this no such thing as "bad press" publicity philosophy in all situations. While he seems to have admired Barnum, he was reticent to test the theory that a showman could gain profitable notoriety from almost any elephantine situation.

After Barnum died in 1891, McCaddon continued on as an on-site manager for Barnum & Bailey shows, but he resented difficult elephants. Like all circus managers, he often disagreed with elephant trainers who knew that their privileged position with the show could be diminished if a dangerous elephant was purged from the herd, or who sincerely sought to protect the lives of their charges, knowing they were innocent of company politics. McCaddon argued with the elephant superintendent, George Conklin, during the Barnum & Bailey European Tour (1898–1902) over an adult bull named Fritz. Conklin related how unpredictable Fritz could be, seemingly an obedient and spectacular performer one instant, a danger the next. "One night in Madison Square Garden I had got the rest of the elephants in line as was going to give Fritz the word to lower his head for me to get on his tusks, when I noticed that he was about to go for me," Conklin explained, with little sense of surprise about one particular episode. Although Conklin escaped by making a swift exit from the performance while a keeper distracted Fritz and the rest of the herd, Conklin knew that the "good punishing" he would give Fritz after the show would be only a temporary solution.

McCaddon, Conklin, the deputy elephant man, George Bates, and the rest of the crew thereafter headed to Europe, where Fritz won "his death warrant" several years later while walking with the herd of twenty-four elephants to the rail yard after a performance in Tours, France. Suddenly, Fritz, who was tethered to two extremely young juvenile elephants, Babe and Columbia, "made for Bates," who "ran for his life" along with the rest of the now terrified handlers, except Conklin and one other crewman, named Deafy. Other circus workers scrambled to locate chains, which were on wagons some distance away and buried "under some other stuff, which had to be unloaded to get at them," Conklin remembered. Deafy finally appeared with chains at the local park, where Conklin had enticed the elephant near a tree by "keeping Fritz's attention on trying to get me," Conklin said with some bravado. Once secured to the tree, the other elephant men rematerialized, and "with tackle pulled him to the ground" for rebreaking. Yet, by then, both James Bailey and McCaddon agreed that Fritz should be killed "where he lay." Conklin lobbied to save Fritz because he believed that the bull was merely suffering from musth, or a "bad spell," and that any elephant could be retrained to be docile.

Although Conklin insisted he could punish the elephant and have him in his rail car on the circus's planned schedule, clearly many on his elephant crew

*Dead elephants are good elephants. Tours 1902.*

Souvenir photograph of the body of Fritz, in Tours, France, in 1902, from the scrapbook of Barnum and Bailey "Greatest Show on Earth" manager Joseph T. McCaddon. G. Dagreau, photographer. Princeton University Library.

wished Fritz gone, and Bailey and McCaddon must have known this when they overruled Conklin. Conklin did acknowledge that handlers commonly "became afraid" of particular elephants and that this was a significant management problem in the circuses, amusement parks, and zoos of the period.[72] Thus, on Bailey and McCaddon's orders, with crowds of onlookers at hand in the park in Tours, Conklin and his crew strangled Fritz by wrapping ropes around his neck that were pulled tight for fifteen minutes by the many dozens of company staff members who stepped forward to help. A photograph of Fritz's corpse, which McCaddon placed in his photo scrapbook of the tour, came labeled with the handwritten words "Dead elephants are good elephants."[73] Today, the last photograph of Fritz also shows the unhealthy and possibly painful condition of his feet—which may have contributed to and been exacerbated by his lack of cooperation with human staff—with their overgrown pads and probable infections produced by many years of chaining and manacling, unsanitary housing conditions (elephants' urine is corrosive to their feet) and confinement on hard surfaces.[74] Although circus people understood the need to manage elephants' feet as one did the horses in a company, often this work was infrequently done if an elephant was resistant or no knowledgeable person was on hand to perform it.

McCaddon was typical of many circus people, holding attitudes about the rights and value of various animals in ways that can appear contradictory. Certainly he resented some of the company's elephants, especially independent-minded males like Fritz. Yet, later in life he would write public rebuttals to those who criticized circus animal use, even if it gave circus staff the power to end an animal's life at their convenience. In one case he complained to the editor of London's *Spectator* about their airing of what he saw as uninformed and facile ideas about "cruelty" emanating from "Jack London clubs" and other early-twentieth-century reform groups. To get across the industry perspective—that no captive animal should live a life of leisure—he quoted William Hornaday's 1922 *Minds and Manners of Wild Animals*: "Who gave to any warm blooded animal that consumes food and requires shelter the right to live without work? No one!" McCaddon signed the letter by naming himself authoritatively as "Late Managing Director Barnum & Bailey's Greatest Show on Earth and Buffalo Bill's Wild West."[75]

At the same time, McCaddon appears to have been a horse lover or at least to have appreciated the SPCA's urban horse euthanasia services, which helped owners of large numbers of horses facilitate collection on equine insurance policies.[76] In the mid-1890s, the rolls of the Pennsylvania Society for the Prevention of Cruelty to Animals (PSPCA) listed McCaddon as a lifetime member. That decade the PSPCA ran an ambulance service and veterinary barn for work horses and addressed many public complaints of animal suffering. Considering the agency's investigations of "abandoning [horses] to die," "impaling live calf on iron hook," "burning rats with a red hot iron," "dashing out a dog's brains" (three cases in 1894), circus people like McCaddon may have believed the privations of their own animals paled in comparison.[77] Regardless of any loose consensus that the public should not critique circus animal management, staff inside each show unit nonetheless disagreed about specific animals and specific situations, all the while keeping these disputes private for the sake of their jobs. The public deaths of these bulls, as ferocious and mean as circus publicity made them out to be, nonetheless drew out the regular "cynics." For instance, Henry Bergh wrote a stern letter to P. T. Barnum in 1883 accusing his elephant men of having turned a bull named Pilot dangerous by being unwilling to accept that "kindness instead of cruelty would have subdued the animal."[78] As critics like Bergh and others sympathetic to animal-use reform perceived it, such complaints were met by "contemptuous chuckling," which revealed a self-satisfied attitude on the part of circus management and animal trainers, who cast the concerns of outsiders as "the sentimental tomfoolery of philanthropic busybodies, [who needed] to supply themselves with commonsense."[79]

However, McCaddon and Conklin actually agreed that it was necessary to limit circus elephant independence for elephants' own good. Elephants thus unknowingly contributed to their own intensive confinement. By going into musth, fighting among themselves, walking away from trainers, or manipulating circus equipment, elephants put workers and managers behind schedule or on edge. Elephant management tactics attempted to control the amount of power animals could exert over the company. An elephant was not like a horse, "just smart enough to be a perfect worker without taking too much initiative." With their curious minds, dexterous trunks, and massive power, circus elephants often took the initiative to make decisions for themselves that complicated human debates over where the power for decision making should lie in a circus organization.[80]

Further, circus company imperatives required staff to modify a given animal's agency to human needs while keeping it alive with minimum inputs of feed, space, and time. It meant that keepers needed an elephant to see the hay and know to put it in his mouth with his trunk, chew, and swallow and to remember a performance routine's movements in correct order on cue (and the consequences of not making those movements on command). At the same time, circuses needed an elephant *not* to use his abilities of learning and self-direction to unfasten his tether, break equipment within trunk reach, attack company horses, tinker with expensive headdresses and other props, throw water or mud on herself or others, kick out the stake to which he was chained, strike with his trunk the particular keeper who was the most injurious with the hook, or walk out of a performance to smash a wagon parked outside, inadvertently panicking the audience.

Thus would circus staff invest considerable energy in developing intensive confinement technologies to limit or suppress elephant action that did not produce value for the company. When not being worked in the ring or rehearsed, elephants spent most of the day and night shackled to posts and stakes in picket lines or crated inside specially reinforced rail cars. When traveling by rail for more than a few hours, most animals were allowed out of their cars only once a day to eat, drink, and recuperate from jostling (no animal could "hold on" when traveling and constant shipping was exhausting to many).[81] Elephants older than

The bull Fritz stepping into his ventilated stock car, which could not be made larger without becoming too high to travel under rail overpass bridges. "Bed Time," ca. 1897–1902. Princeton University Library.

five years of age or so often wore manacle harnesses made of leather straps and runs of chain that bound their chests to shackles on their front ankles or to a headpiece like a horse bridle. If left to their own devices without such fetters, many elephants walked and moved their trunks with such consistency and force that they would be uncontrollable unless one were to employ large teams of men bearing goads and pitchforks to monitor them around the clock. Elephant agency and the profit requirements of the circus companies intersected to necessitate the continued use of elephant hooks and pitchforks as well as the stakes, manacles, and harnesses that became standard equipment in Gilded Age circuses (and, later, many zoos) because they reduced the need for infeasible levels of human labor.

At the same time, the move to continual confinement paradoxically hobbled both elephants and their keepers, giving humans more control but also more responsibility. It was common knowledge in the circus business that an elephant in captivity would "eat all the time, as fancy prompts," consuming at least one hundred fifty pounds of hay per day, and if not distracted by hay might get into

Elephant in head manacle typical of American captivity systems, Lincoln Park Zoo, 1904. Chicago Daily News, Inc., photographer. DN-0003395, Chicago History Museum.

mischief.[82] Attempts to limit those elephant actions that disrupted circuses required companies to supply human labor to provide food for those captives, which elephants had gathered for themselves in the years when menageries and animal exhibitors had driven them from town to town.

George Conklin claimed to have invented "most of the various harnesses, hobbles, tackle, and so on, used in the handling and training of elephants," noting that they served to limit the movement of the trunk or reduce elephant mobility.[83] In fact, many of these techniques originated in Asian practices of tethering and binding elephants during capture and training, practices that men in the animal trades adapted to the overseas transport of elephants. Hoisted on board by a crane, elephants' first introduction to ocean life was "a strange experience to most elephants," witnesses said. While some were kept on the main deck of a ship, smaller elephants would be placed in a sling and "lowered into the hold like bales of cotton" and thereafter confined in crates made of teak or chained to some sturdy beams. One account from India in the 1870s noted that these elephants, mostly juveniles under age five, exhibited signs of great discomfort at being loaded on board: "By watching the eyes of the poor beasts their terror was very manifest. Tears trickled down their mild countenances, and they roared with dread, more especially when being lowered into the hold," it related.[84]

Thereafter, elephants of all types were said by the press and animal men to have "suffered much from a long sea voyage" because confined below decks in "very hot and oppressive" conditions or exposed to the weather while crated above decks. They commonly lost weight and exhibited "bruises and attrition of the skin" as a result of being unable to hold on while the ship rolled in rough seas or from becoming agitated and assaulting their surroundings. They also injured their skins by frequent weaving: "setting to work rocking the ship from side to side, by giving themselves simultaneously, a swinging motion as they stood athwart the ship, the vessel rolling heavily as if in a seaway." Although when hoisted off the ship at the destination, many an elephant appeared to display a "lightness of his heart at being freed from his floating prison," he would regularly sway and rock periodically thereafter.[85] Many in the animal trade and circuses took the habit as a result of elephants' experience in being shipped long distances while confined, which was likewise a necessity for postwar circuses eight months per year. In fact, in Britain, there had been a debate since the early-nineteenth-century death of the elephant Chuny over the degree to which caging and chaining produced elephants whose weaving was evidence that they had become vulnerable to illness and madness. The French naturalist Comte de Buffon had noted elephant weaving as a phenomenon that seemed "as if it were necessary to substitute some exercise for the unrestrained activity of a state of nature" for elephants thus compromised by captivity.[86]

All through the late nineteenth century, people reported seeing elephants who "gave a rocking motion" or were witnessed in the barn "swinging their heads with the incessant . . . motion peculiar to these animals."[87] Some took this to be a habit that pleased elephants and, as the wisdom still goes among some mahouts and circus people, as a positive adaptation to chaining that helped elephants "maintain good blood circulation."[88] Others were focused less on the physiological function of elephant weaving and considered the behavior a window into an elephant's emotional state or experience. Of a juvenile called Baby Ruth, one publicist wrote, "She looked mournful. She stood for hours yesterday morning with her hind feet brought forward between her front feet, so that she looked like a big round ball, swaying her body backward and forward with a very sad expression."[89]

By the turn of the century, people were still reporting elephants "rocking and swaying to and fro."[90] This characteristic had even been absorbed into fictional accounts of wild Asian elephants, like the main character of Ellen Velvin's *Rataplan* (1902): "In a wearied but restless manner . . . every now and then he would lift one of his massive legs and put it down again, or sway his whole body from side to side."[91] At the same time, descriptions of elephant weaving did not appear in the natural-history style booklets and show patter the circuses used to promote and enhance their shows, although these publications often contained a goodly amount of accurate information about what elephants might eat, where they originated, or how they used their trunks.[92] Nor did circus advertisements, color lithograph broadsides, handbills, or show programs depict this action. It was the ubiquitous circus press notices that most often described elephant weaving, probably because elephants were unchained and focused on a handler when performing in the ring or a street parade, and elephant rocking was primarily a backstage aspect of their routines. Yet, this swaying was also a public performance when it appeared in print or was witnessed by a ticket buyer who saw an elephant weaving on his picket line in the menagerie tent. So, when company animal managers, audiences, and the press took elephant weaving as healthy, if ponderous, elephant behavior, they took the instrumentalized elephant of the industrial circus, the normative Gilded Age American elephant, which engaged in such pointless behavior, to be an inherently flawed species.

Nineteenth-century elephant keepers and trainers were correct that elephant weaving was a behavior produced by elephant confinement. Today known as stereotypy or abnormal repetitive behavior (with terminology dependent on the scholar and his or her theory of what is happening in the brains of such animals), elephant weaving is explained as a "repetitive, unvarying behaviour with no obvious goal or function." In other kinds of animals, stereotypies include cage pacing, chronic grooming, repeated jumping or tumbling, which press agents and menagerie and circus customers had been noting for decades. And, like Buffon and other observers of the nineteenth century, researchers over the last thirty years have argued that stereotypic behaviors help animals cope with confinement by reducing the anxiety caused by an inability to "express highly motivated natural behaviors" or by the expectation of a stressful "management event," like being driven from one building or cage to another.[93] Although the complete mechanism behind this phenomenon is not fully documented, today it is known to correlate with suppressed immune systems, unpredictable aggression, and often, early death.[94] Consequently, elephant scientists and many elephant keepers have come to interpret weaving as a highly visible indicator of underfunded management programs employing intimidation and punishment and often poor diet and lack of veterinary care or proper exercise.[95]

Stuart Craven and many an elephant man in the Gilded Age reported all these phenomena with respect to members of their herds. They freely admitted that, although they knew more than their antebellum forerunners, their elephants often ate moldy hay and unhealthy foods fed to them by visitors and were ruled by breaking and pain-aversion training. Furthermore, they publicly explained that some individuals were unpredictable and dangerous, that elephant crews kept their charges confined when not in the ring or a parade, that there were no elephant veterinarians on staff so one had to improvise cures for elephant ills, and that elephants were susceptible to sudden death from chills or other mysterious causes. Inhabiting a radically different human culture than ours, in the early twenty-first century, and having limited financial and manpower resources, elephant men like Craven, Conklin, and Arstingstall approached their elephant herds from a difficult position, in which they would have to balance attempts to remedy these elephant troubles against the demands of their employers, the turnover in their crews, the expectations of critics, and a fickle and often badly behaved circus-going public.

Still, some elephant superintendents and their staff sought to do right by their elephant herds, despite assertively hooking them and resorting to manacles to control their mischievous or "treacherous" acts. At the larger circuses, men worked to refine elephant captivity, believing as they did that bigger circuses were actually better circuses for elephants, a belief that agreed with the industry's fascination with mergers and consolidations. Stuart Craven, the Texan head elephant man for Cooper and Bailey circuses, said of elephants kept by the prewar menageries: "In those days they were subjected in the tenting season to long and fatiguing marches, were insufficiently fed, were exposed to severe weather, which they were ill-fitted to endure, in spring and autumn, and were not given time or suitable surroundings in the winter season to recuperate before their toils recommenced in the spring; so that they never felt really well and fully themselves." For Craven, the advent of the rail circus herd was a way to provide not only a more modern show but also better elephant husbandry: "Knowing what I do of their nervous disposition, and seeing of what affection for each other they are capable, I do not wonder that under the old system the bulls used to get frantic with loneliness in their isolation. Now, they are warmly housed in winter, walk only in street parades, are well fed, kindly treated and kept in herds, so that they have companionship."[96] Did "frantic with loneliness" mean that an elephant was experiencing musth? Certainly that phenomenon continued, although circus crews may have been more adept at managing it than those early menagerie men who first confronted bull elephant maturity in the 1840s.

Today Craven's observations present a jumble of pros and cons for elephants. Confinement and restriction of their legs and trunks was necessary, Craven held. However, other species-typical traits, such as the desire of elephants to be with other elephants, could be accommodated because they happened to coalesce with the mammoth show promotional cultures and bigger-is-better business strategies built on elephant herds. The argument that large rail circuses were more legitimate keepers of elephants than small wagon circuses, which still drove elephants rather than confining them in cramped rail cars, was an ingenious and possibly sincere assertion. Men like Craven had a point in that we know now that juveniles and females probably suffered when kept in singles or in pairs. Solitude has been shown to produce mental distress and such physiological symptoms as a suppressed immune system. On the other hand, the elephant who walked from show to show with one or two others satisfied needs to browse and move that were and are central to elephant physiology and instinct. We also know today that elephants need to walk to avoid foot disease, arthritis, and obesity. So Craven's talk of elephants who walk "only in street parades" was either a well-meaning projection of human preference onto his captives or an attempt to portray industry-standard practices of continuous confinement of elephants, devised for the logistical needs of the business, as an industrial innovation that elephants endorsed as genial celebrity performers in modern rail circuses.

In fact, there would be moments of probable relief for these elephants in the rail circuses when their managers allowed them a chance to do things they could see elephants liked *if* those activities happened to promote the company's work. For instance, elephant handlers sometimes took their herds to bodies of water to bathe. Standing, rolling, and laying in water is something that elephants at large in Asia and Africa do regularly to keep their skins clean, play, and otherwise socialize and promote group survival. This practice might also have saved keepers' labor in washing down company elephants, while creating a promotional spectacle that gave the illusion of elephants' free participation in the shows more generally. Street parades that publicized show openings were just such events, to be sure, although they did not save later labor in preparation for the ring show.

As the herds grew and some handlers continued periodically to take the company elephants, led by the eldest female, for a bath, elephants and their handlers were endangered at risk because members of the public were often ignorant of elephant behavior and saw such a procession as a provocative invitation. For instance, in June 1882, Boston citizens got a free show when Barnum & Bailey elephant handlers drove their herd, including the African elephant Jumbo, to bathe in Frog Pond on Boston Common. Jumbo's handler, William Scott, head elephant man George Arstingstall, and an assistant were riding on horseback to "direct the tractable monsters" when the situation became complicated. Suddenly a barking dog accosted the group and attempted to bite at the elephants' trunks. The elephants became noisy and distracted, and they were further alarmed by a runaway horse who passed by, equally startled by the elephants on the road. Then some "hooting" boys appeared and set off firecrackers around the group of men and elephants to see what would happen. The elephants stuck their ears out perpendicular to their bodies in a sign of alarm and "rushed through the Common Gate" to the water before their keepers could slow them down. The *Boston Journal* surmised, "Owing to the fact that they were rather heated and the younger elephants were frightened, they did not enjoy the bath as much as they would otherwise have done." Nonetheless, the older elephants "rolled and lolled in the water," leaving the pond water muddy when they responded to their keepers and walked back to the circus grounds.[97]

Trainers and keepers knew many of their elephants could be powerfully unpredictable—or at least that *they*, especially if only recently hired or working with an elephant the staff feared, could not predict when an elephant might act out in apparent fear or anger. Additionally, city streets were contexts over which circus people really had no control, and some young citizens saw a parade of elephants as an opportunity to cause trouble. To be sure, many circus workers understood that it was their job to protect skittish elephants from the public as much as it was to protect the public from unpredictable elephants.

As fun as the spectacle at Frog Pond was for people on the sidewalk that day, the truth was that this kind of practice was very dangerous for the circuses. Arstingstall's crew would again lose control of their herd in August at Troy, New York, in an incident in which several members of the public were injured, which newspapers across the country noted.[98] Rail circus elephants who experienced almost continuous confinement and periodic breaking had reduced behavioral repertoires and tended to become dependent upon routines in order to remain calm. "Consistency, consistency, consistency" has long been the tried-and-true method of the elephant manager who knows that routines give elephants some sense of predictability and control over their surroundings.[99] In new situations, they had been trained to look to human handlers for direction, yet often they did not do so. Moreover, although some elephants became worldly and calm over time, for others, their experiences produced an exaggerated startle response that people talked about as "unpredictability." Indeed, Stuart Craven would parrot the conventional wisdom about elephants in captivity when he noted the ease with which they startled: "Big as they are, elephants are timid, nervous and excitable animals. They worry easily, and are frightened sometimes by the most insignificant things."[100] Elephants in public, untethered and handled only by a man with a hook, potentially threatened themselves and public safety. Circus people would persist with pond visits and the celebrated city street parades nonetheless, as they advertised a show cheaply. And, although elephant crews would do everything in their power to prevent disasters in public, they knew that, on some level, many spectators secretly hoped for an elephant panic or a "rampage" in their city or town.

When members of the public considered the ways tragedy seemed to stalk the circuses, it was barn and tent fires or the train wreck disasters that provided the most obvious moments of animal suffering and terror. While the circuses did not publicize these disasters, newspaper accounts often gave reports of "Wild Animals Burned Alive" in circus conflagrations. The *New York Times* reported on one 1879 Detroit blaze in which

> the whole interior of this place was in flames, and the air resounded with the frantic cries of the frightened animals, which sprang fiercely against their bars in a vain effort to escape. . . . In one corner of the barn was securely chained the huge performing elephant . . . now almost frantic with fear, and whose shrill trumpetings rang out loud and clear above the babel of sounds. But the fastenings were secure, . . . leaving the poor brute to roast alive where he stood. . . . Two of the lions fell upon each other and fought desperately, . . . and a dozen men hastily armed themselves to shoot down any poor, maddened, howling beasts should they break through the cordon of flames.[101]

William Coup remembered a similarly tragic train wreck in which, "as the cars had been thrown on their ends, in each horse car twenty horses were thrown into a struggling heap. . . . The elephants were piled up in much the same way as the horses, and in order to extricate them it was necessary to strip the cars completely."[102] Certainly many Americans were saddened by these events and identified with the behavior of the animals caught in such situations—as they had with Horatio when he fell from the bridge and with other elephants over time—assuming that the animals held by circuses were ultimately innocent in these events yet suffered death or injuries that were frightening and painful to them.

As terrible as those public moments of animal misery were, however, elephants probably did not worry about the potential of a train accident or barn fire. Their immediate experience was of the daily grind of confinement, training, discipline, and ill health. Elephants who fared badly in those conditions and became dangerous inspired their handlers—themselves often struggling to cope, drinking heavily or vowing every day to quit and never work on an elephant crew again—to call for more of the hands-on management and behavior-modification practices that drove the precise behaviors that seemed so destructive to circus routines and to elephant and human health.

It was not that more liberal modes of management might not have allowed for better mental and physical health for elephants while increasing worker safety in the circuses. However, the modes of hands-off handling that circus people used with the big cats, for instance (or that many zoo elephant keepers use today) contradicted the very essence of the circus entertainment experience. These businesses were not refuges or conservation societies, who could ask the consumer to fund the lives of elephants to which they did not get close and extraordinary access. There was probably no person in the nation who could manage an elephant without goads, chains, and other modes of containment while meeting circuses' rigorous show and training schedules. Elephants owned by these companies did not collectively and consciously resist circus management as a group of humans might, by sabotaging key equipment, engaging in slowdowns or strikes, or simply quitting. Yet the routines of their captivity caused elephants to behave in species-typical ways that often produced the same effects. Indeed, the transformation of wild-born elephants into such quasi-industrial animals would problematize the marketing icon of the happy elephant performer for modern Americans thereafter.

# CHAPTER SIX

# Going Off Script

In 1900, impresario August Kober pondered the great irony of circus history. The ambition of man's domination over beasts had long been a theme in circus advertising and performance, he said. Yet, circus people everywhere knew they relied upon animal power they controlled only through systematic confinement and an ability to persuade animals to either fear or respect human keepers. Thinking of the fragile human-animal relationships that made the cultural and financial success of his business possible, Kober imagined: "If those twenty-one elephants really woke up one day and became conscious of their strength, how easily they would burst their chains! They could pound their platform to bits, tear down their canvas stable and wreck our entire circus, if they but knew their own power, and every night that I take a promenade through our circus stables with their hosts of elephants, lions, tigers, bears, leopards and buffaloes, I visualize such an awakening of the animal kingdom."[1] Kober's disturbing visions revealed that on some conscious or unconscious level many circus people believed that their animals might smash the means of their captivity if they understood the nature of human power in the world. Kober drew up the metaphor of an elephant tricked by humans into forgetting his own power to indicate to circus fans the mystery at the root of modern circus entertainment: Why did powerful animals like elephants tolerate puny human beings? Why were they so easily intimidated by a person with a chain and a pitchfork? Why did they not band together to defeat their keepers?

Elephants had sentience and an agency that helped them respond to circus managers and trainers and to survive for periods of years in a series of menageries or circuses, but they possessed no human understanding of the cultures that built and organized those ventures, and they had no real power to challenge their captivity in an organized or lasting manner. Kober and others banked on that fact. Day after day, year after year, most captive animals acquiesced to human direction by performing on cue. Most refrained from overpowering the circus staff and running away. Thus circus companies were able to provide the spectacle of potentially dangerous animals performing for human amusement by way of behavior-modification regimes that endeavored to draw out of elephants actions that served the company while suppressing behaviors that did not. It was this skill that gave the circus impresarios and animal trainers their reputation for modernity and efficiency in exploiting and regulating the risks of animal power.

Nonetheless, with their massive herds and mobile containment and behavior-modification practices, the biggest circuses were testing the limits of the elephants in their possession. Their elephants struggled to keep healthy and calm while unable to satisfy hardwired urges to walk, fully explore their surroundings with the trunk, browse for food, or interact freely with other elephants. They weaved and rocked to survive in small spaces, laboring with the resulting suppressed immune systems and undiagnosed infections, chronic foot disease, or stunt-driven arthritis. They tolerated hands-on management out of fear of the men who subdued them, the pain of the goad, and the barn man's pitchfork. Many circus people believed elephants did so only by suppressing their desires to strike out at their keepers and trainers, grab hooks and pitchforks, or simply walk through the tent wall to find the hay pile or the pond they could smell nearby.

The circuses could stretch their elephants like elastic, but only so far. Some elephants reached a point at which, for whatever reason, they could no longer muddle through. By the turn of the twentieth century, to the American public, many modern circus elephants did indeed seem furious at the circuses and liable to destroy them, just as Kober had imagined. An increasing number had been making dramatic news since the 1880s by breaking out of enclosures or walking or running from parades or the circus tent to smash property, injure or kill people, horses, and other menagerie animals, thereby exposing moments when industrial circuses had only a tenuous control over their captives. Certainly, audiences had long seen bull elephants portrayed as destructive brutes. The worrying new reality was that female elephants were instigating just as many destructive events. The circuses would struggle to assuage public fears, but they erred by applying to females the same cliché of elephant as furious brute that they had used to characterize bulls since the 1860s.

Thus, the modern female circus character of "mad elephant" was born. She was a villain like her male counterpart, but seemingly more deceptive and vengeful. Thanks to the mad elephant, many twentieth-century Americans would come to fear all circus elephants as "vicious" "brutes" who "mangled," "crushed," or "attacked" people, as the newspapers sensationalized. That strain of public thought demonized elephants but would ultimately show that the public blamed the circuses for putting ostensibly irrational and uncontrollable animals in places where they became public safety hazards. The 1941 Disney film *Dumbo* would in time speak to sympathetic public knowledge of circus elephant dysfunction:

Dumbo's mother, Mrs. Jumbo, is locked away in a wagon as a "Mad Elephant" when she assaults a circus visitor who ridicules Dumbo's ears. The Disney film portrayed her as sagacious and misunderstood, a fierce defender of her son, who suffers intense grief when separated from him. After *Dumbo*, circuses as a rule hid from the public the disposition of elephants made apparently psychotic by circus management, as well as those injured or accidentally killed in their care.

In the meantime, modern Americans appear to have been subversively fascinated by the turn-of-the century phenomenon of the elephant "rampage." Female elephants run amok, like their male conspecifics, offered observers moments of schadenfreude that enhanced the meaning of "circus" as an entertainment defined by glamour and danger, joy, humor, and fear. Circuses represented the height of human organization and efficiency, backed up by modern technology and managerial expertise, but they could be turned upside down by an "angry" elephant. The mad female elephant particularly coincided with a growing and aging elephant population in the United States. Heavy importation during the Gilded Age had drawn many inexperienced people into the industry as handlers, while the elephants they managed each year grew older, an industry context that multiplied the opportunities for elephantine trouble. "There is a constant demand for 'bull men,' but few will accept the position, and most of us fall into the business by accident," elephant handler John Eck explained in 1907. Although an animal superintendent or head elephant man at a large circus might make "$250 a week," Eck said, most of the men worked with bulls for "small salaries" because of their fascination for elephants and exemption from other, more physical circus labor, such as tent raising.[2]

Consequently, there were two ways elephants and people came to display behavior that generated events of elephant destruction or violent disobedience. First, people or local newspapers at times recast escaped elephants seen walking down the road browsing the trees as "angry" elephants, "rampaging" and causing "a panic," simply because the genre of animal rampage story was a comprehensible and eye-catching old favorite dating back to the days of Hercules, Columbus, and Hannibal.

Second, many female elephants periodically rejected the conditions of their experience because they were stubborn, uninterested, frightened, or aggravated by a particular situation. Overcoming the fear of the bull hook, trainer, or some training not to assault a known person, they walked away from handlers to injure bystanders, frighten audiences and horses, hasten or stroll through towns, evading capture for many hours. Some of these disruptive and often costly escapes stemmed from train accidents that broke open cages, from an elephant's assertive refusal to enter or exit a transfer container, or from an elephant's sudden and apparent desire to destroy his or her rail car from the inside. August Kober had himself advised that, especially when animals were being moved by staff, "it is certainly the most dangerous time, both for the animal itself, as its nerves are liable to be upset, making it scary, and also for the people engaged in the work, as one is much more likely to be bitten, clawed, kicked or squashed, according to the sort of animal one is dealing with."[3]

Other events of elephant non-cooperation occurred during ring performances, which circus men often ascribed to audience noise that seemed to frighten otherwise obedient females into bolting from the circus tent. Later, some might break into nearby tents or buildings, smashing what they found there and sending the people inside running into the street.[4] Observers also said it appeared that the other elephants nearby knew how to raise the alarm when one elephant was being too harshly coerced, was attempting to escape, or had begun inflicting injury on a circus worker. People recalled that elephant trumpeting made these critical situations seem even more stressful. It appeared, they said, that the animals were cooperating at the expense of frantic circus staff, especially when horses, large cats, dogs, and other animals chimed in to the disorienting din.

Of course, circus publicists, owners, and trainers did not wish to encourage public worries that elephant trouble was a damning result of circus management errors and lack of systematic reform—which was exactly how most circus people perceived such breakdowns in animal management. Instead, they often attempted to turn elephants off script into media relations opportunities by explaining these incidents as intriguing evidence of elephant "vagaries" and thus just another incredible mystery of circus life. Americans used that formulation to work through the inscrutability of elephants and the awkwardness of human-elephant communication, explaining elephant vagaries as an indicator of inherent elephant irrationality. For instance, elephants suffering from an exaggerated startle response were commonly described by the circuses and their publicists as having fled the blowing of a locomotive's whistle or the yapping of a small dog, in such cases appealing to the old paradox of the elephant terrified by a mouse.[5]

At one Chicago show, an audience of fifteen thousand watched as one of Barnum's elephants "undertook to vary the entertainment by a special performance not set down on the programme," the *New York Times* sardonically reported in 1883. "Among the features of this performance seems to have been the chasing of the women charioteers around the ring, and as it was probable that this would be followed by an attempt on the part of the elephant, with the support and encouragement of other animals, to 'clean out' the tent, the audience rose and rushed toward the doors."[6] An earlier account argued the whole debacle had been caused by the "baby" elephant in the show, who became distressed for some reason "and went bellowing around the ring, dashing in the procession and smashing one or two of the chariots. To add to the confusion, one of the women charioteers abandoned her horses and they went dashing around the ring. The wild beasts, frightened at the uproar, began bellowing and beating the bars of their cages, and the rest of the elephants became frightened and unmanageable." Worse yet, that report accused, the audience became similarly unmanageable, racing for the exits with such urgency that they threatened to create a stampede in which someone could easily have been trampled.[7]

Whether the audience or the animals were more of a public threat was not clear, but one editorial on the disaster laid the blame on the circus company for assembling unpredictable (to people) animals in an enclosed space containing thousands of spectators: "It is strange that no panic has yet seized upon any one of the vast throngs which Barnum's circus calls together. . . . If Mr. Barnum is as wise as he is popularly supposed to be, he will take every possible measure not only to prevent a panic among his audiences but to enable them when panic-stricken to escape with rapidity and ease."[8] The old showman knew circuses posed dangers, whether tent fire or animal meltdown, and he replied by admitting that his company would take responsibility for any property damage or patron injury caused by its animals. Yet—always looking to turn unexpected situations to his own advantage—he insisted (and so advertised the mammoth industry and thus modernity of his venture), "We also know that these animals are all in perfect subjection to their keepers and are carefully made and kept secure by the 640 men who travel in the interests of the great show." Besides, the young elephant that had started the whole episode had merely "kicked up his heels in childish sport," Barnum said, hoping to persuade.[9] However, real damage to property or life was an enormous risk for Barnum's enterprises or any production that insisted on showing elephants free of cages in the circus tent or city parade, with just a handler there to maintain control.

As with bulls, when female elephants became difficult, circus staff could be equally divided about how to proceed so as to keep them working but limit the potential for accident or rampage. Thus were many females sold off at a discount to a rival company, whose staff would change the elephant's show name in order to obscure her origins and any infamy she bore. Indeed, so often were elephants moved between circuses that their biographers have at times struggled to reconstruct elephant life histories, just as American audiences did at the time.[10]

For modern audiences, the spectacle of an elephant ignoring her handler was becoming increasingly essential to the meaning of the circus. While it was an art form that required regulation and precision, the circus also provided surreal variety through preposterous arrangements of unusual people and extraordinary performances. The circuses reconciled disparate elements in their productions, which seemed especially compelling, haunted as they were by danger and tragedy: glamorous high-wire and trapeze performers who fell to their deaths, beautiful canvas tents that occasionally but dramatically burnt down, decorated circus trains notorious for derailments that killed people and animals. An elephant's rejection of circus routine was costly to the company and any humans or other animals who might be injured, but on another level, for some spectators, it was a fantastic example of the whimsy—or disorder—essential to circus entertainment: during the next show, audiences legitimately knew that anything could happen.

Meanwhile, in the last two decades of the century the middle class would forge new public spaces for debate on show business animal use, tussling with industry insiders over the authority to speak for animals' interests. The older tradition of critique, represented by Henry Bergh's publications or children's stories condemning the treatment of circus elephants, now began appearing in editorials and other comment in major urban newspapers like the *New York Times*. Since the death of Horatio, the newspapers had been fickle allies of the circus companies, ready to promote them one day but report on their disasters the next. With the number of papers and column inches only growing, the largest shows attempted to limit negative coverage by strengthening their partnership with the for-profit popular press. The circuses paid huge sums for newspaper advertising in the latter half of the nineteenth century and accordingly expected that their money and the visual interest their graphics added to newspapers would be acknowledged with friendly editorial policies and polite reviews of circus shows and parades. And, in large part, this was the case. Circus managers also developed cadres of publicists—the famed circus press agents—to make doubly sure that the papers got the circuses' marketing messages over as accurately as possible, often by handing prewritten stories over to editors as free, ready-made content. Only the wealthiest companies could afford the staff for such strategic communications, and by so doing, they often drowned out the publicity and advertising of smaller operations.

The expertise of circus press agents presaged the wisdom of journalist Walter Lippmann. In the 1920s, he would discuss a media maxim the circuses had long known, namely, that the degree to which any company could expect uniform sympathy from every writer and editor—even at a trusted newspaper—was not to be overestimated. "This is the underlying reason for the existence of the press agent," he wrote. "The enormous discretion as to what facts and what impressions shall be reported is steadily convincing every organized group of people that whether it wishes to secure publicity or avoid it, the exercise of discretion cannot be left to the reporter." Of course, newspapers usually acquiesced in the mutually beneficial relationship between circuses and the press; many a hurried newspaper writer understood that "The happy circus elephant is bringing a show to town!" was a story he knew how to tell to readers. Lippmann would later observe that telling familiar stories—even with apparently new facts and events—had long been the stock-in-trade of commercial newspapers. These were organizations that seldom had the luxury of "sufficient space in which even the best journalist could make plausible an unconventional view." The use of common stereotypes (like the happy circus elephant or the vicious bull elephant subdued), which flattered audience knowledge, was not simply a function of capitalism, he cautioned, but was certainly exacerbated by it for commercial media outlets charged with assembling a dependable audience for advertisers.[11]

Some circus people were equally nervous about accepting the consequences of unscripted elephant "violence" and the circuses' portrayal of it as just another aspect of the generally unconventional circus lifestyle. Grounded in daily contact with elephants, many saw elephant-imposed injuries as a sign of the unsafe working conditions bred by hands-on work with enormous wild animals. Many circus workers would never speak publicly of these things. They lived in a nation in which each year thousands of people were bitten, kicked, or thrown to the ground by horses, bitten by dogs, or otherwise injured or killed by animal activities that people certainly regretted but seemed to take in their stride. Louise Montague was famously reputed to have been paid $10,000 by circus impresario Adam Forepaugh to ride in an elephant howdah as "The Lovely Oriental Princess" for his Easternized circus processions. She was unusual in that she later won a $500 award in her suit against the circus company after she was injured, as the papers put it, "by being thrown into the mud by a sore-backed and vicious elephant."[12] Montague would later become notorious for filing various lawsuits against her employers. Therefore, the overall lesson people took away from her case was about American litigiousness rather than whether a woman should be asked by her employer to risk her health by climbing on the back of an elephant.[13]

Either way, circus company managers and owners knew that circus work endangered employees. In 1891, after P. T. Barnum's death, management at the Barnum & Bailey shows was purported to have presented employees with a new contract. Many of its elements were probably common in the industry by then.

The agreement banned such immoral behavior as drinking, gambling, and chasing women ("neither will 'mashing' be tolerated in the tents, streets, hotels or any other public places"). More ominously, it further required employees to bear all responsibility for any injury suffered at work by agreeing, "You promise to . . . waive all claims for damages for losses, injury or death to yourself, horses, beasts, birds, reptiles or any other property or animals that you may take on the road with you."[14]

Common elephant-imposed injuries on circus workers included broken ribs, internal bleeding, peritonitis, paralysis, concussions, and broken limbs.[15] Elephants inflicted such injuries, which often proved fatal, by crushing circus staff members in their haste to move through small spaces or when an inattentive person got too close or poked too forcefully with the pitchfork to indicate that a chained elephant should move over in his space. Being large and powerful, elephants who developed an exaggerated startle response due to their time in captivity certainly became dangerous around new, distracted, or insufficiently trained staff. Many people became objects of apparent elephant aggression when they ignored trainers' warnings not to keep a certain distance from a particular elephant; at times, an employee became the target of elephant retaliation against rough treatment doled out by some other staff member in the company. Such incidents were further fueled by staff overwork and alcohol consumption, whereby men lost their patience, good judgment, or balance in the presence of elephants reputed to be dangerous and seemingly always watching for an opportunity to strike or crush an apparently vulnerable human.

James Sweeny so suffered when one of Barnum's many elephants "struck him with his foot and tusks, and then took him up with his trunk and threw him to the ground, [by which Sweeny was] injured about the breast and shoulders and internally."[16] These kinds of episodes were among the factors that drove the high turnover rates among barn men, keepers, and other circus hands. Even when they stayed on the job, a common anxiety often developed among staff about particular elephants on a tour, which in turn led to more hooking and more consistent confinement of that or all the elephants. Some elephants became meeker in that context; many became more difficult to handle. Of course, officially the circuses would portray elephant-imposed injuries as isolated accidents, although local reporters and elephant trainers would nonetheless often speak publicly of a particular elephant's reputation for violence.

By the late decades of the century, most circus workers understood that female elephants could be just as dangerous as a bull. For instance, a Barnum company elephant named Queen pressed one circus man, Harry Packer, up against a wagon "so heavily as to nearly upset it. The pressure broke several of Packer's ribs and flattened him out almost as thin as a board," the *New York Times* typically alerted readers in 1883.[17] Her keeper said Queen was so clever about planning her attacks that he was positive that when she had crushed Harry Packer it was with patient vengeance in mind by appearing to have only inadvertently pressed Packer up against the wagon: "She is very adroit in her meanness, and quite capable of killing a man without apparently intending to do so."[18] A generation later, even though the circuses had begun purging males from their herds to increase safety and reduce costs, circus performer Connie Clausen remembered that her mother had warned Clausen against going near the elephants on her show. Showing how the frequency of sensational elephant rampage stories would settle into American popular culture, Clausen's mother worried that they were "awful beasts" who "mangled" people. Later Clausen conferred with a co-worker about how to avoid trouble with elephants: "'What makes an elephant go berserk?' I asked. . . . 'Just about anything might. They're very sensitive,' Anne said. 'There are two things to remember about elephants. Never get close to one and never let one get close to you.'"[19]

The new concern over female elephants had been spurred on by newspaper accounts of escaped or "violent" females, which often assured the reader that, having once been found to have injured someone or to have gotten loose, the circuses loudly promised the public that they would discipline those elephants. Such stories became a staple of American journalism and popular culture for sixty years, beginning in the 1880s. Circus managers attempted to use publicity about elephant punishment as a cathartic public means of assuaging public anger over the safety risk circuses caused various communities by transporting elephants around the nation with limited security and often ill-trained or inebriated staff in charge. Female and male elephants would be effectively criminalized by the aggregate body of circus press agent "interviews" with hard-bitten elephant trainers and journalists' accounts of how the circuses dealt with elephant trouble.

The Forepaugh & Sells Brothers' elephant Topsy, for example, found herself manacled with "double chains on her feet and a check chain on her trunk" as "penance for her offense" in the spring of 1902, when she killed Fielding Blount in Brooklyn, New York. The Asian elephant, one of at least five elephants to carry the name Topsy over the decades, probably arrived to the United Sates as a juvenile in 1875 on the order of Adam Forepaugh. She appears to have worked on the various Forepaugh & Sells Brothers' circus units over the years (the companies would combine in 1896) and, as an adult, became known in the industry when she "killed a keeper in Waco, Texas, in 1900; another in Paris, Texas, a month later."[20]

A year later, Topsy was in Brooklyn with the Forepaugh & Sells Brothers Circus. The papers, referring to her as "Tops," said that at five o'clock in the morning that day Fielding Blount had crawled under the edge of the menagerie tent and "staggered into the center of the tent, where the long line of elephants stood, some of them standing stone still asleep, others rocking and swaying to and fro." Blount was reported by a barn man to have offered an empty whisky glass to a number of the elephants, then approached the still sleeping Tops and slapped her on the trunk. Another keeper awoke and from his cot warned Blount, "You better keep away from her. She's ugly." Blount seemed to be ignoring the advice when Topsy suddenly seized the man, threw him down and lowered her front right foot squarely onto his chest. One keeper explained that he then heard "a crushing, crunching noise and [then] everything was quiet." Immediately Tops removed her foot and "quietly rolled the mangled body of her annoyer out toward the ropes" that surrounded the elephant picket line.[21]

As in Mogul's day, many people used the old practices of explanatory anthropomorphism to make sense of Tops's actions by invoking the stereotype of the haughty elephant insulted. Most still believed that sagacious elephant performers should patiently humor humankind, even when people behaved in ways known to irritate or frighten elephants. Hence, some inside and outside the circuses ascribed an illegitimate and inflated sense of self-worth to dangerous elephants like Topsy. By contrast, those opposed to circus elephant captivity were likely to explain Topsy's behavior as the frustrated actions of an honorable being tormented by captivity, rough keepers, and crude visitors to circuses.

The Forepaugh & Sells Brothers' elephant man, Bill Emery, only complicated matters by blaming the incident on the famous vagaries of elephants. Halfheartedly attempting to convince the reporters, who quickly materialized at the scene that morning, that Topsy had not really killed Blount "in anger," he suggested that the elephant had merely been toying with Blount when she accidentally killed him. One local paper recounted his interview: " 'Some men don't know that the only way to play with elephants is with a long handled pitchfork,' said Emery laconically. 'You can never tell when an elephant will do a thing like that.' "[22] Such animal managers were unmoved by the seeming naïveté of critics who did not understand how to contain and manage the species with stern consistency. Give elephants an inch and they will take a mile, trainers like Emery would warn. That is, by heavy-handed discipline and management of circus elephants, these men claimed to be serving the public interest, not offending it. Every elephant rampage proved their case, they thought, by demonstrating that elephants were nasty at heart.

Regardless of why she acted, an elephant like Topsy was put at risk by the chronic understaffing and insufficient security at her circus. Industry-standard, portable but cheap measures like the use of simple "guard chains" to rope off picket lines of elephants and other stock from the public, combined with often-minimal staff surveillance, allowed drunks and other "hangers-on" like Blount access to the menagerie tent.[23] This was a chronic problem for companies with thousands invested in animals but whose security staff was at best limited to a couple of contracted Pinkerton detectives and the diligence of company barn men and roustabouts. Indeed, public talk about female elephant trouble still revealed humans distracted over proximate rather than systemic causes of injuries caused by elephants.

Topsy would ultimately die in Brooklyn, prompting many Americans to consider how wasteful and crass "cruelty inflicted for frivolous purposes" seemed.[24] Sold off from the Forepaugh & Sells Brothers company, Topsy turned up at Coney Island, where her owner placed her in the care of an abusive handler, who was seen prodding her unnecessarily with an elephant hook and feeding her lit cigarettes. People also said the handler was frequently drunk when seen wandering around Coney Island with the elephant, attempting to earn money from tourists by offering rides or petting sessions. He was thrown into jail with some frequency, and locals soon realized that no appropriate handler was available to keep Topsy from injuring people and that her physical condition was deteriorating from improper feeding and housing. City officials and her owner agreed to have Thomas Edison's men electrocute Topsy as a demonstration of the alternating current technology patented by one of Edison's firms. The curious crowded around to watch the event, which was recorded on film as the Edison short *Electrocuting an Elephant* (1903).[25] The *New York Times* concluded that the elephant's killing amounted to "an inglorious affair," though it undoubtedly did not speak for all the New Yorkers who came to witness Topsy's death as a provocative spectacle of modernity.[26]

Nonetheless, goaded on by tough-talking trainers like Bill Emery, by the turn of the century most Americans had the impression that both Asian and African elephants were prone to "murderous" acts.[27] By the most extreme interpretation, these modern elephant villains were the most dangerous kind of "treacherous" animal. The charming favorite of a gullible public, behind the scenes they were wicked brutes planning revenge at every moment. When the time was right such beasts might appear calm until unchained, then they would toss a keeper to the ground and crush him underfoot or throw him into nearby containers of water and hold him under. The famous elephant Empress was reputed to have tossed, pushed, and injured more than a dozen circus workers during the 1880s and to have killed between three and seven men.[28] Although they may not have known it, in learning of such actions Americans were witnessing internal debates spilling out of the circuses and into the public sphere, as the largest of these companies strained under their elephant herds. With their decentralized marketing and advertising plans, the largest circuses seemed to compartmentalize the genial elephant performer as a company mascot on color lithograph broadsides and ring performances, while still exhibiting the "mad" elephant in circus press notices and interviews with newspaper reporters about elephants gone off script. Now, the elephant man or the elephants themselves contradicted the owner, who explained an elephant rampage or injury to a person as an anomaly or a case of innocent elephant fright, an explanation that might have been broadly persuasive if the circuses' managers and trainers had all agreed to stay on message with it, as most elephants did stay calm and contained the vast majority of the time.

A steady supply of circus talk turned gender ideals upside down to demonize female elephants, as well as bulls. At Barnum & Bailey's "Greatest Show on Earth" Brooklyn show grounds in 1896, one press notice had a *New York Times* reporter asking William H. Winner, the superintendent and trainer of the animals, if a boy from the audience might pet any of them safely. Winner replied that the young boy might pet: "Not one. I wouldn't trust one of them. . . . They are all vicious. . . . I don't believe any of the animals have any affection for anyone, and I have been in this business for thirty years. . . . Elephants are treacherous. The males are usually the worst, but sometimes the females are bad. The worse elephant in the country now is Empress, who is with a travelling show. She has killed several men. An elephant will throw you down and trample on you; the males gore you, and the females, which have no tusks, crush you with their heads."[29] Winner was certainly the cliché of the hard-bitten animal trainer, unsentimental and full of wild stories that might impress younger people. He claimed to have had an early career as a race horse jockey, then to have served in the Civil War in the Twelfth United States Infantry. He had also served six months in "Libby Prison," before emerging from those trials to become an authoritarian animal manager.[30] Even if just promotional bravado, this kind of language somehow had to coexist with claims residing on circus bills and embedded in ring performances that circus people and sagacious elephant actors loved one another and that consumers could help animals by paying to be entertained by them.

Hence, the modern elephant rampage and corresponding criminalization of elephants were signs of a crisis among both African and Asian varieties in the United States. Elephants injured and killed many people in those years, and in reply many died at the hands of company managers, zoos (if sold off because dangerous), or local law enforcement. The rising numbers were undoubtedly the combined product of increasing elephant confinement and the aging elephant population in the nation. Imported as juveniles during the elephant wars of the 1880s, many female circus elephants at the turn of the century were nearing adulthood and experiencing an inborn drive for independence from humans that produced their unmanageability. Either way, the numbers showed that circus company owners and senior managers had neither the manpower nor the time to allow their trainers to develop expensive management techniques that might have prevented elephant trouble. Industry leaders had for so long been determined to make the most of their elephants at the least expense that, even if individual elephant handlers had lobbied for particular elephants they knew fondly, owners and managers still took the show business elephant as a transient and perishable inventory. Heavy-handed management and intense confinement may have been productive in the short term, even emotionally satisfying for circus workers injured by elephants, but the experiences of some female elephants in those conditions set the clock ticking to complete intractability, just as that experience, plus musth, did with bulls.

Gypsy was just such an individual. Her story achieved national coverage in papers from California to Kentucky to New York, appearing above the fold or on the front page of city newspapers. Like the other "mad" elephant cases, Gypsy's actions distracted Americans from the circuses' main goal of selling consumers experiences of genial animal celebrity. Her difficult case also shows us how elephants, their trainers, and the public together symbiotically defined the modern female circus elephant as a criminalized evil twin to the happy elephant performer icon. Gypsy came to the United States in 1871, imported by impresario Pogey O'Brien, who originally billed her as Empress. O'Brien sold her to another circus owner, M. L. Clarke, who called her Queen. When his venture "went broke" in 1891, Clarke in turn sold the elephant, possibly to a Forepaugh Circus unit, where she appeared as "Mary" until 1893.[31] Whether she made a stop at the Forepaugh venture or not (the records of her life disagree on the point), by 1896 the elephant was probably around thirty years of age, living and working as Gypsy on the W. H. Harris Circus Company's "Nickel Plate Shows." Among her other services to the company, Gypsy played the part of elephant mother to a juvenile on the show billed as "Her Elephant Baby Barney—The Wonder of the Age."

Gypsy would gain publicity that spring for inflicting fatal injuries on the company's head animal manager, Frank Scott. The show was in Chicago for a series of performances. Early one afternoon, Gypsy curled her trunk around Scott and crushed him, breaking his ribs and pelvis, then pressed his body up against the barn door. Of course, different reports supplied varying details about what precisely had happened in the barn. One set of stories said that company owner Harris confronted the elephant on the spot, and Gypsy turned her attention to him, sure enough. A woman at the scene—Mrs. Harris, some said—saw the elephant now attempting to crush Harris. She grabbed a handy pitchfork and repeatedly stuck Gypsy, until she "soon had blood streaming from the wounds in the elephant's side."[32] With Harris and his woman-rescuer having left the scene, hundreds of people crowded around the alley leading to the barn to see what was happening. Gypsy remained walking freely in the alley behind the structure and in nearby streets as several dozen police officers appeared on the scene to discuss with Harris what caliber of firearm might stop the elephant. A number of officers and bystanders approached Gypsy, scrambling away in haste when she charged them. At that moment, desperate to avoid company liability, Harris apparently offered to pay a large sum to anyone who could think of a way to kill Gypsy.[33]

Editors in distant communities began picking up the story off the news wires, applying their own sensational headlines to the piece for their audiences: "Big Elephant Becomes Unmanageable and Runs Amuck" and "Gypsy Resents Her Keeper's Attempt to Direct Her Movements and Beats Him to Death with Her Trunk." Many papers characterized Gypsy as an "Enraged Brute" who "Beats the Man to Pulp." For local readers, the *Chicago Tribune* account was highly detailed, less formulaic and more critical of Gypsy's situation than most other accounts. Inviting Chicagoans to read between the lines and see the elephant as the victim of an inept manager, their account put the grain of truth in the trope of the "bad elephant trainer." A character that became common to twentieth-century circus memoirs and novels, he is often depicted as mentally ill and/or an alcoholic, a selfish and brutal disciplinarian who physically abuses and terrifies both animals and his fellow circus employees. The *Tribune* recounted how the whole three-hour event with Gypsy had been initiated by Scott. He had ignored orders from Harris to keep the elephant chained, and to "run her around" had taken her into the alley way behind the company's winter quarters barns between Van Buren and Jackson streets.

> Maddened by punishment . . . [Gypsy] did not relish the idea of trotting up and down
> in an alley at the will of a man. Several times its rebellious spirit asserted itself, but

vigorous applications of the elephant hook brought it to time. After ten minutes' work, and as it was about the pass the door of its barn, Gypsy tried to turn in. Scott spoke to it sharply. It shook its head defiantly and continued toward the open doorway. Scott, who was seated on the animal's head, raised his hook and drove its sharp point into one of its ears. With a cry of pain and a "trumpet" the elephant tossed its head and the keeper rolled to the ground. Before he could regain his feet Gypsy was upon him. Seizing the prostrate keeper it wrapped its trunk about him and lifted him high in the air. For a moment the beast held him there, seeming to the few frightened people who saw it to be gloating over the helplessness of the man who a moment before had called himself its master. Again the furious beast screamed and drew the man toward its mouth, as if to crush him. . . . He had not yet become unconscious, and as he saw the people standing about he cried: "Help! Help! She will kill me."

Much of the *Tribune* account came from barn hands and "Mrs. Fred Irwin of No. 796 Jackson Blvd." She claimed to have seen the whole thing from her kitchen window overlooking the alley, exposing the ways circus companies, with their tenuous control over wild animals, existed in close proximity with residential housing.[34] The *Tribune* account also carried a critical tone that described Gypsy as an ambivalent character who might not be entirely responsible for the public hazard she had become. Angry and vengeful to be sure, she was also "Maddened by punishment" and was thus the victim of a cruel trainer, who injured her painfully for the satisfaction of his own ego.

Two hours after inflicting fatal injuries on Scott and nearly fatal injuries on Harris, Gypsy returned to the barn on her own. Yet when circus staff closed and barred the heavy oak doors behind her, she merely pushed through them and walked back out into the alley, where police were trying to keep at bay several hundred curious citizens, a group that now included children walking by on their way home from school. Circus staff brought several baskets of bread loaves from a nearby bakery and lured Gypsy back inside. Claude Orton, apparently the bravest man on the crew, snuck in and attempted to chain the elephant down at the ankle. Each time she saw him, Gypsy gently used her trunk to push the chains aside. Orton persisted and eventually succeeded in chaining her, after which other staff covered the broken barn door opening with canvas and left it guarded by a police officer to keep snooping locals away.[35]

To reporters and police, company owner Harris denied any knowledge of the elephant's past, insisting that he, Scott, and Gypsy's previous trainer, Bernard Shea, had "never had any trouble with her"—a stock explanation used by many circus people, then and since, to deny responsibility for events of this kind. We know that in this case the stock explanation was untrue. The previous August, in that very city, Shea fell down in the ring during a circus performance and sustained serious injuries from the elephant. That day Gypsy, who had just finished "waltzing and creeping on hind legs, and other elephantine tricks," saw Shea go down, attempted to crush him with her foot, then "beat his brains out" with her trunk as he lay in the sawdust. One city paper asserted that when Gypsy realized her trainer was in a compromised position "the desire to kill him beset her." The audience scrambled for the exits as staff rushed in, including Harris, who "prodded her with the hook," to end the incident. Shea suffered multiple bruises and broken ribs, reputedly explaining to journalists that Gypsy was "a treacherous beast." While she would not "permit a horse or dog to be beaten in her presence" and was generally gentle with women and children who visited the circus, she was also diligent about assaulting strangers who attempted to enter her tent. Shea believed Gypsy could turn on him at any time. The lesson Shea took from the event: "I suppose, when I get well, I will have to subdue her anew, and show her again that I am master."[36] Here was both an attempt to reassure the public that the circus could control their elephants, and evidence of continued industry-belief that periodic rebreaking could suppress the presumed desire of female elephants to assault their trainers.

Seven months later, with Shea gone and Scott the target of Gypsy's actions, Harris told the papers that he and Scott had in fact been arguing about how or if to "exercise" Gypsy. Half-admitting to the *Tribune* that he was worried about the elephant, Harris asserted that when Scott freed Gypsy and rode her out into the alley, he had defied Harris's policy that the elephant be kept chained in the barn. A valiant attempt at damage control it may have been, but people in Chicago and elsewhere were already speculating that "Gypsy" was in fact the notorious Empress, an elephant reputed to have killed up to seven people in the 1880s and early 1890s. While she may have inflicted fatal injuries on only one or two men before arriving in Harris's care, it is not clear if Harris knew her complete history when he purchased her from the bankrupt O'Brien or Forepaugh shows in 1893.[37] Regardless, that day in March, Harris made the fateful decision to keep Gypsy on the show and hire back her old trainer, Bernard Shea.[38] Certainly, gossip in the state held that Harris was an ambitious businessman and had bought a circus, "made a mint of money at it, and is living in fine style in Chicago." As such, many observers may have believed that Harris's approach to the problems presented by Gypsy—of simply ignoring the early Chicago episode so as to attribute the later one to Scott's refusal to follow orders—was the least costly measure to take in the short term.[39] It showed Harris operating with a hand-to-mouth mode of management in an entertainment context in which companies could suddenly become insolvent due to a season of rough weather, a downturn in the economy, or financial mismanagement. Indeed, Harris had purchased Gypsy from a rival show that was in just such a situation.

How difficult Gypsy's life was for the next six years we do not know. Was she rebroken by Shea and compliant over those years? Was she generally unpredictable but respectful enough of her immediate trainer's authority that his presence dissuaded her from acting out? Did she kill or injure Harris company staff, perhaps an anonymous roustabout? Such deaths were common but could be easily "hushed up" by the circus; many circus people were estranged from their families or working on a circus precisely in order to stay underground because they were facing criminal charges, family trouble, or debt back home.[40]

We do know that in 1902 the Harris Company "Nickel Plate Shows" were on the rails traveling back to the company's new winter quarters in Valdosta, Georgia. Gypsy was now managed by yet another trainer, James O'Rourke. When they arrived in Valdosta, he and Gypsy disembarked from the Harris train. O'Rourke mounted Gypsy, as Scott had in Chicago, and attempted to direct her through town to the company's barns. Local reports said O'Rourke was "feverish from malaria and had been drinking whisky during the day" and seemed disoriented and unfocused when the elephant turned away from the destination barn. He fell from her back and, observers said, Gypsy calmly "stopped, paused, then kneeled on the motionless O'Rourke, killing him instantly."[41]

Some circus fans later explained Gypsy's behavior as an intriguing case of elephant vagary, contending that Gypsy had started the whole episode, seemingly without cause. As circus memory constructed the tale, near the Georgia destination, Gypsy had become "enraged" inside her rail car, either intentionally or accidentally inflicting fatal injuries on O'Rourke, who was in that small space with her. Thereafter, Gypsy was heard to call and trumpet, then attack the interior of the rail car. Once the train halted, she broke out and made her way into town with Harris, various members of his staff, and crowds of townspeople following. Unbeknownst to Gypsy, the local sheriff was also marching in her direction with a rifle under his arm.

This account seems constructed by circus legend in a typically showman-like way, seeking to divert attention from the industry failures that drove this elephant's mismanagement—kept on a show after she had assaulted people; renamed to cover up her earlier reputation as the dangerous Empress; run through a series of trainers, some of whom drank on the job; confined in rail cars for many hours at a time; driven with use of an elephant hook trainers knew to be painful to her—by discussing the event as a kind of circus world adventure. By this portrayal, O'Rourke's death and Gypsy's probable misery on the Harris show were less important than an industry-friendly interpretation of elephant trouble as just another aspect of the unconventional lives of circus folks. In no way asking entertainment consumers to consider their own role in circus disasters, it posited that circuses were inhabited by mysterious people who should be trusted to manage the basic irrationality of elephants, even if it got the better of them sometimes.

Such was not the consensus in Valdosta. After Gypsy killed O'Rourke, Valdosta Sheriff Calvin Dampier and a group of armed local men pursued Gypsy and fired at her, perhaps at the behest of Harris. Their weapons lodged ammunition in her skin but did not kill her. The next morning the group found Gypsy in a clearing six miles outside Valdosta, and Dampier finally killed her with a shot to the head with a "Krag-Jorgensen military rifle." A curious crowd of men and boys gathered around the corpse as Dampier posed for a photograph sitting on the carcass with his rifle, as though imagining himself a big game hunter. The whole affair exposed a larger lesson about the degree to which people were failing captive elephants, of course. But in Valdosta, Gypsy's death exposed the growing impatience among city and town officials with circuses that put dangerous animals in the hands of incapable men in public spaces.

For Harris and other impresarios presented with local citizens or police who offered or demanded to take matters into their own hands with respect to an elephant gone off script, acquiescence was often a practical matter. When owners and managers ran out of ideas or stamina with an elephant made unmanageable—by former keepers and trainers or her own inability to cope with circus routines—the decision to let local people dispatch her was often equally a function of the exhaustion of being in the middle of a tour that needed to continue at all costs. On the road, circus people lived in trains, on public roads, and in empty lots, where they had no privacy, no immediate place to go and regroup. Thus, they had no choice but to attempt to turn what was inevitably a public event to some kind of advantage as grisly publicity for later shows and (hopefully) reassurance to Americans that the circuses were diligent about protecting public safety. This was an ideal often not achieved. And, on the billboard and in the ring, circuses still sought to give audiences experiences of a sagacious and glamorous celebrity elephant performer. As a group, circus people would remain torn over the issue of elephant management as it only became more obvious in the indus-

Gypsy dead in Valdosta, Georgia. "Photograph of Chief of Police Calvin Dampier atop elephant, 1902, Nov. 22." Courtesy of Georgia Archives, Vanishing Georgia Collection, image low049.

try that in modern America "Audiences Like Accidents," as one satirical book would put it in 1905.[42]

Additionally, there was a sense in Valdosta that November that some whites in the community viewed the hunt for Gypsy as a kind of community justice ritual, that is, a lynching.[43] Certainly, northern and western newspapers on occasion snickered that southerners dealt with elephants who assaulted their trainers or got loose—like Robinson's bull Chief, who killed his trainer John King in North Carolina—in racially grounded ways.[44] While that kind of sectional sniping was a denial of the very serious problems of de facto segregation and racism across the United States, the point was that elephants were addressed in all these contexts as criminals accused of some real or imagined crime. Like the people ritually killed in such ways as community acts of "justice," an elephant's outlaw status was really a reflection of his or her perceived resistance to Anglo-American economic and corporal power. This linkage between circus workers, human and nonhuman, was especially problematic because the circus workers with the least political, social, or economic power, often African American men, were most commonly the ones injured or killed by elephants and their deaths covered up by company owners, managers, and publicists.

Here was a pattern of community justice Americans adapted so easily to elephants that it only reiterates the degree to which people of color, like elephants in the United States, were imagined by many whites to be inferior in a moral as well as a biological sense. As had occurred in formal trials of and community retribution against animals in Europe over the centuries, elephants were at once "personified" in being held accountable for their acts in human moral terms while also being punished in ways reserved for people—blacks and Mexican Americans, primarily—whom many American whites considered inherently immoral.[45] Circus arts had long been premised upon supplying risky and frightening spectacles to the public and on providing performances that reassured audiences of human supremacy over the natural world. The meaning and cultural value of the drama of an elephant punishment or execution as lynching, whether ad hoc (as with Gypsy and Fritz) or carefully planned (as with Chief and Topsy) came from the public nature of that event. Local souveniring of elephant (or human) body parts functioned to provide a cathartic community experience of a violent killing that seemed to resolve the conflict at hand, even if only temporarily, and even if some in the community were disgusted by it.[46] Public elephant killings would function similarly in the North and Midwest as well. Elephants were not people, of course. Unlike human lynchings, which served to warn minorities of the unwritten and often arbitrary rules whites imposed on their conduct, elephant executions did little to prevent other elephants from resisting the conditions of their experience. Moreover, the hunting of an escaped circus elephant connected Gypsy to a longer, transatlantic history of animal sacrifice and ritualized killings dating back centuries.[47]

Elephants' unknowing challenge to Anglo-American manly enterprise propelled both tropes—the treacherous female and the powerful bull brute—neither of which endeared elephants to many people in ways that might have marshaled human political power and resources to prevent further elephant trouble. Until the early twentieth century, well-to-do Americans who embraced conservationism did so to preserve domestic species associated with landscapes the wealthy sought to protect for their own use. The only people who seriously defined themselves by their relation to elephants were circus animal trainers, impresarios, and performers, who had a financial stake in continued elephant imports and captivity in circuses. Neither they nor the critics of elephant use—the newspaper editors and other cultural elites who saw captive elephants as noble beings humiliated by crass and greedy circus folk and their customers, as well as average citizens who believed the circuses created a public safety hazard with their volatile animals—endeavored to protect the home environments of elephants (mostly subject to British colonial rule in those years) or the foreign people and landscapes there. American naturalists and other scientists interested in extinction showed little energy toward protecting foreign species.[48]

Elephants were wild animals, but non-native to the continent and consequently holding no claim to life in the nation roaming at large, like eagles or elk. Still, modern Americans were in effect attempting to define elephants in ways used with respect to the bison a generation earlier. Before their numbers had collapsed in the 1880s and they became an interest of conservationists, bison were derided by proextinction groups seeking Anglo-American control of the West as a predatory "violent species" and competitor with man—almost foreign to the land they had long inhabited. Elephant trainers and many in the public now similarly portrayed elephants as a danger to humankind that should be hunted down and killed if they escaped from the companies that kept them captive.[49] There would be no government action on American elephant killings, as there was for the bison, that would attempt to portray elephants as a precious resource requiring legislated protection. There would be no space made for elephants in Yellowstone National Park, for instance, no national "elephant range" made for them somewhere out West. Elephants were trapped as wild but also captive animals who, regardless of state of mind or intention, would never be sent back to Asia or Africa yet had no right to freedom or even life in the United States.

Why did elephants apparently run amok so often in the latter half of the century? Blanket explanations are hard to prove with great certainty. Certainly today most mahouts in South Asia who are killed by elephants die at the trunk, tusks, or feet of a male in musth.[50] Yet in late nineteenth-century circuses, such males were not the primary killers of people. Circus trainers had had enough experience with elephant bulls by then to know that at the first sign of resistance in his late teens or twenties, an elephant male could be subdued a few times but would soon need to be sold off or killed.

With respect to the many female elephants who ran from keepers or assaulted people, animals, or property there may be a different explanation, one that is grounded in cow elephant experiences of modern management practices. For instance, many "bull men," as modern elephant men began to be known in the early twentieth century, said both males and females could suddenly act out if confronted with a new trainer. Or, they might become unpredictable if in pain due to a health problem (which circus people might or might not be able to accurately diagnose or treat) or to fatigue. Trainer John Eck explained in those days that fatigue seemed to be a factor, saying, "The weather, a hard night ride on the cars, a bruised foot may turn the best natured elephant into a peevish brute, ready to strike and murder in an instant. But these moods pass quickly."[51]

Like musth, the scenario of a good elephant in pain or uneasy around a new handler was an immediate cause of some instances of elephant trouble, but still not the full answer to the puzzle of the elephant psyche. Eck explained (and did so more articulately than most tough-talking elephant men) how the elephant trainer's experience had intersected with elephant experience to produce the character of the treacherous female elephant. Eck described elephants as "the most treacherous, moody, changeable animal in a menagerie" because most similar to humans in logic and emotional range. Yet, not caged at any time—unlike bears and big cats, who could be shut in most of the time to protect worker safety and reduce labor requirements—elephants made the role of bull man the "most dangerous business," he said. "The great danger is when an elephant is just turning into a rogue," he would warn of the often undetected period when a calm elephant began to succumb to destructive internal desires. "Every elephant turns rogue sooner or later, and they never recover. They may be tractable for a time after quieting down, but the attack will return and then probably a keeper will be killed. The female elephants turn rogues earlier and are more violent than the males, and the females are more dangerous in ordinary times."[52]

Over the last twenty-five years, studies have found that confining systems of captivity constitute, for people and elephants, "a volatile combination: keeping powerful, intelligent animals in nonfamily groups, in small spaces, and then subjecting them to abuse and deprivation," which makes elephants more, not less, dangerous because they are under mental and physical duress.[53] Like circus workers one hundred years earlier, more recent researchers have also seen similarities between human and elephant experience, noting that circus (and zoo) elephants are reported by their handlers to show "irritability or outbursts of anger" as well as an "exaggerated startle response," which can make them appear inscrutable, untrustworthy, and frightened by "the most insignificant things," as Stuart Craven had put it.[54]

Just as Americans in 1890 or 1905 said they found some elephants to be "mad," in the sense of being both angry and mentally ill, today some researchers refer to elephants experiencing "neuroethological compromise," or more colloquially, "elephant breakdown."[55] With a lack of inborn affinity for human contact (such as purpose-bred animals like sheep or dogs might demonstrate), combined with an elephantine genetic predisposition to conspecific cooperative hierarchies, elephants can be impaired by individual experience of hands-on dominance training and management. Add on suboptimal physical development from premature removal from their allomother groups and chronic health issues produced by confinement, and such contexts produced elephants who stayed obedient with some effort but were unable to sustain that effort indefinitely.[56] When they reached a moment at which they could no longer remain calm, such elephants had no persuasive or permanent means by which to act on their environments or communicate with humans other than to assault their keepers, trainers, or visitors.[57]

Indeed, the cultural logic in 1895 or 1903, which drove Americans to criminalize female elephants as treacherous individuals deserving of punishment, seems less opaque if we consider that, like prison inmates, they were beings who had no other option for changing the conditions of their experience except powerful and unpredictable behavior, which humans interpreted as "violence." Many trainers saw it just this way, speaking of elephants "spoiled" if handled by complacent men who "allowed [her] to get away with anything she wants to do." [58] A spoiled elephant would become unafraid of human handlers when she discovered that assertive use of her body to crush or strike people and objects was the most powerful means of communicating and maintaining the ability to control her own activities from moment to moment.

Although approaching elephants from different political positions and in contexts separated by a century, early elephant trainers and recent ethological and animal welfare science research basically agree about elephant experience and behavior. Many circus insiders believed it to be precisely the case, as many scientists now argue, that elephants were complicated wild animals, intelligent, emotional, and potentially volatile, and—as circus trainers would have had it— over time turned rogue because they came to hate people. In 1901, an elephant trainer lectured newspaper reporters in Madison Square Garden on this point: "I wouldn't trust one of those damned hides. They're the trickiest beasts that live. They know more than all the rest of the animals in the circus put together. They'll fool you for a long time, by allowing you to think they love you, when they are just waiting for the chance to put you out of business with a swing of that snaky snout of theirs. . . . I long ago considered the elephant as my enemy, and I will always cling to that belief. . . . I hate them, though I'm a trainer. I rule them by brutality and fear, not by kindness."[59] For those who described elephants as untrustworthy, complacent keepers killed by their animals were evidence of how a man would become distracted and weak by his love for an elephant.

In truth, the circuses created the very elephants they most feared. Show business workers operated in contexts of extreme complexity shaped by the hasty transience inherent in mobile shows and the ever-present risk of bankruptcy. Their animal managers and owners created business cultures that put management convenience ahead of worker safety or animal health, even such as it was understood at the time. Under constant public scrutiny, many circus folk developed a stubbornly suspicious view of outsiders that made them resistant to internal cultural change and impatient with critics who had no firsthand experience with elephants. Circus staff, their audiences, and even the few "horse doctor" veterinarians occasionally hired to work with circus animals knew only captive elephants, and they may not have realized how circus captivity shaped elephant behavior and health. Many an American was fooled by elephants' strength and size out of seeing how mentally fragile they actually were.

The typical elephant trainer's bravado not only obscured any sympathy and gentleness with which many circus people addressed their animals when they could; it also created an opening for other entertainers who could use the hard-bitten trainer as a foil. When elephant managers informed Americans that they rejected "kindness" when managing elephants, they created an accumulation of tough talk that aggravated an existing perception that the circuses served crude tastes. Middle- and upper-class audiences had been self-segregating at the symphony and the art gallery for a generation, and they rejected the circuses as low-brow culture for an unintelligent mass. Indeed, the many mad elephant stories coincided at the turn of the twentieth century with a broader controversy over the use of animals in show business, as well as their depiction in literature. The frequent publicity around elephant trainers who punished and ruled elephants with brutality began to color marketing trends across the entertainment industry. Suddenly, those competitors who claimed to use "kindness training" appeared progressive; they asked all Americans to think about whether elephant experience mattered to their perception of circus entertainment. Much of what we know of these debates comes from affluent Americans with privileged access to print media for whom modernity actually implied a vocal concern for animal suffering (as opposed to those modern Americans who relished the mad elephant).[60] Still, it is difficult to imagine that many people from various walks of life did not choose to take a side in the scandals of those years over how animal training for entertainment should take place.

Among the new concerns was one that had been brewing on both sides of the Atlantic over the dogs used on vaudeville stages. It was a concern prompted by many consumers' close-up viewing of small animal acts using dogs, ponies, monkeys, and even "baby" juvenile elephants in popular theater venues like circuses, fairs, medicine shows, and vaudeville. Entertainment writer John Jennings had addressed the contradictions many saw in the trade in which creatures audiences perceived as "pets" lived the lives of disciplined "worker" animals on stage. Vaudeville dog trainers employed cruel force toward their animals in training, he asserted. Yet, some vaudevillians would insist that any corporal punishment of show business canines was performed in the animals' own interest. A dog performer needed "to mind" his director by responding to "firmness" when uncooperative because it saved him trouble later—although if randomly applied, such physical discipline could be considered "cruelty" on a trainer's part. To men like Jennings many a trainer would present confusing explanations of their philosophies, in which they claimed to "rarely [use] the whip upon them [dogs], but endeavored, by properly feeding and speaking kind words to them, to make them obedient to his command." Yet, when Jennings actually observed such work, he concluded that "dogs that hop around on two feet have their little limbs lashed from under them until they almost feel the sting of the rawhide in the tone of the trainer's voice."[61]

Such knowledge was common among entertainers. A 1905 exposé on vaudeville dogs (and later, Jack London's 1917 novel *Michael, Brother of Jerry*) would show that even within the working-class theater community there were men and women who disagreed with the frequent uses of food deprivation and pain to motivate dogs to perform dangerous leaps and acrobatic acts, which resulted in a high death rate among them from injury and compromised health.[62] Concerned observers made a ready audience for contemporary books and periodical stories that encouraged Americans to imagine what nonhuman experiences might be like. Lumped under Theodore Roosevelt's pejorative term *nature fakers*, a series of writers produced fiction that portrayed animals as individual actors with complex human-styled personalities, feelings, and ambitions. These animal characters had both agency and power, since they were written as though they understood in human terms the human contexts and cultures they encountered.

The debate over whether such stories were either legitimate and progressive literature conveying morality and a respect for living things or sentimental dross pitted well-to-do naturalists and conservationists against one another in print. People on both sides agreed especially on the need to protect wild spaces and animal life while disagreeing vehemently about how or if humans should inter- act with the animals in those spaces, especially if it meant doing so as big game hunters. At issue, in part, was the question of whether mankind had a duty to protect animals from suffering, especially if it meant giving up leisure activities— be they hunting or an afternoon of trained dogs and baby elephants at Keith's Theater—and the business opportunities they offered. Like the rough-and-ready circus elephant wrangler, men like Theodore Roosevelt symbolized ostensibly unsentimental but self-described pro-animal thought, believing humankind to be a victor in the competition among species, all of which were tested by the merciless forces of nature. If animals suffered at the hands of humans in the process of humans making their way in the world, this was a fact of life one need not regret or change.[63]

The year 1906 saw the publication of Upton Sinclair's novel *The Jungle*, which became a classic American critique of capitalism and industrialization by depict- ing in shocking and perhaps embellished detail the filthy and inhumane condi- tions in the meat-packing houses that supplied food to markets coast to coast. The pigs, sheep, and cattle in the book were helpless, suffering, and anonymous raw materials for the killing floor, the sausage grinders that sometimes absorbed rats, and the giant kettles that churned out toxic potted hams laced with lye. Al- though his muckraking account was mostly a metaphor for the ways people were digested by the machine of American capitalism, his portrayals of the commodi- fication of animals must have struck many readers who had long believed that the inhumanity of capitalism was directly connected to inhumane treatment of animals in the United States.

All these conversations and the chatter of circus publicity denouncing ele- phants as treacherous and mad contextualized people's understanding of circus elephants and challenged the primacy of the happy circus elephant icon. Thus, it was most galling to the community of Americans involved with or sympathetic to anticruelty activism that show-business animals appeared to endure the worst of human nature in order to produce what many took to be tawdry entertain- ment. Many animal tricks and performances exploited dogs, horses, elephants, *and* people because they capitalized on man's basest interests. "Among educated city dwellers . . . such acts were *not* mere entertainment but rather affronts to basic values," one observer of vaudeville noted. Animals were bad actors to the art critic and many spectators to be sure, but they were also trained to replicate the most "boorish" or unappealing human behavior. So it was with, say, monkeys trained to smoke and play billiards, or any animals whose performances were alternately cute or absurdly satirical of human ways.[64]

Some entertainment entrepreneurs sized up these controversies and sought to employ them in winning the sympathies and dollars of ticket buyers through a marketing fashion for "kindness training." The idea of training through kindness was certainly not new in 1890 or 1905. People who worked with farm horses, pet dogs, or an "educated" pig had long known that one could persuade animals to perform useful labor by offering something a given animal wanted, especially food.[65] Kindness training was premised on this, plus the idea that it was the trainer's duty to discover each animal's individual talents and make the most only of those talents. *Haney's* had already agreed at mid-century: "Public exhibitors are able to show a large array of tricks because of the number of animals they have, each, as a rule, knowing a comparatively few of these tricks, or, in the case of some 'sensation' tricks, perhaps only one." This was most efficient in handling elephants, too, as a trainer could use each individual for particular tricks, rather than attempting to persuade all individuals on hand to perform all tasks.[66]

What was new at the turn of the century in the public talk about kindness training and animal experience was its contextualization: a rising number of people were killing elephants and being killed by elephants, especially previously innocent-seeming cow elephants. With signs of apparent female elephant "anger" and "madness" so obvious in the news coverage and oral tradition around circuses, showmen offering animals prepared only by kindness training attempted to distinguish their productions in the marketplace as more humane. In those years, many showmen would accuse rival trainers of using cruel means, such as starving or severely beating their animals at random, especially "rogue" elephant bulls and big cats, who would thus present more apparently ferocious behavior for audiences.[67] In these moments it was not actual elephant pain or misery that mattered as much as the suspicion of such on the part of the viewer. Without question, over the nineteenth century individual elephants had experienced periodic or chronic hunger, thirst, fatigue, anxiety, pain, frustration, or loneliness, as much as they may have found moments of enjoyment or relief in their daily routines. However, those who styled themselves "kindness trainers" and animal "educators" understood that audiences used their assumptions about show business animal training to determine the entertainment value the performances they watched as alternately edifying, amusing, or horrifying.[68]

The most famous advocate of kindness in the production, distribution, and promotion of animals was the famed German trainer, zoo keeper, and animal trader Carl Hagenbeck. From his headquarters in Europe he quasi-mass-produced internationally branded interspecific performance tools, namely, animals he and his staff had trained for circuses and zoos. Hagenbeck claimed also to have educated dozens of the elephant handlers laboring as trainers for European and American circus companies and to have prepared six hundred large animals for the show trade. Carl Hagenbeck did revitalize the tradition in animal show patter suggesting that one could "educate" animals primarily by patient repetition and positive reinforcement. In this system, pain aversion was used only as a last resort because, he insisted, punishing animals was counterproductive if it made them fearful. Moreover, he claimed not to require his trainees to perform any movement that was not "natural" to them.[69]

People may have looked at the presentations of Hagenbeck-branded animals at Luna Park, the St. Louis Louisiana Purchase Exposition, or Hagenbeck-Wallace traveling circuses and concluded that there was nothing terribly "natural" about what he provided, which seemed like a modern twist on the old "happy family" genre of animal presentation. Hagenbeck-trained animals appeared in astounding shows, photographs, and advertising that had trainers enter an enormous cage to wrangle a polar bear, two cheetahs, two tigers, four lions, and two Great Dane dogs together, arranging them posed on pedestals in pyramid formation.[70] Another display branded with the Hagenbeck name had an elephant in a mountain-themed display enclosure sliding down a chute into a pool of water for the amusement of visitors to the 1905 St. Louis Louisiana Purchase Exposition.[71] The trade journal *Billboard* puffed these presentations as "thrilling, interesting and educational all at once."[72]

Even worldly entrepreneurs like Hagenbeck had limited understanding of the breadth or function of species-typical behaviors for many of the animals they used at the time because many had yet to be systematically observed in the wild. Nonetheless, Hagenbeck's point was that one could devise performances of carefully chosen, seemingly amenable individuals that capitalized on the physical talents and inborn desires of each individual, which could be determined in captivity.[73] A complimentary article in *Scientific American* voiced not only the proposed modernity of these claims but also the astute business sense of the men who used such rhetoric, explaining that the Hagenbeck Trained Animal Company would purchase sixty animals to produce fifteen for a show: "This is where Mr. Hagenbeck scores over his competitors. Being a dealer in wild animals, as well as a trainer, those beasts that are unfit for the stage are sold to zoological gardens and menageries."[74]

Indeed, in the industry many took Hagenbeck's shows largely as a promotional tool for his performing animal production and distribution business, as

well as his trade in zoo stock. He controlled a sophisticated global animal sourcing network based in Germany and sold packages of animals in various combinations of species to American and European animal show operators, investors,
and others. He also sponsored "Carl Hagenbeck's Animal Kindergarten" and
"Hagenbeck's Wild Animal Show," featuring his animals at Coney Island and the
Chicago World's Fair, among other locations, in the 1890s. Some of Hagenbeck's
American buyers may have believed that Hagenbeck animals would be easier to
manage, but they got more than they bargained for when a pretrained but nonetheless disoriented wild animal was delivered to them. Yet without the time,
money, or manpower to prepare the animals themselves, many welcomed the
opportunity Hagenbeck offered.[75]

It was a minor European invasion of sorts that brought these ideas to the
United States, where the promotional context around trained animals was still
influenced by circus staff who had been struggling with elephants since the bull-
crazy 1840s. German showman August Kober—the impresario who imagined
animals smashing his circus—was another who would reassure ticket buyers
that, because of kindness training enacted in his traveling show, "we have no
need of any society for the prevention of cruelty to animals, because the circus or
menagerie animal is just as much a comrade as the human performer."[76] The
Briton Frank Bostock would similarly purport to use kindness training through
a heavy reliance on flexibility in animal selection, consistency, and operant conditioning to develop a mode for the "scientific training of animals."[77] In the
United States, Bostock staged an animal show at the Dreamland Amusement
Park at Coney Island. Although focused primarily on big cats, Bostock claimed to
have developed techniques that purged human emotion from the training process and could gently modify the behavior of a broad variety of species, from dogs
to elephants.

Cast as methods of animal education that people like William Ballou might
have endorsed since they paralleled similar contemporary philosophies of "tough
love" with respect to horses, ostensibly progressive animal management techniques were advocated by trainers like Bostock, Kober, and Hagenbeck, who
were, in effect, reflecting contemporary "civilized manliness" rhetoric.[78] Epitomized by some middling churchmen and other male reformers in the United
States, the modern civilized masculinity challenged the rough and competitive
"strenuous" manhood of the 1880s and 1890s, which undergirded the violent
punishment of captive elephants or the extermination of the bison, for instance.[79]
By contrast, the modern animal wrangler was trying to tame both the beast and
the animal show business. As the argument went, Gilded Age circus elephant

wranglers like Bernard Shea or William Winner exposed their outmoded thinking when they cautioned that elephants were inherently treacherous villains requiring management by "brutality." Kindness trainers insisted that such trainers
were subjective and inefficient. Their animals became belligerent and maddened, not because they were inherently imperfect but because they were imper-
fectly educated. Indeed, to punish uncooperative trainees at times distinct from
an unwanted behavior or out of frustration or a vain need for publicity was unmanly and gratuitous.

Americans seem to have been receptive to the idea. For example, the down-
to-earth *New York Sun* defined the contrast between scientific/kindness and
breaking/carrot-then-stick training for its readers by describing them as "two
kinds of training, mental and muscular."[80] The first was modern and relied on
savvy, experience, and a clear mind; the second was of the nineteenth century and
relied upon a subjective use of force that made the animal as emotional as the
trainer was deemed to be. American consumers truly hoped that the kinds of
cooperation between humans and wild animals that Hagenbeck touted were possible. They hoped that the training of an elephant, bear, or lion could be as gently
executed as the training of a pet dog or family horse (apart from the initial breaking of that horse). Numerous spectators would trust Hagenbeck and others who
presented Hagenbeck-branded animal performers to bring wild animals safely to
customers, with the assurance that their performances were a demonstration of
the most ingenious and benevolent aspects of human nature. Indeed, the marketing mode of kindness training enabled a trainer persona that was brave, powerful, clever, and humane, gladly sharing his celebrity with his animals in order
to help them see that, at heart, they all wanted to be traveling entertainers, just as
he did.

Many circus people must have resented how the marketing vogue for kindness training disparaged their practices during the first decade of the twentieth
century. They noted that operators like Hagenbeck and Bostock often presented
animals at dedicated venues in amusement parks or zoos, and consequently
had the great luxury of reduced logistical costs and complications because they
did not need to move their show from booking to booking every day, eight months
a year. In any event, the proposed ideal of kindness training was a luxury that
most circuses, elephant men, and their crews simply could not afford. If it was
show time, an elephant who was sluggish getting in line for the parade could not
be allowed to delay things while a keeper hunted around for an apple. Nor could
she simply be allowed to refuse an order so as to permanently opt out of that task.
Instead, it was far easier for any trainer or presenter at hand to give her a sharp

pull under a leg or over an ear with a hook so that the elephant stepped back into line quickly, even if she vocalized a bit or was clearly displeased. After the show, one might have time to use a combination of hook and food reward to prepare an elephant for a new act, but this was often simply not feasible. Although in practice circus elephants experienced both, on balance most circus trainers believed that it was not kindness that kept animals tractable, but fear.

In fact, the circuses' elephant managers had a valid argument because they knew that so-called kindness training techniques differed only in degree, not in kind, from the stick-then-carrot approach used in the circuses. Even Carl Hagenbeck still authorized the use of force "in cases of gross disobedience" when working his animals.[81] So, too, did Frank Bostock and other purveyors of this idea concede that they would still strike or tie down their "pupils" as a demonstration of "trainer's firmness." By their formulation, physical punishment and confinement communicated to an animal that it was right to comply with direction and that biting, scratching, or hitting a trainer was absolutely unacceptable (as the trainer had it) and would produce discomfort in the animal trainee (as the animal knew it).[82] Practically speaking, any "kindness" trainer working with elephants, even the more tractable among them, had no choice but to subdue them initially and thereafter to inflict pain at times in order to cope with their sheer physicality and their eventual "boredom" with or inconsistency in performing required movements.[83] What is more, the moments of training were only one portion of the elephant management routine for any herd. Regardless of how "kind" the trainers might intend to be or how much patience the barn men might find in themselves, none of these men was in a position to address the long-term stress on an elephant's psyche or health imposed by captivity and travel.[84]

Regardless, those circus men knew that "kindness" training was historically contingent and specific to the species, and even to the individual animal, because the balance of reward/praise/relief and punishment/pain/deprivation used by a given trainer was a subjective spectrum of action. Thus, the ostensible "revolution in animal training techniques and showmanship" that people called "kindness" in 1900 or 1905 was often simply a shift in promotional presentation.[85] Certainly, kindness training was good public relations strategy for stationary trained animal shows and vaudeville acts, which could thus claim to answer growing public questions about whether wild animals were suited to show business by defining themselves as a modern alternative to circuses. "Willingly they respond to every wish of their master," said one typical advertisement for a vaudeville act, called Miller's Elephants. "Chas. Miller . . . carries them from feat to feat by kindness alone, not once using the goad, which is so offensively in evidence

at every exhibition of Elephantine training in circuses. . . . They play musical instruments, they execute military drills, fight sham battles, die on the field, and
are buried with military honors; they roll ten-pins, tally their own score, and grow
enthusiastic over the game just the same as a local bowling team would do. They
are, without a doubt, the finest, cleverest and quickest performing Elephants in
the world."[86] Many such companies and acts invited ticket buyers to believe that
one could forego the immediate results provided by the hook and heavy-handed
training techniques by instead persuading an animal performer to enjoy various
poses or movements that spoke of the old show business icon of sagacious elephant performer.[87] Still, such trainer's emphasis on food or water rewards may
have caused anxiety for elephants who perceived a connection between hunger
or thirst and training. From an elephant's stance, it is difficult to know if the kindness trainers were more or less rewarding to work for.[88]

When the competition advertised their animal acts as products of kindness
training, they nevertheless performed a service for the circuses. The years marked
by the revitalized kindness training shows in the marketplace corresponded with
a seeming peak in the number of angry elephant stories. Yet, many consumers—
with ample encouragement from some circus advertising—would conflate the
marketing rhetoric of the so-called kindness trainers with the animal presentations provided by the circuses. Not all were persuaded, of course, as the persona
of the hard-bitten elephant man had become so central to the colloquial and commercial cultures of circus entertainment. Even before the concept of kindness
training became especially popular at the turn of the century, some American
observers perceived the idea as more an ideal than a practice, a bit of promotional
patter that presented the best-case scenario as the norm. It was just "the old gag"
fed to reporters, some said, in order to deflect negative attention generated by
people concerned about the experiences—probably miserable, they asserted—of
show business animals.[89]

Thereafter, all sorts of vaudevillians, amusement park animal trainers, and
circus people could be found claiming the kindness technique as their own dis-
covery, even though it was as old as time. As a marketing mode, kindness training allowed showmen to claim novelty while flattering the ostensible good judgment of the audience, conscientious consumers who were open to the idea that
by spending on the right entertainment they could help animals and reform
human-animal relations. To foreground the icon of the happy circus elephant
performer—which implied training by kindness—most circuses began to push
news of elephant disobedience underground in the first decades of the twentieth

century. It took a generation to accomplish and did not mark a dramatic change in training or management regimes behind the scenes.

The circuses gradually ceased to invite their press agents and newspaper contacts to hear about mad elephants "executed," punished, or subdued. To prevent elephants going off script, they quietly killed or sold off individuals who became difficult. They purged males from their collections while obscuring that fact by using the term *bull* to represent all elephants, male and female, in the industry. In the ring and on the billboard and program, the circuses fell back on the older ideal of elephant as sagacious, genial, and glamorous actor in order to promote the hundred or more elephants still in use in the nation's traveling shows. That classic, happy mascot helped many circus fans disavow the consequences of their spending choices. As ticket buyers, most Americans very much wanted access to trained animal shows as affordable entertainment. Ultimately, most would be reassured by circuses that endorsed the idea of training through kindness and a modern hope for animals' consent to their captivity for the enjoyment of a consumer public.[90]

# CHAPTER SEVEN

# Animal Cultures Lost in the
# Circus, Then and Now

The genial circus elephant was a powerful mascot for the menageries and circuses in selling experiences of animal celebrity to a broad consumer audience in the nineteenth century. She was flexible, too, morphing from wonder of nature into various characters, some glamorous, some villainous, which served different venues simultaneously to make circus entertainment all the more meaningful to a diverse audience. Yet, her influence with ticket buyers meant that living elephants were a chronic expense and risk for the animal shows and both joy and terror to their employees. These elephants were uneasy inhabitants of the United States. Although many companies assured their customers that elephants were in essence native to the circus, the practical history of elephant management was more complicated. All the while, the circuses were consumers, not producers, of elephants, just like their customers.

Nonetheless, the dream of a native-born American elephant population would inspire various conjectures about how the elephants that symbolized national ambition in the world of entertainment could be efficiently produced by the circus companies that employed them. As a possible "California Enterprise," one newspaper speculated about Victoria and Albert, the elephant pair shipped to that state in the late 1850s: "Possibly they may be the foundation of a royal line of elephantine live stock, to be classed hereafter among the other products of California. Our climate is sufficiently congenial, . . . why not elephants?"[1]

A quarter of a century later, the situation for elephants in the wild—whether African elephants hunted for ivory or Asian elephants hunted for sport—seemed so dire that the question of American appropriation of these species had become far more urgent. "Are Elephants Dying Out of the World or Not?" asked the *New York Sun* in 1885.

> Only a few years have elapsed since the London Spectator declared it quite likely that
> if Jumbo attained the natural limit of his life, 150 years, he might be the last of his
> race on the globe. The production of the 1,200,000 pounds of ivory used in England
> alone every year necessitates the death of 30,000 elephants and from various causes

the annual death rate of this most interesting of quadrupeds is estimated at not less than 100,000. Breeding in captivity must then be depended on eventually to propagate the species, and how far successful this has been may be inferred from the general rejoicing among show people when at rare intervals a baby elephant is born.[2]

Although the old canard of elephant longevity was in evidence here, the author was right to note that the ivory trade was destroying the African elephant population in those years.

The *Sun* writer also voiced a circus-friendly message, which zoos and animal amusement parks often evoke today with their exotic animal breeding programs. Specifically, he implied that foreign peoples could not be trusted to manage their animal populations and that Westerners should take over this responsibility, even if in practice it was exorbitantly expensive and produced captive animals who would only be shadows of their wild kin, unable to practice or pass on the cultures their ancestors had devised over the centuries to thrive in their native habitats. This was a kind of zoological colonialism by which wild animal keepers, then and since, assured their customers that circuses and zoos could be trusted to deliver wild animals to the public conveniently, safely, and cheaply. Thus did the circuses argue that the privatization of wild animal populations was progressive and that consumers could help animals by paying to see them held in captivity.

Further, read the *New York Sun* story again and note that the author asked readers to consider "the general rejoicing among show people when at rare intervals a baby elephant is born." Here he referred to the two celebrated baby elephants of the late Gilded Age, Columbia and Bridgeport, both female. Columbia was the first elephant born in America, and she arrived on March 10, 1880, at her company's winter quarters in Philadelphia. Her story would captivate the public with the tempting possibility that showmen could create an indigenous American elephant population. She was birthed by a dam called Hebe, a Cooper & Bailey Company elephant in her mid-twenties, and a bull of similar age, known alternately as Mandrei or Mandarin. Animal trader James E. Kelley had imported Hebe and Mandarin together with three others from Ceylon in 1865. He then sold the five as a package to the Cooper & Bailey circus in 1878.[3] According to company publicity, which focused on uncontroversial details evoking the chatter produced when human babies arrive, Columbia weighed 213½ pounds at birth, measuring almost three feet high and four and one-half feet from the tip of her trunk to the end of her tail.[4] "It is thought that little [Columbia] will live to be a full grown young-lady elephant, as she is hardy and healthy . . . [and] the mother is very affectionate toward her offspring, and inclined to regard the approach of strangers with suspicion," Cooper & Bailey publicity assured the public.[5]

Yet, that first night there was a creeping sense that Hebe did not know that this tiny elephant was newborn offspring. None of the extant records give any indication that circus staff recognized the problem, although someone there must have had some indication since they would have been familiar with horse, dog, or cattle reproduction. The various accounts of that moment, although heavily shaped or possibly written by company press agents, nonetheless told the same story. Earlier in the day, when they suspected she might be on the verge of calving, the barn men had chained Hebe to a post in the center of the building. The other elephants they chained in a picket line against one wall so they would not "molest" the expectant cow.

When Columbia emerged from her mother, the lone keeper on duty said the chained elephants began crying, trumpeting, throwing their trunks and moving about with great energy. *Harper's* and other media outlets took this as a sign of "highest glee," although the keeper said it persuaded him to run out of the barn with great urgency.[6] In reply to the noise of the elephants on the picket line, Hebe became agitated and "almost frantic," one account claimed:

> With a terrific plunge she broke the chains and ropes which held her, and grasping up the little baby with her trunk, threw it about twenty yards across the room, letting it fall near a large hot stove—where a fire is always kept burning—then followed with a mad rush, bellowing and lashing her trunk as though she would carry everything before her. . . . Around the stove was a stout timber railing, against which Hebe charged with such effort that she reduced it to kindling-wood in short order. Not stopping here, she struck the stove, and knocked it into the position of the Leaning Tower of Pisa in an instant, and badly smashed the pipe.[7]

Rumor had it the trumpets and cries coming from the elephant barn woke and alarmed the big cats in the building next door, which in turn set off the horses and dogs nearby, who whinnied and barked until every animal in earshot seemed to be making his or her own call.

People on the scene were unclear about the meaning of Hebe's apparent assault on the young Columbia. Some suspected that she was frightened by the behavior of the elephants on the picket line. However, others speculated that Hebe threw her newborn across the barn in an attempt to get her to the other elephants, who might protect Columbia from the humans now streaming into the building. They believed they saw Hebe rolling the calf around like a snowball with her trunk when they burst in the door, so they rushed in with hooks and

## THE EXCITED ELEPHANTS.

"The Excited Elephants." Artist's reimagining of elephant agitation at Columbia's birth. *Harper's Young People* (1880).

chains in hand, and they may have seemed particularly menacing. It appeared to them that Hebe might soon kill her baby, she was so upset and apparently unclear about what was taking place.

Hebe's early moments with Columbia worried bystanders, indicating that the cow did not know how to approach her offspring or even to recognize the tiny elephant as a being to be protected and fed. The crucial elephantine knowledge that elephants had been passing along over the generations had been lost or was at least incomplete for Hebe, who had been removed from her allomother group and shipped to the United States as a youngster. Without having lived in a free group of elephants into adulthood, Hebe had little chance to see and imitate how older elephants managed to feed and socialize newborns.[8] Infanticide was a serious possibility, and company elephant keepers intervened as best they could to salvage the situation, although they were themselves operating more from their experience with cattle calves or foals than elephants.[9] Many of those men may have believed that mothering skills in elephants are purely instinctive (as it is with many kinds of animals who reach adulthood in a year or less), not immediately realizing that, as with humans, the fifteen-year-long youth of elephants shows how much elephant survival practice is actually learned, and over a very long period of time, while an elephant's brain is growing to full size. What they may not have understood is that the strain of captivity on Hebe had suppressed immune system functions and produced a higher tendency for destructive or abnormal behavior, as well as a shorter life span passed on to her calf during gestation. Indeed, because of Hebe's life in America, Columbia's development may have been compromised by her mother's experience even before she was born.[10]

In spite of the confusion at the scene, Columbia settled some long-standing mysteries in the United States around elephant reproduction. Since Hebe was estimated by her keepers to have been pregnant for twenty and a half months, her pregnancy did confirm to the public and scientists alike the approximate length of elephant gestation. Americans discovered that a newborn elephant carries a sparse, fuzzy coat of hairs. When the barn men, trainers, and company managers saw the newborn "run about with its mouth open, very much like a young colt," Columbia also finally settled a lingering debate over the accuracy of Aristotle's ancient description of elephant calves nursing with their mouths rather than their trunks. In fact, Hebe at first refused to nurse the calf. The papers surmised it was because she was weaving too much and would not stand still long enough for Columbia to connect. One of the barn men milked Hebe into a funnel and directed the milk to her baby with a hose, and later that night Columbia was seen to nurse successfully: "Throwing back its trunk, the baby applied its mouth to the mother's breast, and fed itself in a perfectly natural and easy manner," the papers said.[11]

Regardless of any talk in the barn about the implications of Hebe's obvious lack of elephantine mothering skills, James A. Bailey allowed company press agents and journalists to tell to the public that his circus had become an animal husbandry organization. In a company press notice one reporter said Bailey insisted that "nearly all elephants thrive better in this country than in Europe. This he attributed to the care and kindness bestowed upon them and to better understanding of their habits and temperament. 'We are the only people that ever succeeded in breeding elephants.' said [Bailey]; 'climate has nothing to do with it; care, kindness, attention and close study are the only means by which we manage all animals.'"[12] When they talked publicly about having bred Columbia, Stuart Craven, Bailey's head elephant man at the time, and Bailey overestimated their knowledge and power over the situation, taking credit for the resiliency of Hebe and Mandarin in traveling so far and somehow producing offspring.[13] Craven and Bailey had the nerve to speak as though it was they who had impregnated

Adult and juvenile elephant, possibly Hebe and Columbia, with keeper at Barnum, Bailey, and Hutchinson winter quarters barn, ca. 1881–85. Princeton University Library.

Hebe when, really, Mandarin knew more about elephant sex than they did. Indeed, George Conklin would measure everyone's enthusiasm in those days by admitting, "our knowledge of baby elephants is very limited in this country."[14]

Nonetheless, Stuart Craven explained the elephant birth as evidence that the business imperative to gather more and more elephants together was actually a better mode of animal management, while it demonstrated that his circus could be trusted to privately care for these wild animals. Of the birth of Columbia he speculated, "It is possible that all this might have happened years ago if the conditions of elephantine existence in menagerie life had been then what they are now. . . . [Now] they have companionship. The sexes are not unnaturally kept apart as they formerly were, and from all these ameliorations in their condition and approximations to the comforts of a free existence, it seems to me but natural that they should now and here for the first time in the history of the world breed in captivity."[15] Critics would reject this nationalist summation of the implications of the Cooper & Bailey elephant program in various ways, some no doubt pointing out that constant rail travel, chaining, mixing of unrelated individuals male and female, and hooking of elephants was hardly an approximation of a free existence.

Nevertheless, the argument that baby elephants were a sign of elephant happiness must have convinced some observers that the circuses had crossed the line from mere animal traders and showmen to elephant breeders. Certainly, it was an indicator of industrial animal management ideologies at work in show business cultures, whereby industry insiders would assert that reproduction of captive animals was to be taken as a sign of their happiness.[16] Craven had a point in noting that the modern circuses had progressed beyond the old menageries and early Gilded Age rail circuses, certainly technologically, but also in that many animal managers did sincerely strive to give their animals the best possible life (even if that meant harsh discipline at times). The route book commemorating the next season's show would claim that only the newfangled electric lighting at the shows vied with the baby elephant as the most authoritative and popular demonstration of how modern the company had become.[17]

As the first native-born elephant in America, Columbia was soon a darling of the circus world. By telegram P. T. Barnum offered $100,000 for her to Cooper & Bailey. His offer provoked Columbia's owners to exploit the event for a moment of publicity Americans would remember for years. Cooper & Bailey ordered enormous bills and plentiful newspaper advertising that displayed an image of Barnum's telegram and the heading "This is what Barnum thinks of Cooper & Bailey's baby elephant." Seeing a fine opportunity for both parties, within weeks Barnum and Bailey combined their companies to produce the Barnum, Bailey & Cooper Circus.[18] For Cooper and Bailey the costs and potential pitfalls of managing and running a captive breeding program were really an unknown at that point. They persevered nonetheless, and George Arstingstall, the elephant man at Barnum, Bailey & Cooper (who would take over after Stuart Craven), oversaw the birth of a second female two years later at the old Barnum winter quarters in Bridgeport, Connecticut. The dam in this case was Queen, and the sire, Chieftain; both were from Ceylon, having arrived in the United States in 1871 and 1867, respectively. Known first by the name America, this second calf was later renamed Bridgeport by Barnum.[19]

The year 1882 would prove pivotal in American circus history not only for the birth of the second baby elephant. Columbia, Bridgeport, and their parents traveled that year with a highly celebrated company herd of two dozen elephants. Among them was Jumbo, the most famous elephant in world history. (Sold to Barnum by the London Zoological Gardens in 1882, three years later he would be hit and killed by a train after a show in St. Thomas, Ontario.) The group further included the notorious "White Elephant of Siam," an elephant from Thailand over which Barnum and rival impresario Adam Forepaugh would have a media war, each accusing the other in increasingly indignant tones of painting an elephant white or simply calling an elephant carrying normal pigmentation "white" in an attempt to pass him off as a "sacred" and rare white elephant.[20]

The public and the press seemed to be circus crazy throughout the 1880s. The shows were bigger than ever, the audiences larger than ever, and newspapers and magazines carried constant circus advertising and publicity, while it seemed that every fence and wall was regularly plastered with huge tableaux of surreal circus posters featuring animals and people in every kind of incredible pose.[21] The baby elephants Columbia and Bridgeport served as a key feature of the company's shows in that era. They spearheaded productions branded "P. T. Barnum and Great London Shows," billed as the "ONLY TWO BABY ELEPHANTS EVER BORN IN CAPTIVITY" and an exclusive attraction. Cooper had moved on by this point, and the company acquired another partner, James L. Hutchinson, who helped oversee "Nine Monster, Massive, Colossal Shows in One, and each Show increased to astounding proportions, with Everything New and Novel. Capital augmented to $3,500,000 and daily expenses to $6,800," the advertising claimed.[22]

Company agents and the press cooperated to create a cascade of publicity around Columbia and the rest of the herd that elevated their celebrity by elaborating anthropomorphized personalities for each one, as though they were characters in a children's story. One such piece had Columbia and the rest of the herd dressing up in fancy clothes (each outfit emblematic of an individual elephant's temperament) to celebrate Bridgeport's first birthday at "Professor Arstingstall's School for Elephants." There the elephants danced, drank wine (except for the babies, Columbia and Bridgeport), played practical jokes, "stood on their heads and others cut up all sorts of tricks" drawn from their ring routines—just for a lark. The story presented Americans with an elephantine royal family of a sort. "There has perhaps never been a party in Bridgeport [company winter quarters] where there was so much fun in such a short space of time," the publicity assured readers. Barnum, Bailey & Cooper invited spectators to participate vicariously in the famous elephants' celebration, staging the pseudo-event in the circus's private storage barns for the specific purpose of being broadcast to spectators through the newspapers.[23] A private life lived publicly is the essence of celebrity, and these elephants had become parasocial entities that would be known by many Americans, but they would themselves never understand this reality.

Backstage, the situation was, of course, more complicated. At only a few months of age, Columbia had become too much for Hebe to manage. Chained the majority of the time, Hebe could no longer grab the little Columbia with her trunk to prevent her wandering around to explore her surroundings and develop the muscles in her trunk. Within a year Columbia was tethered like the other elephants in the herd. By age two, George Arstingstall and his handlers had taught Columbia to rear up "like a lady," stand on her head, and perform other movements necessary for a life as ring stock.[24] Arstingstall had praised her by saying that, at only two years of age, "she takes the part of a clown in the performance. She sits at a table, uses a napkin, fans herself, and does a dozen little things that make the children, and their elders too, adore her."[25] She probably also underwent the common management practice of being restrained and cast. Like other very young elephants in the business, she submitted to grooming processes for ring stock, which meant having her coat of juvenile elephantine hairs singed off "at the elephant barber's," as one press notice gently broke the news. "Shaving them is a peculiar process. It is done with a lamp, or torch, such as is used to burn paint. They are singed until they have not a hair to their heads, and start out with the show smooth and well groomed."[26]

Just as a sentimental public interest in the original Elephant, Betsey or Horatio, had driven entrepreneurs in the early republic to try their hand at importing and showing elephants, almost a century later, Columbia and Bridgeport unknowingly inspired a speculative market for so-called baby elephants. The public's fascination fostered what one observer called "an artificial value" for extremely young elephants, causing buyers to import younger and younger individuals from south Asia for use in circuses, amusement parks, and the vaudeville stage, sometimes in the hope of passing them off as newborns.[27]

As the shows rumbled along with the seasons in the 1880s and 1890s, the effect of Hebe's inexperience as a mother was exacerbated by the fact that Columbia spent her youth chained to the floor instead of interacting freely with other elephants. Moreover, none of the other elephants in the company herd, even if unchained and left alone with Columbia, would necessarily have known how to take over for Hebe or to demonstrate before her what maternal care in the cultures of *Elephas maximus* looked like. That is, while the desire to mate might have been instinctual, once a calf was born, it is doubtful any individual in the herd would have been able to engage in the kind of cultural transmission of elephantine childrearing and social regulation their ancestors had. The absence of that learning made Columbia very different from wild-caught and broken juvenile elephants, who had nursed and lived free with their allomother groups for some period of months or years before exportation to the United States. Even George Arstingstall, the company's head elephant man in the 1880s, asserted publicly that captive-born Columbia was proving more difficult to manage and train for the ring than imported elephants.[28] Consequently, regardless of her status as "baby," in time he said that, while one should not punish her for no reason (such as humans understood their reasons), Columbia was not to be trusted. Nor could she to be depended upon to work hard for the company without being persuaded to do so "in the severest possible manner," he said.[29]

These efforts were to no avail as, once into adulthood, Columbia began injuring people. In one particularly notorious 1903 case at Madison Square Gardens, Barnum and Bailey press agent Tody Hamilton told the papers that Columbia had struck with her trunk at a company elephant handler, knocked him off his feet, then "dropped on her knees and tried to crush the keeper with her head."[30] She only ceased putting pressure on the man when prodded with a pitchfork, and thereafter other circus staff rushed in and "attacked her with pitchforks and goads and forced her back," a company press notice admitted. "Columbia's shrill trumpetings of rage and pain excited the whole elephant herd, and several tried to break away from their shackles." The official line on Columbia from Hamilton thereafter was that she was "known to the circus men as a bad elephant . . . whose trunk is kept chained," and "the most vicious of the herd." [31]

It is difficult to know if the circus's public demonization of Columbia was carefully planned or inadvertent (for instance, if voiced by company employees made fearful or resentful of her, but aware they were defying company publicity strategies). It is possible there was an informal plan among the elephant men to get the public ready for the regrettable day when the circus would no longer be able to tolerate Columbia's inability to cope with captive life and would give up on her, leaking word to press agents of Columbia's troubling reputation.

Then, in 1907, one hundred and eight years after the first elephant imported into the United States vanished with no trace, Columbia disappeared as well. We actually know what happened to Columbia, though, thanks in part to a circus press agent called Townsend Walsh. He knew the circus for the complicated institution it was. Of course, that complexity was a reality he was not in the business of boosting in the press, but one that nonetheless supplied the raw materials for his promotional writing. Among his papers residing today at the New York Public Library is a set of jottings he made in the early twentieth century on a spare piece of men's club stationary. These handwritten notes show his creative process at work. They also show how traditional narrative modes of drawing emotional identification with circus animals from the public with stories and biographies emphasizing "baby" elephant high jinks and maternal love actually papered over the difficulties of workaday circus life. Walsh's jottings were rough notes for a nostalgia piece on baby elephants that he planned based on a conversation with anonymous circus staff regarding their memories of the elephant crazy 1880s:

Note: <u>Baby</u> elephant

This was the first opportunity that our naturalists had had to determine the elephantine period of gestation which was then found to be twenty months and a half.

The youngster weighted at birth 213 ½ pounds, was 35 inches in height, and measured 4 feet 6 inches from root of tail to root of Trunk, and 3 feet 11 ½ inches around the body. Each of its parents weighed 7,800 pounds. Its papa was 26 years old, its mamma 24. Though elephants use their trunks in drinking, the baby drew her nourishment from her mother by the mouth.

Apropos of the period of gestation being so long, Manager Cooper suggested facetiously that the little elephant might have been sooner if she had not been obliged to stop to pack her trunk.

I regret to record that "Columbia" as she was christened, developed a very ugly disposition. She was quick to learn tricks and was worked in the ring as a performing clown elephant. But like so many other animals born in captivity, she grew more rebellious as she matured than jungle-bred elephants. After crippling one trainer and almost killing another, she was deemed too dangerous to travel with the show and set a bad example to the rest of the herd. So she was quietly sent to oblivion one night at winter quarters.[32]

Columbia's demise was not unknown at the time. The Associated Press (AP) picked up the news of her killing in the company's Bridgeport property, and newspapers across the country reported that circus staff had strangled the elephant to death on November 8, 1907. "Keeper Denman shackled her to one corner of the elephant barn. A rope was thrown around her neck, and with the aid of block and tackle five men pulled the rope around her neck so tightly that the elephant was choked before she had time to make an outcry. Her bulging eyes were the only sign of pain. . . . The body, weighing 9,350 pounds, was dragged out of the barn by Alberta, one of the largest elephants in the herd. It will be cut up into small pieces," the AP noted coolly. People at the scene said it took twelve minutes for Columbia to expire.[33]

A year earlier, the papers had reported that the company saw fit to kill Columbia's father, Mandarin, after he had fought continuously with circus staff and the other males in the herd. Circuses extinguished bulls with great regularity, but Columbia was supposedly a charmed case. She was a native-born female elephant, yet she proved even less equipped for life in show business than many of her wild-born herd mates. Even though the circus was the only world she knew, neither her herd mates nor her human keepers were able to socialize her sufficiently for her to cope with the labor of listening and responding to direction or suppressing her instinctive urges to move and get into mischief with her trunk. Or, perhaps Columbia was just a troubled individual and would have been difficult and belligerent even in the wild with other elephants—although Barnum, Bailey & Hutchinson company staff did not perceive the situation that way, telling Townsend Walsh off the record that captivity and training discipline inspired her intractability.

Many people inside the circuses, too, regretted what a disaster Columbia had become and also the many other elephants the circuses had killed. Still, because few circus owners or elephant trainers could imagine their trade without elephants, some attempted to normalize the death toll in the industry with flippant accounts that whitewashed these internal disputes and outside criticism of elephant management. George Conklin's brief account of Columbia dismissed her mental decline in a single short sentence: "The keepers became afraid of Columbia and she was killed in Bridgeport in 1907." Conklin similarly made light of how disposable elephants had become to the circuses. On the trip back from the company's famous five-year European tour, Conklin told of Mandarin, Columbia's father: "We brought him to New York in a big crate on the upper deck of the boat. On the way over Mr. Bailey decided to have him killed, so instead of unloading him on to the pier Mr. Bailey had a big seagoing tug come alongside, and the crate, elephant and all, was swung down to the deck of the tug, which then put out to sea. When far enough outside the crate was loaded down with pig iron, swung out over the water and let go. And so ended Mandarin."[34] Mandarin and Columbia both died at the peak of circus business history, and the height of the cultural-criminalization of elephant trouble, also publicly exposed by the deaths of Topsy, Gypsy, Fritz, Pilot, Albert, Prince, and many others whom circus fans knew by name.[35]

Columbia's death in 1907 additionally marked the peak of animal-based circuses in the United States, when about one hundred companies operated in the country. Throughout the previous century, the United States had been on balance a consumer nation, drawing into itself money, materials, and life from around the world—including animals. By the turn of the twentieth century, although the nation still traded reciprocally with the outside world, it had shifted from a net consumer of workers, capital, and information to a net producer, a pattern that extended to the nation's first mass entertainment as well.[36] Several companies would experiment with touring schedules to Canada, Mexico, Europe, Australia, and beyond.[37] Beginning in 1899, Barnum & Bailey circus managers accordingly founded a wholly independent show branded "Greatest Show on Earth," based in Britain and offered as a publicly held limited liability venture with £400,000 initial capital.[38] To boost sales of shares in the company, the financial papers reminded readers in London and beyond that the Barnum & Bailey brand was defined by, among other things, the fame around the "baby elephant" Columbia.[39] The shares sold very well to Britons, Europeans, and some American investors, and the "Greatest Show on Earth" circus toured Europe over four years and kept an inventory of animals, rolling stock, tents, and more at winter quarters in Olympia, just outside London. The venture ultimately returned to the United States, but it showed off the global aspirations of the world's most ambitious impresarios. Barnum & Bailey was perhaps the biggest and most valuable entertainment brand in the world at that point. Managers of the elephant herds on the largest circus shows still rightly claimed the elephant as a unique American innovation to circus arts and an entertainment mascot that encapsulated the globalized ambitions of Anglo-Americans. However, they never bred enough elephants to export, only lending them out to foreign audiences who paid to buy a ticket.

Many Americans inside and outside the circus community hoped that elephants could be dependable, easily reproduced workers like horses, cattle, dogs, pigs, and other domestic animals. But elephants were too powerful and too self-directed. They could not simply be put out to pasture in a paddock with a simple fence, expected to walk back to the barn from the field, left alone on the porch to bark when a wagon passed by, or allowed to roam the streets in packs eating garbage until rounded up by the man who would sell them to the butcher. Although Americans hoped to domesticate elephants, they remained net consumers of wild-caught elephants.

Nonetheless, elephants in captivity still needed to learn things from one another—things that no human could train them for—in order to survive in America and measure up to industry hopes. In captivity to humans, wild and powerful elephants were chained and manacled, held in picket lines, and otherwise constrained from the autonomous social interactions by which elephant culture—how to recognize and care for baby elephants or relate to an amorous conspecific—is transmitted to younger herd members. In truth, circus animal management routines actually prevented the interelephant interactions that might have allowed elephants to sustain and perpetuate themselves as a population native to the circuses.

By contrast, the self-reliance of domestic animals gave them the power to feed themselves, avoid predators, keep their skins and coats healthy, communicate among themselves, find mates, reproduce, and raise young.[40] Americans would speak about their productive relationships with those animals by saying that they were "easy to work with," not because they were necessarily docile or friendly but because there was a "reciprocal dependence" between humans and domesticated animals; those creatures could be trusted to take on much of the labor of their own care and reproduction with minimal constraints—perhaps just a collar, brand, or fence.[41] By contrast, American circus people found themselves dependent upon wild populations of elephants thousands of miles away and impossible to replicate in North America, with their own elephants likely to die at a young age due to accident, illness, or death by company order. Circus people perceived the self-directed activities of elephants as destructive to profit-making, and as a result, the elements of elephant autonomy that might have served circuses were interrupted.

So it was that the animal show business was fatefully built on the back of an elephant, as circus owner John Robinson had insisted.[42] Although elephants facilitated the development of animal celebrity as a cultural product and human concept, it is fair to ask if the American circus became a smashing success as entertainment and industry because of elephants or in spite of them.[43] In the nineteenth-century circus business there appears to have seldom been systematic bookkeeping complex enough to enumerate the total cost of obtaining, feeding, staffing, and transporting elephants. There were no cost-benefit analyses performed with respect to each company's elephant program, and thus no way to measure the actual earnings accrued from elephants versus the costs of managing them. Indeed, circus impresarios had long had a reputation for making ill-advised business decisions on gut instinct, not hard financial data.

Still, the thinking went that, if any menagerie or circus had elephants, they all had to have them. Circus men puffed their elephant herds with industry elites like George Arstingstall, insisting, "No circus, however small, could hope to exist without an elephant."[44] Animal trainer Lucia Zora agreed at the turn of the century, "Circuses are in the business of giving audiences what they want."[45] But menagerie and circus producers routinely leased, hired, and fired acts and show elements from season to season to keep spectators engaged. Not all companies had a big cat act or monkey horse jockeys or a cannonball man each season, and many circuses prospered without elephants. To keep elephants on a show was not inevitable, but a human choice in which both the circuses and their customers participated.

The dream of domesticating elephants was a hubristic moment in the broader story of "zoological manifest destiny" in U.S. history.[46] While sheep, horses, cattle, and other creatures became abundant successes, in the nineteenth century no menagerie or circus species was self-supporting, although companies did breed a few ungulates and (in the twentieth century) big cats, produced mostly in winter quarters barns. Although the genial elephant icon was native to the American circus, the living elephants failed to be genial because they were powerful yet fickle beings unable to flourish in captivity and because American entertainers and audiences were unable to come to terms with elephantine limits. In spite of the high cost of containing, controlling, and restocking elephants, circuses pressed on with the species because not to have done so would have required cultural changes in the entertainment business that most people were not prepared to make.

"It hurts to be beautiful," so the show business wisdom goes. Circus people have known this truism for generations. They took it in stride when they joined a traveling show and confronted the paradox of the American circus, an entertainment defined by both glamour and grime, exciting energy and exhausting labor, human triumph and human failure. Circus folk coped with these contradictions by romanticizing the complexity of their experience, not least with the elephants so central to the genre, at times gentle and obedient, at other times frightening and dangerous. "There is something about its size and shape that defeats the imagination," circus "elephant girl" performer Connie Clausen would explain. "That piercing primeval trumpet; that wrinkled, seemingly elastic trunk, and those solid squared-off legs, tiny eyes, and long, siren lashes are all strictly elephant, nothing else. Certainly their intelligence, ability to respond to affection and that hint of jungle violence that lies beneath their apparently easy-going natures made them fascinating."[47] If indeed stardom is produced by individuals who can reconcile such contrasts, then truly the circus elephant was the original American star.[48]

The first decade of the twentieth century represented both the height of the circus business and the peak of human-elephant conflict in the United States. The decade also marked a turning point in U.S. entertainment history, when live entertainments of all kinds would begin to contend with various new media. Decade by decade, Americans would gain access to a growing universe of commercial entertainment options with which Betsey and Mogul and Romeo had never vied, including cheap illustrated magazines, amusement parks, vaudeville and rodeos, and—importantly—cinema and television.[49]

Those venues would take up the methods and theories pioneered by menagerie and circus showmen to present American consumers with imagined chimps, dogs, horses, bears, and more, all performing human stories. The circuses would be in direct competition with compelling animal characters like Felix the Cat, Mickey Mouse, Bugs Bunny, Lassie, Yogi the Bear, Flipper, Elmo, Nemo, and dozens of others. Indeed, the 1941 Disney film *Dumbo* portended troubles ahead for the circuses, not only for its portrayal of the "Mad Elephant" but also for its logistically clean simulations of anthropomorphized elephants and other animals voiced by human actors who flattered and amused spectators with better "acting" than a living elephant could ever produce. These were characters that needed no training, food, or confinement, never aged or became sick, never went off script, never killed anyone—unless the script called for it. They in turn prepared the cultural landscape for a series of famous living animals, like Koko the gorilla, Knut the polar bear, Alex the parrot, and Zenyatta the thoroughbred horse. Those animals' lives are more complicated, but onto those creatures millions have projected their hopes and assumptions as consumers of living animals made figurative and symbolic. In this crowded media and animal character context, the circuses would increasingly offer only the older ideal of the circus elephant as sagacious and genial performer, downplaying and covering up circus workers who punished or killed off elephants who broke down under circus management, explaining away elephants who went off script as temporarily confused or anomalous.

Along the way, various people have predicted the death of the traveling circus, and yet, the form and the Americans who produce it are more resilient than they often appear. Certainly, there was the terribly difficult season of 1936, when the Great Depression crushed ticket sales as consumers spent what they still had on frequent trips to the movies. There was the infamous Ringling Brothers Circus Fire of 1944 in Hartford, Connecticut, which exposed the precariousness of the trade. The company's big top circus tent burnt down, resulting in 168 persons either burnt or trampled to death, almost 500 more injured. Many dozens of children were among the victims, and liability for the company was a crippling $4 million. Although many companies folded then and thereafter, including the much-loved Clyde Beatty Circus in 1956, the industry did not evaporate. The surviving circuses found ways to coexist with more episodic entertainments, like movies and carnivals, but would stumble again in the face of television's constant broadcasts, and they gradually became smaller and more thinly capitalized. By 1960 only about a dozen remained.[50]

With little capital available for reform of animal management procedures in the business, the living conditions of circus elephants have changed only minimally. They are still confined the majority of the time, now in box vans as often as rail cars. They are still broken for training and steered with elephant hooks. The profession of elephant handler remains a dangerous one, subjecting many elephant trainers to destructive or unpredictable elephants and additionally to

Audience view of elephants in performance, Standard Oil Company Wild Animal Show, Century of Progress Exposition, Chicago, 1933. Keystone-Mast Collection, UCR / California Museum of Photography, University of California, Riverside.

open criticism by fellow circus workers and the public. A few hardy souls continue on with the work, seeing it as an honorable tradition of interaction with elephants grounded in human relations with captive elephants that goes back centuries. "Bull men" can still present an intimidating presence on a circus lot, persevering even though they know they may eventually be killed on the job. They are somehow compelled to be with elephants nonetheless.

Over the twentieth century, American circuses remained material consumers of elephants, just as their customers were figurative consumers of the reflections of those elephants offered by traveling shows. Bought, sold, and leased frequently, elephants became incredibly inexpensive at mid-century, sold in want ads in the *Billboard* and through matter-of-fact animal dealers who could get a person "anything, anytime," no questions asked. By 1952, one "Elephant Census" commissioned by the trade journal *Billboard* had the total number of elephants in American circuses at 126, and 142 more in stationary animal shows, zoos, carnivals, and exotic animal dealers' inventories.[51] By then circus people referred to all adult elephants, male and female, by the masculinized term *bull*, which distracted from the fact that the circuses held fewer than ten male elephants.[52]

In the 1970s the circuses confronted a new challenge as public perceptions—especially among the middle and upper classes—of elephants in the wild shifted dramatically. The cause was new celebrities, created by "the rise to stardom of both researchers and their extended elephant families." Famous elephant researchers like Iain Douglas-Hamilton, Cynthia Moss, and others began gaining great media attention describing elephants doing elephant things in Africa, rather than human-directed things in America. To fund their research and to publicize threats to elephants, they studied and explained the personalities of individual wild elephants, asking audiences to identify with elephants as "mothers" and "daughters" and "families."[53] In using the media and public interest to cultivate consumer experiences of wild elephants as intrinsically valuable beings—in autonomous behavior and physicality—in order to halt the ivory trade and find funding for their research teams, these scientists produced popular natural history and ethological theories emphasizing animal individuality, sentience, and subjectivity, which audiences have accepted with great relish. Once it was the circuses that provided the public with popular depictions of animal life in the wild. Yet, with the advent of wildlife film, the National Geographic Society, and these celebrity animal scientists and their famous elephant research subjects, the circuses have become less authoritative in that respect.

Since the 1970s, circus elephant men and marketing teams have also coexisted with a large network of zoos, which similarly claim to speak for elephants and assure visitors that as privileged consumers they may serve as benevolent stewards of animals by their consumption of them. Certainly, contemporary zoo and animal amusement parks are, like the antebellum museum, meeting places of science and show business. Many of them have undergone a kind of Disneyfication, by creating presentations that try to construct wild or exotic animals as untouchable but still "friends," with names and personalities visitors can know.

Conservation-themed presentations and parks, inspired by Disney theme amuse-ment parks, allow visitors to enact that knowing by viewing and possibly feeding animals, speaking to handlers and tour guides, reading "information" plaques, paying for themed souvenir merchandise or food, attending "birthday party" pro-motional events for juvenile zoo animals, and visiting zoo websites, among other practices.[54]

What, then, has the circus become at the beginning of the twenty-first cen-tury? Productions offering no animals (save humans and maybe some horses) actually appear to be once again at the center of this age-old entertainment art form, just as they were in the day of John Rickett's circuses of trick horse riding and human acrobats two hundred years ago. By refusing exotic animal acts and offering more abstract performances, these circuses have been able to serve a broad, if wealthier and less family-oriented, audience while avoiding the scrutiny of the many animal reform organizations that gained notoriety in the 1990s. Cirque du Soleil—the "Montreal Monster,"[55] as some refer to the company in the trade—is one of many companies that have considered the expense and logistical complications of working with live elephants and decided to go a different way.[56] In 2001, a writer in *Time* magazine exposed the cultural divide in circus arts when he commented that the more traditional animal-act circuses seemed "un-comfortably out of place in today's entertainment market. It's the interspecies version of a minstrel show," whose speciesist caricatures of animals like tigers and elephants seem dated and in bad taste.[57]

Still, to many people, "circus" still means "elephant." This is wonderful or ter-rible, depending upon one's perspective. Newly volatile debates rage today over how or if circuses can continue to make their money by providing consumers (mostly families with small children) with shows of live animals. For instance, two dueling elaborations of the old happy circus elephant have appeared in the last few years, with advocates for each hoping to convince a critical mass of con-sumers that their interpretation tells the truth about the circus elephant life: the baby elephant Barack, born at the Ringling Brothers, Barnum & Bailey breeding farm in Florida, versus the icon of the "sad circus elephant," offered by People for the Ethical Treatment of Animals (PETA).

The young elephant Barack was born the day before President Obama's Inau-guration, in January 2009. He had been conceived using artificial insemination (AI) almost two years earlier by staff at the Center for Elephant Conservation (CEC), a five-million-acre elephant breeding and training compound founded in 1995 by Ringling Brothers' parent company Feld Entertainment.[58] Barack was the twenty-second of twenty-three (and counting) claimed Asian calves born there. The CEC is the most productive elephant breeding program on the continent, although it is not clear how many of those offspring survive today, and captive-bred elephant calves have only about a 50 percent chance of surviving to age ten.[59] Nonetheless, a few weeks after he arrived, Ringling Brothers staff posted on the center website a birth announcement ("It's a Boy!") and photographs of the newborn that replicate once more the genial circus elephant icon.[60] Presented in profile, Barack's promotional image shows his baby trunk raised in a salute to the viewer.

The company website touts the property as a conservation center, rightly noting that Asian elephants are in crisis in the wild because of loss of habitat. It proposes that the compound will serve as a "state-of-the-art facility . . . dedicated to the conservation, breeding and understanding of these amazing animals."[61] Outsiders are divided on what the center means for elephants, calling it "wonderful" and "the leading elephant-breeder in the Americas" or, alternately, "expensive acres of concrete and reinforced steel pipe [that] suggest a gulag for pachyderms" and "elephant puppy mill."[62] Ringling Brothers has not released any elephants back into the Asian or African countryside, as the CEC's name seems to promise. However, it has certainly solved two management problems for the circus: overcoming an effective elephant-importation ban created by the 1973 Endangered Species Act and adapting the Ringling Brothers brand to current zoo marketing trends, which emphasize captive animals as tools for the study and preservation of wild species and their shrinking habitats. Here the circus has revisited the old nineteenth-century educational role of traveling shows by appropriating the "zoo ark" concept from contemporary zoo, aquarium, and animal amusement park industries. The CEC claims to act out of a responsibility for stewardship over elephant species threatened by poaching and human encroachment in Asia and Africa.[63] Like the old-time circuses, it also continues to shape its brand to a broad, working-class audience. Company President Kenneth Feld routinely tells the press that, by delivering the essential American circus experience defined by elephants, the Greatest Show on Earth brand functions as a cultural "security blanket" and the "Walmart of show business," "priced . . . for the masses."[64]

Conservation ambitions aside, within a year baby Barack was on the road with the circus. He and the rest of the staff traveled the continent in eighteen-wheel trucks, sometimes also by rail, playing in hockey arenas and agro-domes. In the early 1960s, when the circus business was reeling from the combined effect of automobiles, cinema, and television, many famous companies stopped touring. The rebirth of the Ringling Brothers, Barnum & Bailey "Greatest Show on Earth" back then happened thanks to an intervention by live event production company

Feld Entertainment. Led by brothers Irvin and Israel Feld, the company slashed costs on the Ringling shows, laying off more than a thousand workers and leasing space and equipment, people and animals only when needed. They sold the company tents and created an eleven-month indoor show schedule. Irvin Feld also reinvented the Ringling Brothers brand with fresh advertising and prime-time specials, getting the show over with a new generation of kids tuned into TV.[65] Other companies similarly streamlined their operations and engaged television, and by the late 1990s there were thirty animal-act circuses operating in the United States, although fewer than ten still tour today.[66] Ringling Brothers is probably the preeminent company among them, and one reason that they, along with the Carsen & Barnes Circus, the Royal Hanneford Circus, and a few others, now take pride of place in the market is that they *do* have elephants.

Thus, when Barack entered show business it was in an era when the duties of circus elephants had been scaled back from the all-purpose heavy labor and acrobatic work of the nineteenth and early twentieth centuries. In contemporary circuses, elephants perform pyramids or sit up and salute the crowd with their trunks as they have for decades. Yet, at smaller companies, elephants seen in the ring show serve largely as an advertisement for the elephant rides that take place on the arena floor. This elephant labor constitutes the main source of direct circus-elephant income, and the ever-longer "intermissions" during performances—to provide time for people to pay for a trip—have not gone unnoticed.[67] For Ringling Brothers, even though Barack did not perform any real tricks and no one was allowed to ride him, his novel presence consisted of appearing in the show and "waving and extending his trunk into the air" to salute the crowd.[68] His owners hope that he, like the many performing elephants that preceded him, will reassure customers that he, in essence, endorses the business venture he inhabits and that they can learn about his species and help elephants by paying to be entertained by them. However, the basic conflict between a conservation ethic and the demands of show business was revealed when it became public in February 2010 that Ringling was abruptly pulling the baby elephant off the tour. Barack had come down with endotheliotropic herpesvirus (EEHV), a difficult-to-detect virus that is nonfatal to African elephants but one that has spread through circuses and zoos to Asian elephants, killing one third of zoo-bred Asian elephants in North America over the last twelve years.[69]

Around that time, people dressed as elephant mascots began appearing outside elementary schools, just as classes let out, in cities where the Ringling Brothers, Barnum & Bailey Circus was scheduled to perform. For instance, in Albany, New York, and Beaumont, Texas, the costumed people came across as sad circus elephants wearing a bloody head bandage and T-shirts bearing the words "Circuses are no fun for me!" The costumed people confronted bewildered six- and nine-year-olds headed onto school busses with posters and coloring book handouts decrying circus elephant captivity. The author of the sad circus elephant character was PETA, the much celebrated and much despised animal protection organization. Since the early 1980s, the group, founded by Ingrid Newkirk and Alex Pacheco, has publicized whistleblower accounts and citizen's complaints over just about every form of animal use. Many observers find their activities unconvincing or offensive because they often use provocative publicity campaigns to expose broad audiences to their views in ways that are brash and emotionally manipulative. Most people commenting on the online editions of the news stories documenting these events were irked by the demonstrations that bypassed parents' role as gatekeepers of their children's exposure to controversial information.[70] Regardless, PETA staff disseminated various depictions of bleeding or weeping cartoon elephants, plus secretly acquired photographs of actual juvenile elephants being bound, cast, hooked, and shocked with cattle prods during training at the Ringling Conservation Center. (The company responded by explaining that indeed the photos were genuine and represented the ways baby elephants are trained and managed in preparation as ring stock.) PETA staff posted these images on their website, in one public square in South Carolina (as a fiberglass sculpture), on freeway billboards, and in a full-page ad in the *New York Times*.

Indeed, the presence of protesters is today as routine a part of "circus day" as clowns and popcorn.[71] The bulk of public opinion against elephant use in circuses is not so much represented by PETA's particular activities as by the broader context of other animal-use reform groups. Less dramatic in their tactics, they carry the same basic message: the hands-on training and rigorous travel and performance schedules of mobile circuses make elephants miserable and represent a frivolous misuse of endangered and threatened species. They include the Humane Society of the United States, the American Society for the Prevention of Cruelty to Animals, the Animal Liberation Front, In Defense of Animals, the Animal Welfare Institute, the Captive Animal Welfare Society (which also uses a weeping circus elephant as a logo), the Performing Animal Welfare Society, and others. Indeed, the number of organizations publicly challenging elephant use in entertainment today may, in fact, outnumber the American circus companies still employing elephants. Understanding that the entertainment value of a given event is all in the mind of the consumer, those groups and PETA present competing icons of a sad circus elephant to challenge the happy circus elephant mascot, especially as implied by baby elephants like Barack. They share a belief that if circuses lose the authority to give meaning to the animals they offer ticket buyers, the very nature of their entertainment product and its assertion of the benevolence of human control of other species is no longer valid.

Consider how the tables have turned over the last few generations. How many in the industry and in the stands these days must resent the idea that circuses may not simply serve the segment of the public sympathetic to animal acts but must also be obligated to answer to the segment that would *never* visit a circus featuring animal acts? In the nineteenth century, circuses ignored or made promotional hype out of the complaints of detractors because, as providers of the dominant mass entertainment in the nation, most consumers and most newspaper men needed the circuses, for fun or for advertising revenue. So did the many thousands of people whose communities gained revenue from contracts to circuses or increased traffic at restaurants and hotels on circus day. Now, by contrast, the circuses operate, not at the center, but on the margins of the entertainment world, and far fewer people are directly financially dependent on companies like Ringling Brothers or the industry as a whole. The press now gives more attention to PETA publicity stunts against circuses than to circus publicity and seems less interested in innocently puffing circus parades or reprinting (now highly censored) interviews with circus elephant handlers.

So, since the 1990s, a number of circuses have been pushing back forcefully against protest and reform groups in order to retain control over the public meaning of their shows. Along with all the typically glamorous show advertising, their publicists have produced various "Our Commitment to Animal Welfare"-style statements for their websites. A few have "founded" retirement and breeding farms like the CEC by renaming their winter quarters properties as such. Various circuses also fund a lobby group called the Center for Consumer Freedom, which attempts to paint industry animal-use critics as hypocrites and "radicals."[72] And circus people know they will be protected by state and federal government law and practice. Most circuses featuring elephants, for example, have had little significant trouble from the United States Department of Agriculture (USDA), the underfunded federal agency charged with promoting animal-based commerce by gentle regulation of only its worst excesses.[73]

Considering the preschool and preteen demographic group to which the circuses cater, the most formidable challenge to the happy circus elephant ideal—and Barack as a living example of it—may come from the double-pronged challenge represented by the Internet (aided by the handheld video recorder) and related generational cultural changes. Not only does cyberspace offer a world of competing activities and entertainments to children and parents that make the postwar challenge of television seem mild by comparison, but it also fosters reform group organizing and the publicizing of such groups' evidence and claims. For people born after 1995—the circuses' future consumer base of working-class parents with young children—the Internet is the first stop when looking for information on any topic. Those who offer the most ubiquitous and youth-centered interpretations of the circus elephant online may become the most persuasive with potential ticket buyers of the future.[74]

To see how this is so, try an Internet search of "circus elephant" or "circuses and elephants." In early 2010, among the first ten hits of such a search, five were sites run by groups seeking to discredit and abolish elephant use by circuses. PETA's Circuses.com and RinglingBeatsElephants.com came up at numbers three and seven, respectively, demonstrating where users have clicked in previous searches. None of the top ten was a promotional or business website run by the circuses themselves. Here the crying and bleeding sad circus elephant has crowded out Ringling Brothers' Barack.

The phenomenon we call "Web 2.0" privileges audience-authored understandings of circuses because the online world is dominated by user-generated content, not commercial broadcasts of marketing messages to clients. There we find diverse amateur records of circus life, some cheerful and joyous, some not. One notorious item shows a decorated circus elephant in Honolulu being shot many times, collapsing against the side of a car, struggling to get up, bleeding and, eventually, dying there. This was Tyke, a now famous—she has her own Wikipedia page—female African elephant killed in August 1994. Tyke assaulted three men, one of them fatally, then ran from keepers into a nearby parking lot. Pursued by police and circus staff, she was shot dozens of times to the horror of bystanders. In the amateur video capturing the event, a disgusted onlooker can be heard yelling at one of the riflemen, "You fucking asshole!"[75]

This study was able to access such evidence of nineteenth-century circus life only in static print and graphic form in old newspapers, line drawings, memoirs, and photographs, sources that are far less emotive. Recent citizen-authored evidence is found in abundance on Youtube.com, where PETA even has its own channel as a collection point for whistleblower undercover footage of elephant life in various circuses. As color moving images with sound, these sources show signs of elephants trouble, like weaving or trumpeting while being forcefully hooked, that for some will lend credence to the idea that circus captivity really does make elephants "sad."

The volume and shocking nature of such online evidence also hints that the icon of the happy circus elephant may be definitively, if quietly, reinterpreted for the public by another set of challengers, namely, animal-themed amusement parks and zoos. In 2004, a meeting of the Association of Zoos and Aquariums (AZA) responded to the growing body of scientific research on African and Asian elephants showing that zoos (and certainly circuses) are not even minimally meeting elephants' needs. They suggested some dramatic changes in zoo elephant management meant to preserve zoos as much as elephants. It was a well-meaning start at keeping abreast of public sentiment, even if the changes still prove insufficient, as many scientists not employed by zoos insist.[76] Nonetheless, the change occurred in part after a committee of AZA members wrote a protocol for the care of African and Asian elephants in captivity, the first species-specific standard of animal care ever produced. It was grounded in industry knowledge of the risks of injury and death to staff and elephants using older modes of hands-on management, a reality with which circus folk labored for two hundred years.[77] The new codes demand, among other things, hands-off management (in which handlers do not enter pens with elephants or employ ropes, elephant hooks, cattle prods, or punishment), larger spaces, and better substrate surfaces to attempt to preserve elephant foot health, since foot disease causes approximately half of zoo elephant deaths. The new hands-off strategy clearly has an elephant welfare component and a zoo worker safety component, but it also creates a publicity benefit because it protects zoos from any injury to elephants or people, as well as from the attending negative media attention, due to staff malpractice. (In 2000, for instance, people across the industry revisited this truism of captive elephant management when the famed Portland Zoo elephant program was fined $10,000 and publicly scrutinized for an exceptionally violent hooking of the juvenile elephant Rose-Tu by a drunken handler).[78]

In light of the new standards, at least ten zoos decided to close down their elephant programs entirely, sending their elephants to other zoos or to one of the three elephant sanctuaries in North America. Many others embarked upon a proposed spending and building spree financed by debt, bond issues, and donations. At those zoos, industry trends see facilities keeping groups of a minimum of six animals and spending millions on AI breeding programs in a risky (and yet unproven) plan to take conservation out of the hands of Asians and Africans running wild game preserves and put it into the hands of North American zoo keepers.

The underlying premise of the move to hands-off and breeding-centered management of elephants is to showcase elephants as representatives of threatened and endangered species whose means of self-preservation and self-direction need to be understood and promoted by staff and public alike. Animal parks that present conservation-themed animal displays assert to visitors that captive animals have and should exercise their preference with respect to how, when, or if to engage with humans who contain them. New elephant enclosures will assert that elephants require larger spaces in which to engage in greater mobile exercise and elephant socialization practices like allomothering. Barack the baby elephant may turn out to be ill-suited as an ambassador for this message since the Ringling Brothers Florida breeding and training farm is not open to the public, does not employ a hands-off management policy, and is not accredited by the Association of Zoos and Aquariums.[79]

If the volume of online citizen-authored depictions of circuses, flattering and not so flattering, does not come to define the meaning of circus elephant for the next generation of potential circus-goers, family and school trips to the zoo may do so. The zoos are elephant-management insiders, but they inadvertently bolster PETA's sad circus elephant mascot by promoting hands-off management while striving to tell a story about how elephant cultures and practices, not just elephant bodies, are necessary for the survival of those species. Circus trainers who use hands-on management imagine elephants as inherently in need of training, as raw material to be shaped and modified to a circus's needs and sold off when unproductive. Especially for the endangered Asian species, with somewhere between only 35,000 and 60,000 members left alive on earth, circus folk will ultimately need to decide if the living circus elephant performer is a sign of cruel folly or honorable artistic tradition that can be adapted for the next generation of circus fans.

## Introduction • Turning the Circus Inside Out

1. In 1966 the U.S. Postal Service also issued a circus stamp, this one depicting a grinning clown, to commemorate the 1866 birth of famous circus owner John Ringling. That year they also issued—without irony—a stamp commemorating Henry Bergh's founding of the American Society for the Prevention of Cruelty to Animals, an organization that was a constant irritant to the circuses.

2. Harriet Ritvo, *The Animal Estate: The English and Other Creatures in the Victorian Age* (Cambridge, MA: Harvard Univ. Press, 1987), 3.

3. James W. Cook, *The Arts of Deception: Playing with Fraud in the Age of Barnum* (Cambridge, MA: Harvard Univ. Press, 2001), 73–117.

4. Graham Huggan and Helen Tiffin, *Postcolonial Ecocriticism: Literature, Animals, and the Environment* (New York: Routledge, 2010), 136; Jennifer Mason, *Civilized Creatures: Urban Animals, Sentimental Culture, and American Literature, 1850–1900* (Baltimore: Johns Hopkins Univ. Press, 2005), 122–25; Marjorie Spiegel, *The Dreaded Comparison: Human and Animal Slavery*, 3rd ed. (New York: Mirror Books / I.D.E.A., 1997), 33–38.

5. See, e.g., Alan Beardsworth and Alan Bryman, "The Wild Animal in Late Modernity," *Tourist Studies* 1, no. 1 (2001): 83–104; John Berger, "Why Look at Animals?" in *About Looking* (New York: Pantheon Books, 1980), 13; Derek Bousé, *Wildlife Films* (Philadelphia: Univ. of Pennsylvania Press, 2000); Chilla Bulbeck, *Facing the Wild: Ecotourism, Conservation, and Animal Encounters* (London: Earthscan Books, 2005); Jonathan Burt, *Animals in Films* (London: Reaktion, 2002); Cynthia Chris, *Watching Wildlife* (Minneapolis: Univ. of Minnesota Press, 2006); Susan Davis, *Spectacular Nature: Corporate Culture and the Sea World Experience* (Berkeley: Univ. of California Press, 1997); Donna Haraway, *Primate Visions: Gender, Race, and Nature in the World of Modern Science* (New York: Routledge, 1989); Walter Hogan, *Animals in Young Adult Fiction* (Lanham, MD: Scarecrow Press, 2008); Margaret J. King, "The Audience in the Wilderness," *Journal of Popular Film and Television* 24, no. 2 (1996): 60–68; Randy Malamud, "Famous Animals in American Culture," in *A Cultural History of Animals in the Modern Age*, ed. Randy Malamud (London: Berg, 2007), 1–26; Gregg Mitman, *Reel Nature: America's Romance with Wildlife on Film* (Seattle: Univ. of Washington Press, 1999); Nigel Rothfels, *Savages and Beasts: The Birth of the Modern Zoo* (Baltimore: Johns Hopkins Univ. Press, 2002).

6. Reuel Denney, *The Astonishing Muse* (1957; New Brunswick, NJ: Transaction,1989), lv–lxix.

7. Haraway, *Primate Visions*, 19–25; Davis, *Spectacular Nature*, 14.

8. Nicole Shukin, *Animal Capital: Rendering Life in Biopolitical Times* (Minneapolis: Univ. of Minnesota Press, 2009), 11–12.

9. See, e.g., Jerry Apps, *Ringlingville USA: The Stupendous Story of Seven Siblings and Their Stunning Circus Success* (Madison: Wisconsin Historical Society Press, 2005); Roland Auguet, *Histoire et Légende du Cirque* (Paris: Flammarion, 1974); George L. Chindahl, *A History of the Circus in America* (Caldwell, ID: Caxton, 1959); Rupert Croft-Cooke and Peter Cotes, *Circus: A World History* (London: Elek, 1976); John Culhane, *American Circus: An Illustrated History* (New York: Holt, 1990); Janet M. Davis, *The Circus Age: Culture and Society under the American Big Top* (Chapel Hill: Univ. of North Carolina Press, 2002); John Durant and Alice Durant, *Pictorial History of the American Circus* (New York: A. S. Barnes, 1957); Mildred Sandison Fenner and Wolcott Fenner, comp. and ed., *The Circus: Lure and Legend* (Englewood Cliffs, NJ: Prentice-Hall, 1970); Charles Philip Fox, *Circus Parades: A Pictorial History of America's Greatest Pageant* (Watkins Glen, NY: Century House, 1953); Isaac John Greenwood, *The Circus: Its Origin and Growth Prior to 1835* (New York: B. Franklin, 1970); Earl Chapin May, *The Circus: From Rome to Ringling* (New York: Duffield & Green, 1932); Joe McKennon, *Horse Dung Trail: Saga of the American Circus* (Sarasota, FL: Carnival, 1975); Jennifer L. Mosier, "The Big Attraction: The Circus Elephant and American Culture," *Journal of American Culture* 22, no. 2 (1999): 7–18; Marian Murray, *Circus! From Rome to Ringling* (New York: Appleton, 1958); Tom Ogden, *Two Hundred Years of the American Circus: From Aba-Daba to the Zoppe-Zavatta Troupe* (New York: Facts On File, 1993); Gregory J. Renoff, *The Big Tent: The Traveling Circus in Georgia, 1820–1930* (Athens: Univ. of Georgia Press, 2008); George Speaight, *History of the Circus* (London: Tantivy Press, 1980); Helen Stoddart, *Rings of Desire: Circus History and Representation* (Manchester: Manchester Univ. Press, 2000).

10. In this broad genre, see, e.g., these twentieth-century examples: "Elephant Census," *Billboard*, Apr. 12, 1952; Charlie Campbell, "Elephants, Good and Bad," *Circus Review* 5, no. 1 (1957): 2–3; Bob Cline, *America's Elephants* (n.p.: T'Belle LLC Productions, 2009); Ann Colver, *Old Bet* (New York: Knopf, 1957); John "Chang" Reynolds Papers, Robert L. Parkinson Library and Research Center, Circus World Museum, Baraboo, WI; Morton Smith, "Elephant Census of the United States," *Hobbies: The Magazine for Collectors*, Mar. 1942, 33, item #6, box (vol.) 6, Leonidas Westervelt Circus Collection, New-York Historical Society, New York; Phil Stong, *A Beast Called an Elephant* (New York: Dodd, Mead, 1955); Jay Teel, "True Facts and Pictures: The Crime and Execution of Diamond the Insane Elephant" (1930), item #15, box 14, "Pamphlets and minor books, [18—]-c1947," Leonidas Westervelt Circus Collection. See also the valuable online database at www.elephant.se/. An important exception in this case: William Johnson, *The Rose-Tinted Menagerie* (London: Heretic, 1990).

11. Christine Meisner Rosen and Christopher C. Sellers, "The Nature of the Firm: Towards an Ecocultural History of Business," *Business History Review* 73, no. 4 (1999): 577–600; also see the other articles in this special environmental history issue of *Business History Review*; Ted Steinberg, "Down to Earth: Nature, Agency, and Power in History," *American Historical Review* 107, no. 3 (2002): 800.

12. Stephen Pemberton, "Canine Technologies, Model Patients: The Historical Production of Hemophiliac Dogs in American Biomedicine," in *Industrializing Organisms: Introducing Evolutionary History*, ed. Susan R. Schrepfer and Philip Scranton (New York: Routledge, 2004), 195.

13. Susan McHugh, "Readings That 'Work Every Time' in the Animal Studies Classroom," e-mail message delivered on H-Animal Discussion Network, Apr. 16, 2009; David Nibert, *Animal Rights, Human Rights: Entanglements of Oppression and Liberation* (New York: Rowman & Littlefield, 2002), 3.

14. Chris Philo and Chris Wilbert, "Animal Spaces, Beastly Places: An Introduction," in *Animal Spaces, Beastly Places: New Geographies of Human-Animal Relations*, ed. Chris Philo and Chris Wilbert (New York: Routledge, 2000), 2–3.

15. Virginia DeJohn Anderson, *Creatures of Empire: How Domestic Animals Transformed Early America* (New York: Oxford Univ. Press, 2004); Jon T. Coleman, *Vicious: Wolves and Men in America* (New Haven, CT: Yale Univ. Press, 2006); Ann Norton Greene, *Horses at Work: Harnessing Power in Industrial America* (Cambridge, MA: Harvard Univ. Press, 2008); Helena Pycior, "Together in War and Memory: Fala and President Franklin Delano Roosevelt," paper presented at the Organization of American Historians Annual Meeting, Mar. 29, 2008; Clay McShane and Joel A. Tarr, *The Horse in the City: Living Machines in the Nineteenth Century* (Baltimore: Johns Hopkins Univ. Press, 2007). See also various chapters of Sarah E. McFarland and Ryan Hediger, eds., *Animals and Agency: An Interdisciplinary Exploration* (Leiden: Brill, 2009); Peter Benes, ed., *New England's Creatures, 1400–1900, Dublin Seminar for New England Folklife Annual Proceedings* 18 (Boston: Boston University, 1995). For a notable example with respect to Britain, see Robert Malcolmson and Stephanos Mastoris, *The English Pig: A History* (London: Hambleton & London, 2001). On African elephants who "might have actively influenced their history," see Bernhard Gissibl, "The Nature of Colonialism: Being an Elephant in German East Africa," paper presented at the "Animals in History: Examining the Not So Human Past" Conference, German Historical Institute, Cologne, Germany, May 2005.

16. McShane and Tarr, *Horse in the City*; Greene, *Horses at Work*.

17. Susan D. Jones, *Valuing Animals: Veterinarians and Their Patients in Modern America* (Baltimore: Johns Hopkins Univ. Press, 2003), x.

18. Jason Hribal, "'Animals Are Part of the Working Class': A Challenge to Labor History," *Labor History* 44, no. 4 (2003): 435–36. On the attempts to shape nonhuman bodies and their habits to market requirements, see also Barbara Noske, *Beyond Boundaries: Humans and Animals* (Montreal: Black Rose Books, 1997), 1–39; Ritvo, *Animal Estate*, 3.

19. Erica Fudge, "The History of Animals" *Ruminations* 1 (May 25, 2006), www.h-net.org/~animal/ruminations_fudge.html; Michael S. Roth, "Ebb Tide," *History and Theory* 46, no. 1 (2007): 66; Gabrielle M. Spiegel, "The Task of the Historian," *American Historical Review* 114, no. 1 (2009): 9.

20. For some recent discussions of this debate see, e.g., Philip Armstrong, *What Animals Mean in the Fiction of Modernity* (New York: Routledge, 2008), 3–4; Jason Hribal, *Fear of the Animal Planet: The Hidden History of Animal Resistance* (Oakland, CA: AK Press, 2010), 26–30; Huggan and Tiffin, *Postcolonial Ecocriticism*, 190–93; Sarah E. McFarland and Ryan Hediger, "Approaching the Agency of Other Animals: An Introduction," in McFarland and Hediger, *Animals and Agency*, 1–20; Philo and Wilbert, "Animal Spaces," 5; Sandra Swart, *Riding High: Horses, Humans and History in South Africa* (Johannesburg: Wits Univ. Press, 2010), 197–205.

21. Erica Fudge has addressed this thorny issue in the introduction to her collection, *Renaissance Beasts: Of Animals, Humans, and Other Wonderful Creatures*, ed. Erica Fudge (Urbana: Univ. of Illinois Press, 2004), 7, 15n25: "Social and cultural history, perhaps recognizing a progression from the study of the working class, women, ethnic minorities, and homosexuals to the study of animals, has begun to pay attention to the nonhuman in new and productive ways. . . . [Yet] this is always an unfortunate formulation that seems to reiterate some particularly unpleasant concepts of a chain of being that placed white, Western man at the top and animals at the bottom, with various human and 'subhuman' groups in between.". See also Jon T. Coleman, "Two by Two: Bringing Animals into American History," *Reviews in American History* 33, no. 4 (2005):

489. For some recent arguments in favor of subaltern, or "working class," status for animals seen as engaging in a kind of "infrapolitics," see Hribal, "Animals are Part of the Working Class," 435–36; Swart, *Riding High*, 201–2.

22. In fact people often operate as historical actors while living with little knowledge of or control over the broader ramifications of their actions. In an interview in the early 1980s Michel Foucault plainly explained: "People know what they do; they frequently know why they do what they do; but what they don't know is what what they do does." Quoted in Hubert L. Dreyfus and Paul Rabinow, *Michel Foucault: Beyond Structuralism and Hermeneutics*, 2 ed. (Chicago: Univ. of Chicago Press, 1982), 187.

23. Etienne Benson, "Historiography, Disciplinarity, and the Animal Trace," paper presented at the "Animals: Past, Present, and Future Conference," Michigan State University, East Lansing, MI, Apr. 16, 2009; see also Anderson, *Creatures of Empire*, 1–3; McFarland and Hediger, "Approaching the Agency of Other Animals," 9–10; John Simons, *Animal Rights and the Politics of Literary Representation* (Houndsmills, UK: Palgrave Macmillan, 2002), 5–6, 85–87.

24. Lucy A. Bates and Richard W. Byrne, "Creative or Created: Using Anecdotes to Investigate Animal Cognition," *Methods* 42, no. 1 (2000): 14; Robert W. Mitchell, "Anthropomorphism and Anecdotes: A Guide for the Perplexed," in *Anthropomorphism, Anecdotes, and Animals*, ed. Robert W. Mitchell, Nicholas S. Thompson, and H. Lyn Miles, (Albany, NY: SUNY Press, 1997), 407–28.

25. Shana Alexander, *The Astonishing Elephant* (New York: Random House, 2000); Silvio A. Bedini, *The Pope's Elephant: An Elephant's Journey from Deep in India to the Heart of Rome* (New York: Penguin, 1997); Richard Carrington, *Elephants: Their Natural History, Evolution, and Influence on Mankind* (New York: Basic Books, 1958); J. C. Daniel, *The Asian Elephant: A Natural History* (1998; repr. Dehra Dun: Natraj, 2009); Oliver Goldsmith, *The Asian Elephant: A Natural History* (1955); Donald F. Lach, "Asian Elephants in Renaissance Europe," *Journal of Asian History* 1/2 (1967–68): 133–76; Ivan T. Sanderson, *The Dynasty of Abu: A History and Natural History of the Elephants and Their Relatives, Past and Present* (New York: Knopf, 1962); Eric Scigliano, *Love, War, and Circuses: The Age-Old Relationship between Elephants and Humans* (New York: Houghton Mifflin, 2002); H. H. Scullard, *The Elephant in the Greek and Roman World* (London: Thames & Hudson, 1974); Dan Wylie, *Elephant* (London: Reaktion, 2008).

26. For definitions and methodologies, see, e.g., David Fraser, *Understanding Animal Welfare: The Science in Its Cultural Context* (London: Wiley-Blackwell, 2008); Richard D. Ryder, "Measuring Animal Welfare" *Journal of Applied Animal Welfare Science* 1, no. 1 (1998): 75–80; or see the publications of the Universities Federation for Animal Welfare,www.ufaw.org.uk/index.php. Some relevant recent titles incorporating animal welfare science research with respect to elephants include: G. A. Bradshaw, *Elephants on the Edge: What Elephants Teach Us about Humanity* (New Haven, CT: Yale Univ. Press, 2009); Ros Clubb and Georgia Mason, *A Review of the Welfare of Zoo Elephants in Europe: A Report Commissioned by the RSPCA* (Oxford: University of Oxford, Animal Behaviour Research Group, 2002), 4–7; Debra L. Forthman, Lisa F. Kane, David Hancocks, and Paul F. Waldau, eds., *An Elephant in the Room: The Science and Well-Being of Elephants in Captivity* (North Grafton, MA: Tufts Center for Animals and Public Policy, 2009); John Webster, *Animal Welfare: Limping towards Eden* (London: Blackwell, 2005), 1–23; Chris Wemmer and Catherine A. Christen, eds., *Elephants and Ethics: Toward a Morality of Coexistence* (Baltimore: Johns Hopkins Univ. Press, 2008).

27. Michael Hutchins, "Variation in Nature: Its Implications for Zoo Elephant Management," *Zoo Biology* 25, no. 3 (2006): 161–71.

28. G. A. Bradshaw goes so far as to explain the difference as one between " 'elephants in captivity' rather than 'captive elephants' " to acknowledge that the animals are not defined by their circumstances but affected by them." G. A. Bradshaw, *Elephants in Circuses: Analysis of Practice, Policy, and Future,* Animals and Society Institute Policy Paper (Ann Arbor, MI: Animals and Society Institute, 2007), 4.

29. Joanna Latimer and Lynda Birke, "Natural Relations: Horses, Knowledge, Technology," *Sociological Review* 57, no. 1 (2009): 1–27; Lynda Birke, Joanna Hockenhull, and Emma Creighton, "The Horse's Tale: Narratives of Caring for/about Horses," *Society and Animals* 18, no. 4 (2010): 331–47. The classic work on this point, of course, is Clifford Geertz, *The Interpretation of Cultures* (New York: Basic Books, 1973).

30. In general, this study makes use of syntheses of the scientific literature on Asian and African elephants by Raymond Sukumar and others, but it refers to individual research papers and studies when relevant findings are of recent vintage and thus available only as journal articles.

31. Edmund Russell, "Introduction: The Garden in the Machine: Toward an Evolutionary History of Technology," in Schrepfer and Scranton, *Industrializing Organisms*, 2.

32. Americans first began to use this metaphor with respect to veterans of the war with Mexico over Texas, explaining that one had thus seen the elephant "in the figurative sense." The concept was also applied in that sense to 1850s California gold rush prospectors, and in the next decade as a metaphor for having seen battle in the Civil War. See, e.g., "Glimpse of the Elephant," *Pittsburgh Morning Post*, May 7, 1845; "California Enterprise," *Daily Alta California*, May 21, 1859; George P. Hammond, *Who Saw the Elephant? An Inquiry by a Scholar Well Acquainted with the Beast* (San Francisco: California Historical Society, 1964). My thanks go to Richard Reid for his advice on this point with respect to the Civil War.

33. August H. Kober, *Circus Nights and Circus Days*, trans. Claud W. Sykes (London: Sampson Low, Marston, 1928), 7.

## Chapter 1 · Why Elephants in the Early Republic?

1. Robert McClung and Gale McClung, "Captain Crowninshield Brings Home an Elephant," *American Neptune: A Quarterly Journal of Maritime History* 18, no. 2 (1958): 141.

2. The numbers of Old Bet stories are far too numerous to list here. For a sampling across many decades and several genres of publication, beyond what I cite elsewhere in the introduction to this chapter, see "First Imported Elephant," *Atlanta Constitution*, Sept. 7, 1896; "An Elephant's Monument," *Youth's Companion* 71, no. 28 (1897): 336; "Elephant to Start 53-Mile Hike Today," *New York Times*, Apr. 9, 1922; Ann Colver, *Old Bet* (New York: Knopf, 1957); Phil Stong, *A Beast Called an Elephant* (New York: Dodd, Mead, 1955); Tracy Garrity, "The 'Cradle' of the US Circus," *Christian Science Monitor*, Feb. 13, 1979; "Step Right Up! Bob Brooke Presents the History of the Circus in America," *History Magazine*, Oct.–Nov. 2001, Nov. 30, 2007, www. history-magazine.com/circuses.html.

3. W. S. Adams, "Elephants and Their Keepers," *Wilkes' Spirit of the Times*, Mar. 18, 1865, 38; see also George G. Goodwin, "The Crowninshield Elephant: The Surprising Story of Old Bet," *Natural History: The Journal of the American Museum* 60 (Oct. 1951): 359.

4. "The Queerest Pilgrimage," *Washington Post*, Mar. 31, 1912.

5. "First Imported Elephant," *Atlanta Constitution*, Sept. 7, 1896; "An Elephant's Monument," *Youth's Companion* 71, no. 28 (1897): 336.

6. Gil Robinson, *Old Wagon Show Days* (Cincinnati: Brockwell, 1925), 33.

7. "First Imported Elephant," *Atlanta Constitution*, Sept. 7, 1896.

8. Raymond Sukumar, *The Living Elephants: Evolutionary Ecology, Behaviour, and Conservation* (New York: Oxford Univ. Press, 2003), 52.

9. Goodwin, "Crowninshield Elephant," 358; McClung and McClung, "Captain Crowninshield," 137–38.

10. McClung and McClung, "Captain Crowninshield," 138.

11. See, e.g., "Curious Account of Capturing Elephants," *American Universal Magazine* 3, no. 6 (1797): 423; Richard Carrington, *Elephants: Their Natural History, Evolution, and Influence on Mankind* (New York: Basic Books, 1958), chap. 12; Peter Kolb, *A Collection of Voyages and Travels* (Philadelphia: Spotswood, 1787), 122–30; Charles Nordhoff, "Peep at Elephant," *Harper's New Monthly Magazine* 20, no. 118 (1860): 455–67; Ivan T. Sanderson, *The Dynasty of Abu: A History and Natural History of the Elephants and Their Relatives, Past and Present* (New York: Knopf, 1962), 181–85; Sujit Sivasundaram, "Trading Knowledge: The East India Company's Elephants in India and Britain," *Historical Journal* 48, no. 1 (2005): 39–41.

12. Asian critics countered that such tactics were not effective and indicated either a lack of civilization or incompetence on the part of the human handler. Sivasundaram, "Trading Knowledge," 40–41.

13. In various parts of Asia today, people still use elephants for ceremonial purposes and also for construction and logging operations, since the elephant is less expensive and less damaging to the environment than heavy machinery.

14. Silvio A. Bedini, *The Pope's Elephant: An Elephant's Journey from Deep in India to the Heart of Rome* (New York: Penguin, 1997); Brian Cummings, "Pliny's Literate Elephant and the Idea of Animal Language in Renaissance Thought," in *Renaissance Beasts: Of Animals, Humans, and Other Wonderful Creatures*, ed. Erica Fudge (Urbana: Univ. of Illinois Press, 2004), 173; Donald F. Lach, "Asian Elephants in Renaissance Europe," *Journal of Asian History* 1/2 (1967–68): 133–76; Sivasundaram, "Trading Knowledge," 27–63.

15. Louise E. Robbins, *Elephant Slaves and Pampered Parrots: Exotic Animals in Eighteenth-Century Paris* (Baltimore: Johns Hopkins Univ. Press, 2002), 19.

16. Raman Sukumar, *Elephant Days and Nights: Ten Years with the Indian Elephant* (New York: Oxford Univ. Press, 1996), 106.,

17. Ros Clubb, Marcus Rowcliffe, Phyllis Lee, Khyne U. Mar, Cynthia Moss, and Georgia J. Mason, "Compromised Survivorship in Zoo Elephants," *Science* 322, no. 598 (2008): 1649; Jeheskel Shoshani and John F. Eisenberg, "Elephas maximus," *Mammalian Species* no. 182 (June 18, 1982), 1–8; Sukumar, *Elephant Days and Nights*, 102–7.

18. Goodwin, "Crowninshield Elephant," 357–59.

19. Nathaniel Hathorne, quoted in McClung and McClung, "Captain Crowninshield," 139, and Goodwin, "Crowninshield Elephant," 357.

20. McClung and McClung, "Captain Crowninshield," 139–40.

21. Robbins, *Elephant Slaves*, 11–17.

22. Ibid., 32.

23. "The America," *Argus; or, Greenleaf's New Daily Advertiser*, Apr. 18, 1796. This piece also appeared verbatim a day later in *Greenleaf's New York Journal and Patriotic Register*, Apr. 19, 1796.

24. Ibid.; see also, Goodwin, "Crowninshield Elephant," 358; McClung and McClung, "Captain Crowninshield," 140.

25. Richard R. John, *Spreading the News: The American Postal System from Franklin to Morse* (Cambridge, MA: Harvard Univ. Press, 1995), 25–31.

26. Brett Mizelle, "Contested Exhibitions: The Debate over Proper Animal Sights in Post-Revolutionary America," *Worldviews: Environment, Culture, Religion* 9, no. 2 (2005): 226–27.

27. Julia Rowland Myers, "Robert Wylie: Philadelphia Sculptor, 1856–1863," *Archives of American Art Journal* 40, no. 1/2 (2000): 4–17; John Frederick Walker, *Ivory's Ghosts: The White Gold of History and the Fate of Elephants* (New York: Atlantic Monthly Press, 2009).

28. T. H. Breen, *The Marketplace of Revolution: How Consumer Politics Shaped American Independence* (New York: Oxford Univ. Press, 2004), 129–32, 136.

29. Ibid., 94–96; Fred Somkin, *Unquiet Eagle: Memory and Desire in the Idea of American Freedom, 1815–1860* (Ithaca, NY: Cornell Univ. Press, 1967), 11–54; William Earl Weeks, "American Nationalism, American Imperialism: An Interpretation of United States Political Economy, 1789–1861," *Journal of the Early Republic* 14, no. 4 (1994): 485–95.

30. Antonello Gerbi, *The Dispute of the New World: The History of a Polemic, 1750–1900* (Pittsburgh: Univ. of Pittsburgh Press, 1973), 3–27, 157–58; Gilbert Chinard, "Eighteenth-Century Theories on America as Human Habitat," *Proceedings of the American Philosophical Society* 91, no. 1 (1947): 30–57; Paul Semonin, *American Monster: How the Nation's First Prehistoric Creature became a Symbol of National Identity* (New York: New York Univ. Press, 2000), 6, 13–14.

31. William Winterbotham, *An Historical, Geographical, Commercial and Philosophical View of the American United States, and of the European Settlements in American and the West-Indies*, 4 vols. (London: J. Ridgway, H. D. Symonds & D. Holt, 1795), 3:523.

32. Semonin, *American Monster*, 265–66, 286, 300.

33. As a war measure, federal law banned "all horse racing, and all kinds of gaming cock fighting, exhibitions of shows, plays, and other expensive diversions and entertainments" between the outbreak of rebellion against Britain and 1780. Breen, *Marketplace of Revolution*, xii; Richard W. Flint, "Origin of the Circus in America," *Bandwagon* 25, no. 2 (1981): 18; Mizelle, "Contested Exhibitions," 221.

34. See also, e.g., "The Elephant," *City Gazette*, Jan. 25, 1799.

35. William Bentley, *The Diary of William Bentley, D.D., Pastor of the East Church, Salem, Massachusetts*, 4 vols. (Gloucester, MA: P. Smith, 1962), 2:34, 247, 356–57; "Male Moose," (ca. 1798–1808), 49473, Early American Imprints Series 1 and 2, American Antiquarian Society, Worcester, MA.

36. Robert McClung and Gale McClung, "Tammany's Remarkable Gardiner Baker: New York's First Museum Proprietor," *New-York Historical Society Quarterly* 42 (1958): 150, 159, 165; Robert I. Goler, " 'Here the Book of Nature Is Unfolded': The American Museum and the Diffusion of Scientific Knowledge in the Early Republic," *Museum Studies Journal* 2 (spring 1986): 16.

37. Bentley, *Diary*, 4:400–402.

38. Robert M. McClung and Gale S. McClung, "America's First Elephant," *Nature Magazine* 50 (Oct. 1957): 403; James W. Shettel, "The First Elephant in the United States," *The Circus Scrapbook*, July 1929, 7–8, vol. 6, item #3, Leonidas Westerveld Circus Collection, New-York Historical Society, New York.

39. "The Elephant," *City Gazette* (Charleston, SC), Dec. 27, 1798.

40. Ros Clubb and Georgia Mason, *A Review of the Welfare of Zoo Elephants in Europe: A Report Commissioned by the RSPCA* (Oxford: University of Oxford, Animal Behaviour Research Group, 2002), 16.

41. "The Elephant," *City Gazette* (Charleston, SC), Dec. 27, 1798; "To the Curious," *Argus; or, Greenleaf's New Daily Advertiser*, Apr. 23, 1796.

42. Brett Mizelle, " 'Man Cannot Behold It without Contemplating Himself': Monkeys, Apes, and Human Identity in the Early American Republic," *Explorations in Early American Culture: A*

*Supplemental Issue of Pennsylvania History* 66 (1999): 148; Brett Mizelle, "'I Have Brought My Pig to a Fine Market': Animals, Their Exhibitors, and Market Culture in the Early Republic," in *Cultural Change and the Market Revolution in America, 1789–1860*, ed. Scott C. Martin (Lanham, MD: Rowman & Littlefield, 2005), 190.

43. "To the Curious," *Argus; or, Greenleaf's New Daily Advertiser*, Apr. 23, 1796.

44. Bentley, *Diary*, 2:235. Male elephants also have teats, but Bentley was probably correct about the sex of the elephant, assuming that this information was relayed accurately from seller to buyer back in India, then from handler to handler as the elephant traveled to and around the United States. It can be very difficult to tell male from female because the male's testes are internal and both sexes have similar folds of skin between their hind legs, and one needs some instruction to identify the sex correctly in juveniles.

45. Mizelle, "Contested Exhibitions," 223.

46. Bentley, *Diary*, 2:235.

47. *Murder of the Elephant: An Accurate Account of the Death of the Noble Animal* (Boston: Coverley, 1816), 8. See also Mizelle, "Man Cannot Behold It," 151.

48. "The Museum," *The Youth's Companion* 21, no. 43 (1848): 1; see also, e.g., "The Elephant," *Farmer's Cabinet*, Apr. 24, 1835.

49. Sukumar, *Living Elephants*, 195–96, 214.

50. L. E .L. Rasmussen and B. L. Munger, "The Sensorineural Specializations of the Trunk Tip of the Asian Elephant, Elephas maximus," *Anatomical Record* 246, no. 4 (1996): 127–34.

51. Lach, "Asian Elephants," 137–39; Cummings, "Pliny's Literate Elephant," 166–69; Nigel Rothfels, "Elephants, Ethics, and History," in *Elephants and Ethics: Toward a Morality of Coexistence*, ed. Chris Wemmer and Catherine A. Christen (Baltimore: Johns Hopkins Univ. Press, 2008), 106–8; Sivasundaram, "Trading Knowledge," 48.

52. Ralph R. Acampora, *Corporal Compassion: Animal Ethics and Philosophy of Body* (Pittsburgh: Univ. of Pittsburgh Press, 2006), 112. More recently, Chilla Bulbeck has found people expressing a similar desire for egalitarian experiences of other species, in her case study, by touching dolphins. In these interactions people do not imagine themselves dominating another species but sharing with them, as friends of a sort, explaining that tactile contact with animals is more satisfying on some level than viewing them from a distance or seeing pictures of them. Chilla Bulbeck, *Facing the Wild: Ecotourism, Conservation, and Animal Encounters* (London: Earthscan, 2005), xviii-xxii, 31–40; D. F. Lott, "Feeding Wild Animals: The Urge, the Interaction, and the Consequences," *Anthrozoös* 1, no. 4 (1988): 255–57.

53. This remark appeared in numerous announcements thereafter. See, e.g., "Elephant," *Argus; or, Greenleaf's New Daily Advertiser*, June 3, 1796.

54. "The Elephant," *Salem Gazette*, Sept. 1, 1797.

55. *Murder of the Elephant*, 11.

56. Bentley, *Diary*, 2:261.

57. Mizelle, "Man Cannot Behold It," 146.

58. Stuart Thayer, "The Keeper Will Enter the Cage: Early American Wild Animal Trainers," *Bandwagon* 26, no. 6 (1982): 38.

59. Robbins, *Elephant Slaves*, 66–67; Sivasundaram, "Trading Knowledge," 54.

60. Harriet Ritvo, "The Order of Nature: Constructing the Collections of Victorian Zoos," in *New Worlds, New Animals: From Menagerie to Zoological Park in the Nineteenth Century*, ed. R. J. Hoage and William A. Deiss (Baltimore: Johns Hopkins Univ. Press, 1996), 47; Robbins, *Elephant Slaves*, 9–36.

61. Zoological Institute, *A Delineated Description and History of the Beasts, Birds, and Reptiles Contained Therein* (New York: Zoological Institute / J. W. Bell Printers, 1837), 16, McCaddon Collection of the Barnum and Bailey Circus, Manuscripts Division, Department of Rare Books and Special Collections, Princeton University Library.

62. Bentley, *Diary*, 2:34, 247, 356–57.

63. Exhibitors often displayed animals in such combinations "to draw custom" (higher earnings) by making the most of whatever animals they could acquire. Thus they also minimized expenses, such as show space rental fees, while drawing in people who might visit in order to see one particular animal among those in the show or who were intrigued by the novelty of juxtaposing different kinds of beings. Bentley, *Diary*, 2:75–76.

64. Mizelle, "Man Cannot Behold it," 144–173.

65. "The Elephant," *City Gazette* (Charleston, SC), Dec. 27, 1798; Bentley, *Diary*, 2:409.

66. "The Majestic Animal Columbus," *Massachusetts Spy* (Worcester), July 15, 1818.

67. Elizabeth Drinker, *The Diary of Elizabeth Drinker: The Life Cycle of an Eighteenth-Century Woman*, ed. Elaine Forman Crane (Boston: Northeastern Univ. Press, 1994), 176.

68. "Elephant," *Argus*, June 3, 1796.

69. Bentley, *Diary*, 2:235.

70. Mizelle, "Contested Exhibitions," 223; Mizelle, "I Have Brought My Pig," 202.

71. Foster Rhea Dulles, *America Learns to Play: A History of Popular Recreation, 1607–1940* (New York: D. Appleton-Century, 1940), 22–43; Lawrence Levine, *Highbrow/Lowbrow: The Emergence of Cultural Hierarchy in America* (Cambridge, MA: Harvard Univ. Press, 1988), 178–98; Len Travers, *Celebrating the Fourth: Independence Day and the Rites of Nationalism in the Early Republic* (Amherst: Univ. of Massachusetts Press, 1997); Peter W. Cook, "Cockfighting in North America and New England, 1680–1900," in *New England's Creatures, 1400–1900: Dublin Seminar for New England Folklife: Annual Proceedings, 1993*, ed. Peter Benes (Boston: Boston University, 1993), 175–82.

72. Drinker, *Diary*, 159, 166, 176, 280.

73. Ibid., 249.

74. "The Elephant," *City Gazette* (Charleston, SC), Jan. 25, 1799; ibid., Dec. 27, 1798.

75. Flint, "Origin of the Circus," 18; Mizelle, "Contested Exhibitions," 221; Jennifer L. Mosier, "The Big Attraction: The Circus Elephant and American Culture," *Journal of American Culture* 22, no. 2 (1999), 9; R. W. G. Vail, "This Way to the Big Top," *New-York Historical Society Quarterly Bulletin* 29, no. 3 (1945): 138–42.

76. Jennifer Mason, *Civilized Creatures: Urban Animals, Sentimental Culture, and American Literature, 1850–1900* (Baltimore: Johns Hopkins Univ. Press, 2005), 12.

77. Rod Preece, *Awe for the Tiger, Love for the Lamb: A Chronicle of Sensibility to Animals* (Vancouver: Univ. of British Columbia Press, 2002), 165–66; Robbins, *Elephant Slaves*, 88–89.

78. Robbins, *Elephant Slaves*, 97–99.

79. For other examples of this broadside, see, e.g., "The Elephant," *Salem Gazette*, Sept. 1, 1797.

80. Scott C. Martin, *Killing Time: Leisure and Culture in Southwestern Pennsylvania, 1800–1850* (Pittsburgh: Univ. of Pittsburgh Press, 1995), 5–8; Elizabeth Lehuu, *Carnival on the Page: Popular Print Media in Antebellum America* (Chapel Hill: Univ. of North Carolina Press, 2000), 14–35, 126–40.

81. Joseph J. Ellis, *After the Revolution: Profiles in Early American Culture* (New York: W. W. Norton, 1979), 36–37, 43.

82. Goler, "Here the Book of Nature Is Unfolded," 10–21.

83. Drinker, *Diary*, 239.

84. Bentley, *Diary*, 2:261.

85. Mizelle, "Man Cannot Behold It," 145–73.

86. Anderson, *Creatures of Empire*, 86; Ruth Wallis Herndon, "Breachy Sheep and Mad Dogs: Troublesome Domestic Animals in Rhode Island, 1750–1800," in Benes, *New England's Creatures*, 61–72; Martin V. Melosi, *Garbage in the Cities: Refuse, Reform, and the Environment*, rev. ed. (Pittsburgh: Univ. of Pittsburgh Press, 2005), 11, 18.

87. Mizelle, "I Have Brought My Pig," 182–207; Mizelle, "Contested Exhibitions," 220.

88. "The Economist; *from the (Boston) Mercury*," *Carey's United States Recorder* (Philadelphia), Feb. 27, 1798.

89. Bentley, *Diary*, 2:235.

90. "Address Spoken by Mr. Hodgkinson," *Argus; or, Greenleaf's New Daily Advertiser*, Oct. 3, 1796.

91. Baker quoted in McClung and McClung, "Tammany's Remarkable Gardiner Baker," 160.

92. "Circus," Knickerbocker 13 (Jan. 1839): 67.

93. Peter Benes, "Itinerant Entertainers in New England and New York, 1687–1830," in *Itinerancy in New England and New York: Dublin Seminar for New England Folklife: Annual Proceedings, 1984*, ed. Peter Benes (Boston: Boston University, 1986), 112–30; Dulles, *America Learns to Play*, 39–40; Walter A. Friedman, *Birth of a Salesman: The Transformation of Selling in America* (Cambridge, MA: Harvard Univ. Press, 2004), 4–33.

94. John Davis, *Travels of Four Years and a Half in the United States of America* (London: E. Edwards, 1803), 61–63.

95. I cite these authors here in the order in which I discuss the elephant's keepers: Shettel, "The First Elephant in the United States," 7–8; McClung and McClung, "America's First Elephant," 403; Goodwin, "Crowninshield Elephant," 357–59.

96. Davis, *Travels of Four Years and a Half*, 61–63.

97. Sukumar, *Living Elephants*, 192–200, 210–18.

98. Clubb et al., "Compromised Survivorship," 1648; Sukumar, *Elephant Days and Nights*, 107.

99. "Speculation," *Boston Gazette*, May 31, 1803.

100. "A Living Elephant," broadside reprinted in Peter Benes, "To the Curious: Bird and Animal Exhibitions in New England, 1716–1825," in Benes, *New England's Creatures*, 155.

101. See, e.g., "Infernal Transaction--Death of the Elephant," *Boston Gazette*, July 29, 1816; "Infernal Transaction," *Columbian* (New York), Aug. 1, 1816; "Death of the Elephant," *Eagle* (Maysville, KY), Sept. 13, 1816.

102. David Jaffee, "Peddlers of Progress and the Transformation of the Rural North, 1760–1860," *Journal of American History* 78, no. 2 (1991): 512–17.

103. *Murder of the Elephant*, 10.

104. Ibid., 3–4.

105. Mizelle, "I Have Brought My Pig," 191, 200–201. The informal practices by local men (rubes) of picking fights or otherwise disrupting staff would later come to be known under the circus phrase "Hey Rube," in which show workers would holler the words to bring the company's resident Pinkerton detectives and largest workmen over to take control of the situation.

106. Bentley, *Diary*, 4:400–402.

107. "The Elephant. Notice," *Commercial Advertiser* (New York, NY), Apr. 8, 1817; "Notice," *Evening Post* (New York, NY), Apr. 10, 1817.

## Chapter 2 · Becoming an Elephant "Actor"

1. "Three Great Natural Curiosities," *Centinel of Freedom* (Newark), Apr. 11, 1820; "Great Natural Curiosity," *Woodstock Observer* (VT), Aug. 29, 1820; "Shocking Calamity," *Massachusetts Spy* (Worcester), Sept. 27, 1820. The first female elephant in America was never taller than seven and a half feet. Robert McClung and Gale McClung, "Captain Crowninshield Brings Home an Elephant," *American Neptune: A Quarterly Journal of Maritime History* 18, no. 2 (1958): 138; see also "The Elephant," *City Gazette* (Charleston, SC), Dec. 27, 1798; "To the Curious," *The Argus; or, Greenleaf's New Daily Advertiser*, Apr. 23, 1796.

2. E. Lakin Brown, "Autobiographical Notes," in *Historical Collections, Made by the Michigan Pioneer and Historical Society*, vol. 30 (Lansing, MI: Wynkoop Hallenbeck Crawford, 1906), 433–35.

3. The male elephant known as Columbus arrived sometime before the summer of 1818, and another known as Tippoo Sultan arrived in June 1821 on a ship called the *Bengal*. "The Majestic Animal Columbus," *Massachusetts Spy* (Worcester), July 15, 1818; Richard W. Flint, "Entrepreneurial and Cultural Aspects of the Early-Nineteenth-Century Circus and Menagerie Business," in *Itinerancy in New England and New York: Dublin Seminar for New England Folklife: Annual Proceedings, 1984*, ed. Peter Benes (Boston: Boston University, 1986), 135n15; Stuart Thayer, "One Sheet" *Bandwagon* 18, no. 5 (1974), 23. Some authors also write of another female in this period, "The Learned Elephant," reputedly killed in Chepachet, Rhode Island, on May 26, 1825, or sometime in 1822. "Little Bet, The Learned Elephant, 198," Box 1, John "Chang" Reynolds Papers, Robert L. Parkinson Library and Research Center, Circus World Museum, Baraboo, WI ; Stuart Thayer, "The Elephant in America Before the 1840s," *Bandwagon* 31, no. 1 (1987): 20–26.

4. "Three Great Natural Curiosities," *Centinel of Freedom* (Newark), Apr. 11, 1820.

5. "Shocking Calamity," *Massachusetts Spy* (Worcester), Sept. 27, 1820; "Shocking Calamity!" *New Hampshire Sentinel* (Keene), Sept. 23, 1820.

6. "Great Natural Curiosity," *Woodstock Observer* (VT), Aug. 29, 1820; "Three Great Natural Curiosities," *Centinel of Freedom* (Newark), Apr. 11, 1820.

7. "Shocking Calamity!" *New Hampshire Sentinel* (Keene), September 23, 1820. See also, "Inaccurate Statement," *Vermont Intelligencer* (Bellows Falls), Oct. 9, 1820.

8. "The Elephant Horatio," *Albany Advertiser* (NY), Sept. 29, 1820.

9. "Shocking Calamity!" *New Hampshire Sentinel* (Keene), Sept. 23, 1820.

10. James Rennie, *The Menageries: Quadrupeds, Described and Drawn from Living Subjects* (London: Charles Knight, 1831), 2:11–16.

11. "Horse and Elephant," *Portsmouth Oracle* (NH), Jan. 1, 1820; "Summary," *Brookville Enquirer* (KY), Jan. 7, 1820; "The Elephant, Horatio," *National Messenger* (Georgetown, DC), Oct. 9, 1820; "The Skin of the Elephant Horatio," *New Hampshire Sentinel* (Keene), Jan. 20, 1821.

12. Flint, "Entrepreneurial and Cultural Aspects," 140.

13. "Circus," *Knickerbocker* 13 (Jan. 1839): 70; Maurice Willson Disher, *Greatest Show on Earth* (London: Bell & Sons, 1937); James S. Moy, "Entertainments at John B. Ricketts' Circus, 1793–1800," *Educational Theatre Journal* 30, no. 2 (1978): 186–202; A. H. Saxon, *Enter Foot and Horse: A History of Hippodrama in England and France* (New Haven, CT: Yale Univ. Press, 1968); A. H. Saxon, *The Life and Art of Andrew Ducrow and the Romantic Age of the English Circus* (Hamden, CT: Archon Books, 1978).

14. David Carlyon, *Dan Rice: The Most Famous Man You've Never Heard Of* (New York: Public Affairs, 2001), 159; James S. Moy, "John B. Ricketts' Circus, 1793–1800" (Ph.D. diss., University of Illinois, Urbana-Champaign, 1977), 79–80.

15. George L. Chindahl, *A History of the Circus in America* (Caldwell, ID: Caxton, 1959), 30–33; Flint, "Entrepreneurial and Cultural Aspects," 140–42.

16. "Arrival of the Britannia," *Pittsfield Sun* (MA), May 2, 1845.

17. See, e.g., "Two Great Exhibitions United in One!" *Essex County Republican* (Keeseville, NY), 4 May 1849.

18. Ann Norton Greene, *Horses at Work: Harnessing Power in Industrial America* (Cambridge, MA: Harvard Univ. Press, 2008), 45; Ian R. Tyrrell, *Transnational Nation: United States History in Global Perspective since 1789* (New York: Palgrave Macmillan, 2007), 27.

19. Ronald J. Zboray, *A Fictive People: Antebellum Economic Development and the American Reading Public* (New York: Oxford Univ. Press, 1993), 4.

20. Carlyon, *Dan Rice*, 189.

21. Ronald J. Zboray, "Literary Enterprise and the Mass Market: Publishers and Business Innovation in Antebellum America," *Essays in Economic and Business History* 10 (1998): 169; see also Richard R. John, *Spreading the News: The American Postal System from Franklin to Morse* (Cambridge, MA: Harvard Univ. Press, 1995), 32–33.

22. Scott A. Sandage, *Born Losers: A History of Failure in America* (Cambridge, MA: Harvard Univ. Press, 2005), 26–27; "Circus," *Knickerbocker*, 76.

23. "Speculation," *Boston Gazette*, May 31, 1803.

24. John C. Kunzog, *One Horse Show: The Life and Times of Dan Rice, Circus Jester and Philanthropist* (Jamestown, NY: John C. Kunzog, 1961), 24.

25. Richard W. Flint, "American Showmen and European Dealers: Commerce in Wild Animals in Nineteenth-Century America," in *New Worlds, New Animals: From Menagerie to Zoological Park in the Nineteenth Century*, ed. R. J. Hoage and William A. Deiss (Baltimore: Johns Hopkins Univ. Press, 1996), 98–99; Penelope M. Leavitt and James S. Moy, "Spalding and Rogers' Floating Palace, 1852–1859," *Theatre Survey* 25, no. 1 (1984): 15–27, 21–23. See also Phineas Taylor Barnum, *The Life of P. T. Barnum: Written by Himself* (New York: Redfield, 1855), 219–220; "Zoological Institute Articles of Association, 1835," www.westchesterarchives.com/HT/muni/wchs/zoological.html, June 9, 2007.

26. E.g., a patchwork of laws and taxes excluded circuses and, to a lesser degree, menageries from performing in Connecticut from 1773 to 1840 and in Vermont, effectively, from 1824 to 1933. Flint, "Entrepreneurial and Cultural Aspects," 144–45.

27. "Circus," *Knickerbocker*, 67–76.

28. Scott C. Martin, *Killing Time: Leisure and Culture in Southwestern Pennsylvania, 1800–1850* (Pittsburgh: Univ. of Pittsburgh Press, 1995), 5–8; Elizabeth Lehuu, *Carnival on the Page: Popular Print Media in Antebellum America* (Chapel Hill: Univ. of North Carolina Press, 2000), 49.

29. "Disembarkation of Elephants at Calcutta from Burmah," *Littell's Living Age*, Mar. 27, 1858.

30. Flint, "American Showmen," 97–98; Brett Mizelle, "'I Have Brought My Pig to a Fine Market': Animals, Their Exhibitors, and Market Culture in the Early Republic," in *Cultural Change and the Market Revolution in America, 1789–1860*, ed. Scott C. Martin (Lanham, MD: Rowman & Littlefield, 2005), 186–89.

31. Elbert Bowen, "The Circus in Early Rural Missouri," *Missouri Historical Review* 47 (1952): 1; Leavitt and Moy, "Spalding and Rogers' Floating Palace," 15–27.

32. Popularly called Simón Bolívar, he was a Spaniard turned South American independence leader who figured into the political struggles of Venezuela, Bolivia, Peru, and Colombia. Jenny Lind was the Swedish opera singer famous in those years in Europe. She would achieve enormous celebrity touring America under the promotion of P.T. Barnum in 1850–51. Bluford Adams, *E Pluribus Barnum: The Great Showman and the Making of U.S. Popular Culture* (Minneapolis: Univ. of Minnesota Press, 1997), 41–74; Carlyon, *Dan Rice*, 165–66.

33. Katherine C. Grier, *Pets in America: A History* (Chapel Hill: Univ. of North Carolina Press, 2006), 67–69, 205.

34. Margaret Derry, *Bred for Perfection: Shorthorn Cattle, Collies, and Arabian Horses since 1800* (Baltimore: Johns Hopkins Univ. Press, 2003), 23–63. On parallel ideologies and practices in Britain, see Harriet Ritvo, *The Animal Estate: The English and Other Creatures in the Victorian Age* (Cambridge, MA: Harvard Univ. Press, 1987), 45–69.

35. Richard White, "Animals and Enterprise," in *The Oxford History of the American West*, ed. Clyde A. Milner II, Carol A. O'Connor, and Martha A. Sandweiss (New York: Oxford Univ. Press, 1994), 240.

36. Mark V. Barrow, *A Passion for Birds: American Ornithology after Audubon* (Princeton, NJ: Princeton Univ. Press, 1998), 107–111.

37. Nancy C. Carlisle, "The Chewed Chair Leg and the Empty Collar: Momentos of Pet Ownership in New England," in *New England's Creatures, 1400–1900: Dublin Seminar for New England Folklife: Annual Proceedings, 1993*, ed. Peter Benes (Boston: Boston University, 1993), 130–46; Katherine C. Grier, "Buying Your Friends: The Pet Business and American Consumer Culture," in *Commodifying Everything: Relationships of the Market*, ed. Susan Strasser (New York: Routledge, 2003), 44–48; Ruth Wallis Herndon, "'Breachy Sheep and Mad Dogs': Troublesome Domestic Animals in Rhode Island, 1750–1800," in Benes, *New England's Creatures*, 68–69.

38. Yi-Fu Tuan, *Dominance and Affection* (New Haven, CT: Yale Univ. Press, 2003).

39. "Replies, Gen. John P. Boyd," *Historical Magazine, and Notes and Queries Concerning the Antiquities* 3, no. 6 (1859): 191.

40. Dan Wylie, *Elephant* (London: Reaktion, 2008), 117–19. The apparent historicity of the faux–South Asian elements in a menagerie parade was loosely based in centuries of global war, trade, and diplomatic history. People in West and South Asia had used elephants in battle from at least the Alexandrian age (330s CE). Although incompletely documented by historians, these elephants probably did not actually engage in combat but were possibly meant to unnerve the enemy, and certainly his horses, by their presence on the battlefield. Others pulled large chariots or carried riders and cargo along supply routes. Later drivers would use Asian elephants to charge and either trample or disperse the opponent's troops. Robin Lane Fox, *The Classical World* (New York: Penguin, 2006), 232–47, 256–57; John M. Kistler, *War Elephants* (Lincoln: Univ. of Nebraska Press, 2007), 97–100; H. H. Scullard, *The Elephant in the Greek and Roman World* (London: Thames & Hudson, 1974), 64–66, 75–100, 154–81. In China people had trained elephants for war beginning around 1600 BCE, while for millennia in Africa and across Asia people used elephants for peacetime labor, political and religious ceremony, and as currency by way of living animals or their ivory tusks. Silvio A. Bedini, *The Pope's Elephant: An Elephant's Journey from Deep in India to the Heart of Rome* (New York: Penguin, 1997); Carl W. Bishop, "The Elephant and Its Ivory in Ancient China," *Journal of the American Oriental Society* 41 (1921): 290–306; Edward H. Schafer, "War Elephants in Ancient and Medieval China," *Oriens* 10, no. 2 (1957): 281; Nicholas Tarling, *The Cambridge History of South East Asia*, Vol. 1, pt. 1, *From Early Times to c. 1500* (Cambridge: Cambridge Univ. Press, 2000), 241, 249–65.

41. Susan Nance, *How the Arabian Nights Inspired the American Dream, 1790–1935* (Chapel Hill: Univ. of North Carolina Press, 2009), 26–33.

42. "The Grand Entre," *Baltimore Patriot*, Nov. 25, 1834.

43. "For 3 Days!—Van Amburgh & Co's Menagerie," *Brooklyn Eagle and Kings County Democrat*, Nov. 11, 1846.

44. "A Strong Team," *Brooklyn Eagle and Kings County Democrat*, Oct. 26, 1848. This article also claimed that Sands and "Lind" company had three elephants, but the company's advertising of the period shows no evidence of this.

45. Brian Cummings, "Pliny's Literate Elephant and the Idea of Animal Language in Renaissance Thought," in *Renaissance Beasts: Of Animals, Humans, and Other Wonderful Creatures*, ed. Erica Fudge (Urbana: University of Illinois Press, 2004), 168–73; Thomas Veltre, "Menageries, Metaphors, and Meanings," in Hoage and Deiss, *New Worlds, New Animals*, 20–21, 26–27.

46. Michael A. Osborne, "Zoos in the Family: The Geoffroy Saint-Hilaire Clan and the Three Zoos of Paris," in Hoage and Deiss, *New Worlds, New Animals*, 36–37.

47. T. H. Breen, "Narrative of Commercial Life: Consumption, Ideology, and Community on the Eve of the American Revolution," *William and Mary Quarterly*, 3rd ser., vol. 5, no. 3 (July 1993): 480–81; Walter A. Friedman, *Birth of a Salesman: The Transformation of Selling in America* (Cambridge, MA: Harvard Univ. Press, 2004), 28–31; Joseph T. Rainer, "The 'Sharper' Image: Yankee Peddlers, Southern Consumers, and the Market Revolution," in Martin, *Cultural Change and the Market Revolution*, 96–102; Fred Somkin, *Unquiet Eagle: Memory and Desire in the Idea of American Freedom, 1815–1860* (Ithaca, NY: Cornell Univ. Press, 1967), 11–54; William Earl Weeks, "American Nationalism, American Imperialism: An Interpretation of United States Political Economy, 1789–1861," *Journal of the Early Republic* 14, no. 4 (1994): 485–95.

48. Hans Bergmann, "Panoramas of New York, 1845–1860," *Prospects* 10 (1985): 119–37; Miles Orvell, *The Real Thing: Imitation and Authenticity in American Culture, 1880–1940* (Chapel Hill: Univ. of North Carolina Press, 1989), 21; Len Travers, *Celebrating the Fourth: Independence Day and the Rites of Nationalism in the Early Republic* (Amherst: Univ. of Massachusetts Press, 1997), 15–23, 123–34.

49. Moy, "Entertainments at John B. Ricketts' Circus," 186–202.

50. "Sands, Lent & Co's Hippoferaean Arena," *Brooklyn Eagle and Kings County Democrat*, Oct. 25, 1848.

51. George P. Hammond, *Who Saw the Elephant? An Inquiry by a Scholar Well Acquainted with the Beast* (San Francisco: California Historical Society, 1964), 5.

52. "News of the Day," *New York Times*, Dec. 3, 1858.

53. Carlyon, *Dan Rice*, 123.

54. Stuart Thayer, *Annals of the American Circus: 1793–1829*, 2 vols. (Manchester, MI: Rymark Printing, 1976), 2:116.

55. "The Menagerie of Waring, Tufts & Co.," *New Hampshire Sentinel*, Sept. 11, 1834.

56. See, e.g., "Caravan. Consisting of the Great India Elephant," *Haverhill Gazette* (MA), Apr. 7, 1827.

57. "The Great India Elephant," Broadside, Newburyport, MA (1832), American Broadsides and Ephemera Series, 4393, American Antiquarian Society. Studies of twentieth-century elephants, captive and wild, have reported a common height for male adults of about nine feet and a weight of around 8,500 pounds, with full grown females usually a foot shorter and somewhat lighter. As such Siam may have been significantly underweight or simply not as tall as claimed.

D. Mariappa, *Anatomy and Histology of the Indian Elephant* (Oak Park: MI: Indira Publishing, 1986), x.

58. "From the American Traveller: The Elephant," *Farmer's Cabinet* (Amherst, NH), Apr. 24, 1835; "Dreadful Disaster at Sea," *Pittsfield Sun* (MA), Nov. 3, 1836; "Another Great One Gone," *Picayune* (New Orleans), Mar. 25, 1837.

59. "Menagerie. Macomber & Co.'s New Collection of Live Animals," *Connecticut Mirror* (Hartford), July 30, 1831.

60. On historical confusion over this issue, see Bishop, "Elephant and Its Ivory," 293; Scullard, *Elephants in the Greek and Roman World*, 60–63.

61. Ernst Mayr, *The Growth of Biological Thought: Diversity, Evolution and Inheritance* (Cambridge, MA: Belknap Press of Harvard Univ. Press, 1982), 251–85; Harriet Ritvo, The *Platypus and the Mermaid: And Other Figments of the Classifying Imagination* (Cambridge, MA: Harvard Univ. Press, 1997), 51–59; Wylie, *Elephant*, 13. For example, American schoolbooks did not discuss contemporary theories on evolution or species--and would not do so even after the release of Darwin's most important works in the 1850s and 1860s--but explained nature from a position of biblical literalism. Ruth Miller Elson, *Guardians of Tradition: American Schoolbooks of the Nineteenth Century* (Lincoln: Univ. of Nebraska Press, 1964), 16–17.

62. John Rickards Betts, "P. T. Barnum and the Popularization of Natural History," *Journal of the History of Ideas* 20, no. 3 (1959): 353–68; Ritvo, *Animal Estate*, 15–42.

63. Edmund Flagg, *The Far West; or, A Tour beyond the Mountains*, 2 vols. (New York: Harper & Bros., 1838), 1:225.

64. Tyrrell, *Transnational Nation*, 32.

65. Laura Browder, *Slippery Characters: Ethnic Impersonators and American Identities* (Chapel Hill: Univ. of North Carolina Press, 2000), 53–58; Patricia Cline Cohen, Timothy J. Gilfoyle, and Helen Lefkowitz Horowitz, *The Flash Press: Sporting Male Weeklies in 1840s New York* (Univ. of Chicago Press, 2008); Jim Cullen, *Art of Democracy: A Concise History of Popular Culture in the United States*, 2nd ed. (New York: Monthly Review Press, 2002), 33–86; Lehuu, *Carnival on the Page*; Guy Reel, *The National Police Gazette and the Making of the Modern American Man, 1879–1906* (London: Palgrave Macmillan, 2006).

66. Flint, "Entrepreneurial and Cultural Aspects," in Benes, *Itinerancy in New England and New York*, 132.

67. "For 3 Days More!" *Brooklyn Eagle and Kings County Democrat*, Nov. 16, 1846.

68. W. G. Vail, "This Way to the Big Top," *New-York Historical Society Quarterly Bulletin* 29, no. 3 (1945): 137–59.

69. "Places of Public Amusement," *Putnam's Monthly Magazine* 3 (Feb. 1854): 152.

70. Ritvo, *Animal Estate*, 25–26, 37.

71. Franz B. M. De Waal, *The Ape and the Sushi Master: Cultural Reflections by a Primatologist* (New York: Basic Books, 2001), 76–78. Randall Lockwood, for one, calls this practice "personification," a mode by which people, "superimpose their own desires on animals . . . [with] no recognition of the true biological needs and adaptations of the animals involved." Randall Lockwood, "Anthropomorphism Is Not a Four-letter Word," in *Perceptions of Animals in American Culture*, ed. R. J. Hoage (Washington, DC: Smithsonian Institution Press, 1989), 46–47. Indeed, Nigel Rothfels suggests that such human-mediated representations of other species, or even actual interactions with them, can mistakenly persuade the viewer he or she understands the true nature of that species in spite of the limitations of human perception. Nigel Rothfels, "Introduction," in *Representing Animals*, ed. Nigel Rothfels (Bloomington: Indiana Univ. Press,

2002), x-xi. For a broader survey of the opportunities and consequences of anthropomorphism, see also Tom Tyler, "If Horses Had Hands," in *Animal Encounters*, ed. Tom Tyler and Manuela Rossini (Leiden: Brill, 2009), 13–26.

72. Disher, *Greatest Show on Earth*, 73–87, 249–72; Michael Dobson, "A Dog at All Things: The Transformation of the Onstage Canine, 1550–1850," *Performance Research* 5, no. 2 (2000): 119; Moy, "Entertainments at John B. Rickett's Circus," 192.

73. Dobson, "A Dog at All Things," 116.

74. Matthew Bliss, "Property or Performer: Animals on the Elizabethan Stage," *Theatre Studies* 39 (1994): 45–59; Dobson, "A Dog at All Things," 116–24.

75. *American Daily Advertiser*, quoted in James W. Shettel, "The First Elephant in the United States," *Circus Scrapbook*, July 1929, 8; William B. Wood, *Personal Recollections of the Stage, Embracing Notices of Actors, Authors, and Auditors during a Period of Forty Years* (Philadelphia: Henry Carey Baird, 1855), 415; see also David Grimsted, *Melodrama Unveiled: American Theater and Culture, 1800–1850* (Berkeley: Univ. of California Press, 1988), 86–87.

76. William Dunlap, *History of the American Theatre*, 2 vols. (London: Richard Bentley, 1833), 1:351–52. On this phenomenon in Britain, see Disher, *Greatest Show on Earth*, 80.

77. Wood, *Personal Recollections of the Stage*, 171, 186, 228.

78. Ibid., 359.

79. Cummings, "Pliny's Literate Elephant," 164–86; Nigel Rothfels, "Elephants, Ethics, and History," in *Elephants and Ethics: Toward a Morality of Coexistence*, ed. Chris Wemmer and Catherine A. Christen (Baltimore: Johns Hopkins Univ. Press, 2008), 106–8.

80. Vail, "This Way to the Big Top," 145. See also Brett Mizelle, " 'Man Cannot Behold It without Contemplating Himself': Monkeys, Apes and Human Identity in the Early American Republic," in *Explorations in Early American Culture: A Supplemental Issue of Pennsylvania History* 66 (1999), 150; "Grand National Menagerie," *Traveller* (Boston), Jan. 27, 1832, reproduced in Richard J. Reynolds, "Circus Rhinos," *Bandwagon* 12, no. 6 (1968): 5.

81. "The Elephant Caravan," *Salem Gazette*, Mar. 30, 1827.

82. Moy, "Entertainments at John B. Rickett's Circus," 192–99.

83. "Grand Caravan of Living Animals," *American Mercury* (Harford, CT), Apr. 5, 1825.

84. "The Elephant," *City Gazette and Daily Advertiser* (Charleston, NC), Mar. 5, 1799.

85. "Circus," *Knickerbocker*, 68–69.

86. Ibid.

87. "The National Caravan," *Farmer's Cabinet*, Aug.t 8, 1829.

88. "For 3 Days!--Van Amburgh & Co's Menagerie," *Brooklyn Eagle and Kings County Democrat*, Nov. 11, 1846.

89. "The Elephant," *Farmer's Cabinet*, Apr. 24, 1835.

90. Ibid.

91. John Stokes, " 'Lion Griefs': The Wild Animal Act as Theatre," *New Theatre Quarterly* 20, no. 2 (2004): 145.

92. Joanne Carol Joys, *The Wild Animal Trainer in America* (Boulder, CO: Pruett Publishing, 1983), vii, xiv, 1, 11, 17–21; see also Stokes, "Lion Griefs," 140.

93. Jesse Haney, *Haney's Art of Training Animals* (New York: J. Haney, 1869), 123.

94. George Conklin, *The Ways of the Circus: Being the Memoires and Adventures of George Conklin, Tamer of Lions* (New York: Harper & Bros., 1921), 124.

95. "Raymond & Co. and Van Amburgh's Long Established Menageries," *Franklin Gazette* (NY), Aug. 15, 1850. Although new to Americans, these acts had been seen before, although

rarely. Some Roman circus elephants had performed acts humans took to be trained performances. One contemporary observer wrote: "I myself have seen an elephant clanging cymbals,
and others dancing; two cymbals were fastened to the player's forelegs, and one on his trunk,
and he rhythmically beat with he struck the cymbal on either leg in turn; the dancers danced in
a circle, and raising and fending their forelegs in turn moved rhythmically, as the player with the
cymbals marked the time for them." Arrian quoted in Scullard, *Elephants in the Greek and Roman
World*, 57; James Rennie, *The Menageries: Quadrupeds, Described and Drawn from Living Subjects*
(London: Charles Knight, 1831), 2:12.

96. "Circus," *Knickerbocker*, 76.

97. Reynolds, "Circus Rhinos," 12–13.

98. Kunzog, *One Horse Show*, 116–17.

99. James Knowles, "'Can ye not tell a man from a marmoset': Apes and Others on the Early
Modern Stage," in Fudge, *Renaissance Beasts*, 140; see also Suzanne Laba Cataldi, "Animals and
the Concept of Dignity: Critical Reflections on a Circus Performance," *Ethics and the Environment*
7, no. 2 (2002): 104–26; Mizelle, "Man Cannot Behold It," 151; Michael Peterson, "The Animal
Apparatus: From a Theory of Animal Acting to an Ethics of Animal Acts," *TDR: The Drama Review* 51, no. 1 (2007): 35.

100. "Raymond & Waring's Immense Menagerie," *Brooklyn Eagle and Kings County Democrat*, Oct. 31, 1846.

101. Bowen, "Circus in Early Rural Missouri," 4–5; see also Nigel Rothfels, *Savages and
Beasts: The Birth of the Modern Zoo* (Baltimore: Johns Hopkins Univ. Press, 2002), 158–60.

102. Horace P. Batcheler, *Jonathan at Home; or, A Stray Shot at the Yankees* (London: Collin-
gridge, 1865), 77.

103. "Van Amburgh & Co's Mammoth Menagerie," *New York Times*, Nov. 24, 1863.

104. Jane Desmond, "Displaying Death, Animating Life: Changing Fictions of 'Liveness'
from Taxidermy to Animatronics," in Rothfels, *Representing Animals*, 159–63; Mizelle, "I Have
Brought My Pig," 194.

105. Rothfels, "Why Look at Elephants?" *Worldviews: Global Religions, Culture, and Ecology* 9,
no. 2 (2005): 168.

106. "Sands, Nathan's & Co.'s American Circus," *Columbia Spy* (PA), Aug. 22, 1857; "The
Elephants in San Francisco," *Sacramento Daily Union*, May 25, 1859.

107. "The Elephants in San Francisco," *Sacramento Daily Union*, May 25, 1859; "The Elephants," *Daily Alta California*, May 26, 1859.

108. "American Theatre," *Daily Alta California*, Dec. 6, 1859.

109. John Berger, "Why Look at Animals?" in *About Looking* (New York: Pantheon Books,
1980), 13.

110. Yoram S. Carmeli, "'Cruelty to Animals' and Nostalgic Totality: Performances of a Travelling Circus in Britain," *International Journal of Sociology and Social Policy* 22, no. 11/12 (2002):
82–83.

111. W. H. P., "Our London Correspondence," *Spirit of the Times* 26, no. 3 (1856): 26; *Daily
Advertiser* (Detroit), Apr. 30, 1856, quoted in Chindahl, *History of the Circus*, 50.

112. Yoram S. Carmeli, "The Sight of Cruelty: The Case of Circus Animal Acts," *Visual Anthropology* 10 (1997): 1–15.

113. "The Elephants," *Daily Alta California*, Oct. 4, 1859.

114. "From California," *New York Times*, July 31, 1860.

115. "California Enterprise," *Daily Alta California*, May 21, 1859; "Exhibition of the Elephants,"

*Daily Alta California*, May 24, 1859; "Grand Circus and Elephant Exhibition!" *Daily Alta California*, Oct. 3, 1859.

116. *Daily Advertiser* (Detroit), Apr. 30, 1856, quoted in Chindahl, *History of the Circus*, 50.

117. "Elephant Exhibition," *Daily Alta California*, Oct. 3, 1859; "The Elephants," *Daily Alta California*, Oct. 4, 1859; "American Theatre," *Daily Alta California*, Dec. 6, 1859.

118. "Rule That Won't Work Both Ways," *Brooklyn Eagle and Kings County Democrat*, Apr. 17, 1844.

119. Bowen, "Circus in Early Rural Missouri," 5, 16; Chindahl, *History of the Circus*, 3–4; Thayer, *Annals of the American Circus*, 1:68–71; Mizelle, "Man Cannot Behold It," 58–59; Mizelle, "I Have Brought My Pig," 201–2.

## Chapter 3 · *Learning to Take Direction*

1. Horace Townsend, "Animals and Their Trainers," *Frank Leslie's Popular Monthly* 26, no. 6 (1888): 733. See also "Horace Townsend Dies as Gas Escapes," *New York Times*, May 10, 1922.

2. John Stokes, "'Lion Griefs': The Wild Animal Act as Theatre," *New Theatre Quarterly* 20, no. 2 (2004): 148. On this point, see also Richard Schechner, *Performance Theory* (London: Routledge, 1988), 225, 248; David Williams, "The Right Horse, the Animal Eye—Bartabas and Théâtre Zingaro," *Performance Research* 5, no. 2 (2000): 34–35.

3. Some circus fans will find disagreeable the argument that elephants did not share the actor's trust and sense of theatricality with human circus workers. Jennifer Mosier explains that because the elephant is so powerful, it "would never have become integral to the circus if it were not willing to perform and be seen." Certainly, other people have claimed to be *positive* that they witnessed dogs, horses, elephants, and other animals glory in proud performance for human audiences, stepping lively with bright eyes to the sound of applause. Jennifer L. Mosier, "The Big Attraction: The Circus Elephant and American Culture," *Journal of American Culture* 22, no. 2 (1999): 14; see also Marthe Kiley-Worthington, "The Training of Circus Animals," in *Animal Training: A Review and Commentary on Current Practice, Proceedings of a Symposium Organized by the Universities Federation for Animal Welfare*, ed. UFAW (Potters Bar, UK: Universities Federation for Animal Welfare, 1990), 72–73; Jeffrey Moussaieff Masson and Susan McCarthy, *When Elephants Weep: The Emotional Lives of Animals* (New York: Dell/Bantam, 1995), 118–20. So, to be clear, in arguing that elephants did not understand or endorse the human stories people told in menageries and circuses, we need not argue that they had no experience of that context. We can argue that, with limited access to human language and cultures, they had their own experiences of the circus and menagerie context, some of which caused them to exhibit behaviors that people *interpreted* as an endorsement of the show, the trainer, the audience's applause, or circuses in general. There does not appear to be evidence to suggest that elephants have ever acted to deceive an audience in a menagerie show in the way human players do with a wink and a nod as a sign of shared suspension of disbelief. Maxinne D. Morris, "Large Scale Deception: Deceit by Captive Elephants?" in *Deception: Perspectives on Human and Nonhuman Deceit*, ed. Robert W. Mitchell and Nicholas S. Thompson (Albany, NY: SUNY Press, 1986), 183; Stokes, "Lion Griefs," 151.

4. Douglas Gilbert, *American Vaudeville: Its Life and Times* (1940; repr. New York: Dover, 1963), 324.

5. G. A. Bradshaw, *Elephants on the Edge: What Elephants Teach Us about Humanity* (New Haven, CT: Yale Univ. Press, 2009), 14–15.

6. Vicki Hearne, *Adam's Task: Calling Animals by Name* (New York: Knopf, 1982), 19.

7. Nicole Shukin, *Animal Capital: Rendering Life in Biopolitical Times* (Minneapolis: Univ. of Minnesota Press, 2009), 5.

8. Frank C. Bostock, *The Training of Wild Animals,* edited by Ellen Velvin (1903; repr. New York: Century, 1920), 24; Treva J. Tucker, "Early Modern French Noble Identity and the Equestrian 'Airs above the Ground,'" in *The Culture of the Horse: Status, Discipline, and Identity in the Early Modern World,* ed. Karen Raber and Treva J. Tucker (London: Palgrave, 2005), 273–310; Nigel Rothfels, *Savages and Beasts: The Birth of the Modern Zoo* (Baltimore: Johns Hopkins Univ. Press, 2002), 160; Stokes, "Lion Griefs," 145–46.

9. Donald M. Scott, "The Popular Lecture and the Creation of a Public in Mid-Nineteenth-Century America," *Journal of American History* 66, no. 4 (1980): 795–96.

10. "An Exhibition" and "A Living Elephant," broadsides reprinted in Benes, "To the Curious: Bird and Animal Exhibitions in New England, 1716–1825," in *New England's Creatures, 1400–1900, Dublin Seminar for New England Folklife: Annual Proceedings, 1993,* ed. Peter Benes (Boston: Boston University, 1995), 54–55.

11. See, e.g., "Van Amburgh & Co's Mammoth Menagerie," *New York Times,* Dec. 4, 1863. On graphic uses of print in early newspapers, see David Henkin, *City Reading: Written Words and Public Spaces in Antebellum New York* (New York: Columbia Univ. Press, 1999), 101–35.

12. "The Museum," *Youth's Companion* 21, no. 43 (1848): 1.

13. "Lion Theatre," Boston 1836, American Broadsides and Ephemera Series, 4978, American Antiquarian Society.

14. Bluford Adams, *E Pluribus Barnum: The Great Showman and the Making of U.S. Popular Culture* (Minneapolis: Univ. of Minnesota Press, 1997), 86–89.

15. Olive Logan, *The Mimic World, and Public Exhibitions: Their History, Their Morals, and Effects* (Philadelphia: New-World Publishing, 1871), 336.

16. On the changing ideology of dominionism see, e.g., Erica Fudge, *Animal* (London: Reaktion, 2004), 12–18.

17. Elbert Bowen, "The Circus in Early Rural Missouri," *Missouri Historical Review* 47 (1952): 4–5; Rothfels, *Savages and Beasts,* 158–60.

18. Adams, *E Pluribus Barnum,* 85; Rothfels, *Savages and Beasts,* 153; Stokes, "Lion Griefs," 152–53.

19. *Kalamazoo Gazette* (MI), Sept. 11, 1846, quoted in Stuart Thayer, *Annals of the American Circus: 1793–1829,* 2 vols. (Manchester, MI: Rymark Printing, 1976), 2: 115. See also David Carlyon, *Dan Rice: The Most Famous Man You've Never Heard Of* (New York: Public Affairs, 2001), 192–93.

20. Jane Desmond, "Displaying Death, Animating Life: Changing Fictions of 'Liveness' from Taxidermy to Animatronics," in *Representing Animals,* ed. Nigel Rothfels (Bloomington: Indiana Univ. Press, 2002), 159–63; Penelope M. Leavitt and James S. Moy, "Spalding and Rogers' Floating Palace, 1852–1859," *Theatre Survey* 25, no. 1 (1984): 15–27.

21. *Kalamazoo Gazette* (MI), Sept. 11, 1846, quoted in Thayer, *Annals of the American Circus,* 2:115.

22. Richard W. Flint, "Entrepreneurial and Cultural Aspects of the Early-Nineteenth-Century Circus and Menagerie Business," in *Itinerancy in New England and New York: Dublin Seminar for New England Folklife: Annual Proceedings, 1984,* ed. Peter Benes (Boston: Boston University, 1986), 136.

23. Logan, *Mimic World,* 344–45.

24. Carlyon, *Dan Rice,* 162.

25. John J. Jennings, *Theatrical and Circus Life; or, Secrets of the Stage, Green-Room and Sawdust Arena* (St. Louis: Dan Linahan, 1882), 598.

26. Logan, *Mimic World*, 347. See also, Flint, "Entrepreneurial and Cultural Aspects," 146–47; Thayer, *Annals of the American Circus*, 2:88.

27. Ronald J. Zboray, "Literary Enterprise and the Mass Market: Publishers and Business Innovation in Antebellum America," *Essays in Economic and Business History* 10 (1998): 168–82.

28. Logan, *Mimic World*, 340.

29. See, e.g., Edward L. Stevenson, "Some Hints on Dog Teaching," *Harper's Young People* 3, no. 156 (1882): 823–24.

30. Jesse Haney, *Haney's Art of Training Animals* (New York: J. Haney, 1869), 12, 123.

31. Ibid., 14.

32. Ibid., 122.

33. George A. Palmer, *Secrets and Mysteries of Educating Animals: A Practical Experience of Twenty Years* (Portland, IN: Sun Job Department, 1890), 7; see also Haney, *Haney's Art of Training Animals*, 15.

34. "The Destruction of the Elephant at Exeter Change, March 1, 1826," vol. 16, folder 19, Leonidas Westerveld Circus Collection, New-York Historical Society, New York; "Destruction of the Elephant at Exeter Change, *Times* (London), Mar. 2, 1826; "Exeter Change Elephant," *Times* (London), Mar. 3, 1826.

35. Haney, *Haney's Art of Training Animals*, 118.

36. See, e.g., George Conklin, *The Ways of the Circus: Being the Memoires and Adventures of George Conklin, Tamer of Lions* (New York: Harper & Bros., 1921), 121; Palmer, *Secrets and Mysteries*, 6–9.

37. Michael Dobson, "A Dog at All Things: The Transformation of the Onstage Canine, 1550–1850," *Performance Research* 5, no. 2 (2000): 121.

38. "Educated Horse Creates Stir in Berlin," *Billboard*, Sept. 24, 1904.

39. Heini K. P. Hediger, "The Clever Hans Phenomenon from an Animal Psychologist's Point of View," in *The Clever Hans Phenomenon: Communication with Horses, Whales, Apes, and People*, ed. Thomas A. Sebeok and Robert Rosenthal (New York: New York Academy of Sciences, 1981), 1–17.

40. "Princess Trixie," *Billboard*, Dec. 2, 1905; Mim Eichler Rivas, *Beautiful Jim Key: The Lost History of a Horse and Man Who Changed the World* (New York: Morrow, 2005).

41. Gilbert, *American Vaudeville*, 322–23.

42. Carlyon, *Dan Rice*, 32–34, 227–30; Richard J. Reynolds, "Circus Rhinos," *Bandwagon* 12, no. 5 (1968): 12; John C. Kunzog, *One Horse Show: The Life and Times of Dan Rice, Circus Jester and Philanthropist* (Jamestown, NY: John C. Kunzog, 1961), 116–17.

43. Michael Peterson, "The Animal Apparatus: From a Theory of Animal Acting to an Ethics of Animal Acts," *TDR: Drama Review* 51, no. 1 (2007): 36. See also Michael Kirby, "On Acting and Non-Acting," *Drama Review* 16, no. 1 (1972): 3–15.

44. Dobson, "A Dog at all Things," 123.

45. Haney, *Haney's Art of Training Animals*, 14.

46. Ibid., 13.

47. Conklin, *Ways of the Circus*, 122; Haney, *Haney's Art of Training Animals*, 15.

48. Conklin, *Ways of the Circus*, 122; Haney, *Haney's Art of Training Animals*, 123.

49. Robert Boakes, *From Darwin to Behaviorism: Psychology and the Minds of Animals* (Cambridge: Cambridge Univ. Press, 1984).

50. Trevor B. Poole, "Principles in Training Animals," in UFAW, *Animal Training*, 5.

51. Haney, *Haney's Art of Training Animals*, 14; Palmer, *Secrets and Mysteries*, 6–7.

52. For more on this idea, see Stephen Budiansky, *The Truth about Dogs: An Inquiry into the Ancestry, Social Conventions, Mental Habits, and Moral Fiber of* Canis familiaris (New York: Penguin, 2001), 124–58.

53. W. C. Coup, *Sawdust and Spangles: Stories and Secrets of the Circus* (Chicago: Herbert S. Stone, 1901), 28.

54. Garrard Tyrwhitt-Drake, *Beasts and Circuses* (Bristol: Arrowsmith, 1936), 103–4.

55. Peter Kolb, *A Collection of Voyages and Travels* (Philadelphia: Spotswood, 1787), 127. See also, e.g., "The Female Elephant," *Mackenzie's British, Irish and Canadian Gazette*, June 9, 1838.

56. Haney, *Haney's Art of Training Animals*, 115–22.

57. Kiley-Worthington, "The Training of Circus Animals," in UFAW, *Animal Training*, 77.

58. Dan Wylie, *Elephant* (London: Reaktion, 2008), 30.

59. Conklin, *Ways of the Circus*, 112–14.

60. Eric Ames, *Carl Hagenbeck's Empire of Entertainments* (Seattle: Univ. of Washington Press, 2008), 29.

61. Conklin, *Ways of the Circus*, 115.

62. "Forepaugh's Wild Beasts," *New York Times*, June 3, 1882; "Attacked by an Elephant," *New York Times*, July 19, 1885; "Empress Kills a Man," *New York Times*, Oct. 12, 1885.

63. Conklin, *Ways of the Circus*, 121.

64. Joyce Poole and Petter Granli, "Mind and Movement: Meeting the Interests of Elephants," in *An Elephant in the Room: The Science and Well-Being of Elephants in Captivity*, ed. Debra L. Forthman, Lisa F. Kane, David Hancocks, and Paul F. Waldau (North Grafton, MA: Tufts Center for Animals and Public Policy, 2009), 5–6.

65. "Good Taste among the Mighty," *St. Paul Daily Globe* (MN), Sept. 3, 1882.

66. Conklin, *Ways of the Circus*, 116.

67. Marcello Truzzi, "The American Circus as a Source of Folklore: An Introduction," *Southern Folklore Quarterly* 30, no. 4 (1966): 295.

68. Benjamin L. Hart, Lynette A. Hart, and Noa Pinter-Wollman, "Large Brains and Cognition: Where Do Elephants Fit In?" *Neuroscience and Biobehavioral Reviews* 32, no. 1 (2008): 86–98.

69. Bostock, *Training of Wild Animals*, 171–72.

70. Rivas, *Beautiful Jim Key*, 139.

71. "In a Circus Menagerie," *New York Times*, Apr. 30, 1882.

72. "Among the Elephants," *New York Times*, Feb. 22, 1884.

73. Donna J. Haraway, *When Species Meet* (Minneapolis: Univ. of Minnesota Press, 2008), 205; Vicki Hearne, *Adam's Task: Calling Animals by Name* (New York: Knopf, 1983), 42; Joanna Latimer and Lynda Birke, "Natural Relations: Horses, Knowledge, Technology," *Sociological Review* 57, no. 1 (2009): 2.

74. This argument is often employed by the marketing programs of contemporary zoos, animal amusement parks, and circuses, but here I quote, Marthe Kiley-Worthington, *Animals in Circuses and Zoos: Chiron's World?* (London: Little Eco Farms Publications, 1990), 11, 64, 135.

75. Townsend, "Animals and Their Trainers," 733.

76. Bostock, *Training of Wild Animals*, 185.

77. "In a Circus Menagerie," *New York Times*, Apr. 30, 1882. See also, e.g., Bostock, *Training of Wild Animals*, 167–73.

78. Palmer, *Secrets and Mysteries*, 21–22.

79. Lucia Zora, *Sawdust and Solitude* (Boston: Little, Brown, 1928), 58.

80. See, e.g., Carol Joys, *The Wild Animal Trainer in America* (Boulder, CO: Pruett Publishing, 1983), 4–18; Kiley-Worthington, "Training of Circus Animals," 69.

81. Zora, *Sawdust and Solitude*, 77.

82. "Cruelty to Animals," *Saturday Evening Post*, Apr. 11, 1868.

83. Palmer, *Secrets and Mysteries*, 6.

84. Haney, *Haney's Art of Training Animals*, 15.

85. Gilbert, *American Vaudeville*, 322–23.

86. Flint, "Entrepreneurial and Cultural Aspects," 145; Logan, *Mimic World*, 290.

87. Logan, *Mimic World*, 291, 338–41.

88. "The Little Circus Boy, Leo Lawrence's Complaint against Mr. Forepaugh, Jr.," *New York Times*, Nov. 14, 1883.

89. Brian Cummings, "Pliny's Literate Elephant and the Idea of Animal Language in Renaissance Thought," in *Renaissance Beasts: Of Animals, Humans, and Other Wonderful Creatures*, ed. Erica Fudge (Urbana: Univ. of Illinois Press, 2004), 174. See also, e.g., Bostock, *Training of Wild Animals*, 160.

90. Bradshaw, *Elephants on the Edge*, 103.

91. S. L. Bensusan, "The Torture of Trained Animals," *English Illustrated Magazine* 13, no. 151 (1896): 26.

92. Bostock, *Training of Wild Animals*, 159–60; see also Haney, *Haney's Art of Training Animals*, 15.

93. Stokes, "Lion Griefs," 150–51.

## Chapter 4 · Punishing Bull Elephants

1. Harriet Ritvo, *The Animal Estate: The English and Other Creatures in the Victorian Age* (Cambridge, MA: Harvard Univ. Press, 1987), 5–6. Other authors have also written about how people have long believed that nonhuman animals are, as Claude Lévi-Strauss famously wrote, "bonnes à penser," namely, good to think with. Claude Lévi-Strauss, *Le Totemisme Aujourd'hui* (1962; repr. Paris: Presses Universitaires de France, 1974), 132. See also, e.g., Eileen Crist, *Images of Animals* (Philadelphia: Temple Univ. Press, 2000); Steve Baker, *Picturing the Beast: Animals, Identity, Representation* (Manchester: Manchester Univ. Press, 1993); Chilla Bulbeck, *Facing the Wild: Ecotourism, Conservation, and Animal Encounters* (London: Earthscan, 2005), xiii-xv; Lorraine Daston and Gregg Mitman, eds., *Thinking with Animals: New Perspectives on Anthropomorphism* (New York: Columbia Univ. Press, 2006).

2. Michael Peterson tells us that this particular rhetorical animal appeared in England (one would assume it has some history in Asia as well). After an elephant exhibition at the St. Bartholomew Fair in London in 1675, an enterprising publisher produced and sold a pamphlet purporting to contain the elephant's message to the nation. In it the elephant called the English population "the great Beast called the Rabble" and likened himself to them, "when we are mad, we are hard to be tam'd; there is nothing that will govern us but an Iron Hook thrust into my Pole, and an Iron Hook thrust into thy Nostrils; yet thou seest at other times how tame and gentle we are. And truly, Brother, take this from me, that I never finde my self better at ease, than when I am obedient to my Keepers." *The Elephant's Speech to the Citizens and Countrymen of England*, quoted in Michael Peterson, "The Animal Apparatus: From a Theory of Animal Acting to an Ethics of Animal Acts," *TDR: The Drama Review* 51, no. 1 (2007): 36.

3. Norval Luxon, *Niles' Weekly Register: News Magazine of the Nineteenth Century* (Baton Rouge: Louisiana State Univ. Press, 1947).

4. Scott A. Sandage, *Born Losers: A History of Failure in America* (Cambridge, MA: Harvard Univ. Press, 2005), 28–31.

5. "How to Tame an Elephant!" *Niles' Weekly Register*, Apr. 24, 1819; See also Brett Mizelle, " 'I Have Brought My Pig to a Fine Market': Animals, Their Exhibitors, and Market Culture in the Early Republic," in *Cultural Change and the Market Revolution in America, 1789–1860*, ed. Scott C. Martin (Lanham, MD: Rowman & Littlefield, 2005), 192.

6. Graham Huggan and Helen Tiffin, *Postcolonial Ecocriticism: Literature, Animals, and the Environment* (New York: Routledge, 2010),147; Nigel Rothfels, *Savages and Beasts: The Birth of the Modern Zoo* (Baltimore: Johns Hopkins Univ. Press, 2002), 77; Bernhard Gissibl, "The Nature of Colonialism: Being an Elephant in German East Africa" (paper presented at the "Animals in History: Examining the Not So Human Past" Conference, German Historical Institute, Cologne, Germany, May 2005).

7. "Training Elephants," *Harper's Weekly*, June 18, 1881, 393.

8. "The Elephant," *Farmer's Cabinet*, Apr. 24, 1835.

9. Ibid.

10. "With this approach," Randall Lockwood explains, "we are inclined to offer circular definitions and explanations of animal behavior, thinking that by naming a behavior we have explained its basis." Randall Lockwood, "Anthropomorphism Is Not a Four-Letter Word," in *Perceptions of Animals in American Culture*, ed. R. J. Hoage (Washington, DC: Smithsonian Institution Press, 1989), 48–49.

11. "The Elephant," *Farmer's Cabinet*, Apr. 24, 1835.

12. Stuart Thayer, *Annals of the American Circus: 1793–1829*, 2 vols. (Manchester, MI: Rymark Printing, 1976), 2: 310–15; "Siam, 8360, Asiatic Male," Box 1, John "Chang" Reynolds Papers, Robert L. Parkinson Library and Research Center, Circus World Museum, Baraboo, WI (hereafter CWM).

13. "A Man Nearly Killed by an Elephant," *Brooklyn Daily Eagle*, Dec. 7, 1842; "All about Elephants," *Ohio Farmer* 14, no. 31 (1865): 243.

14. "All about Elephants," *Ohio Farmer* 14, no. 31 (1865); Thayer, *Annals of the American Circus*, 2:114.

15. "Great Natural Exhibition," *New Hampshire Sentinel* (Keene), Oct. 7, 1840; "A Man Killed by an Elephant," *Brooklyn Daily Eagle*, Mar. 20, 1845; "Pizarro, 760, Asiatic Male," Box 1, John "Chang" Reynolds Papers, CWM.

16. "Pizarro, 760, Asiatic Male," and "Virginius, 9850, Asiatic Male," Box 1, John "Chang" Reynolds Papers, CWM; Thayer, *Annals of the American Circus*, 2:116; John Fanning Watson, *Annals of Philadelphia and Pennsylvania*, 3 vols. (Philadelphia: Parry & M'Millan, 1879), 3:374.

17. See, e.g., Melvin J. Olsen, "Newspaper Story Places John Robinson on Raymond & Wahring Show of 1839," *Bandwagon* 4, no. 4–5(1954): 3–4; Shana Alexander, *The Astonishing Elephant* (New York: Random House, 2000), 24; Marian Murray, *Circus! From Rome to Ringling* (New York: Appleton, 1958), 260–61; Esse Forrester O'Brien, *Circus: Cinders to Sawdust* (Naylor, 1959), 18.

18. "Columbus, 284, Asiatic Male," Box 1, John "Chang" Reynolds Papers, CWM.

19. Thayer, *Annals of the American Circus*, 2:78.

20. "Raymond & Co. and Van Amburgh's Long Established Menageries," *Franklin Gazette* (NY), Aug. 15, 1850.

21. Thayer, *Annals of the American Circus*, 2:78.

22. "Awful Scene in a Menagerie," *Democratic Banner* (Clearfield, PA), Jan. 8, 1848; "Sagacity of the Elephant," *Essex County Republican* (Keesville, NY), Feb. 12, 1848. See also Elbert Bowen, "The Circus in Early Rural Missouri," *Missouri Historical Review* 47 (1952): 6–7.

23. Thayer, *Annals of the Circus*, 2:110–20; "Columbus, 284, Asiatic Male," Box 1, John "Chang" Reynolds Papers, CWM.

24. "Tobacco's Doings," *Brooklyn Daily Eagle*, Aug. 28, 1846.

25. Raymond Sukumar, *The Living Elephants: Evolutionary Ecology, Behaviour, and Conservation* (New York: Oxford Univ. Press, 2003), 100–103.

26. Ritvo, *Animal Estate*, 228.

27. "Destruction of the Elephant at Exeter Change, *Times (London)*, Mar. 2, 1826; "Exeter Change Elephant," *Times* (London), Mar. 3, 1826.

28. Sujit Sivasundaram, "Trading Knowledge: The East India Company's Elephants in India and Britain," *Historical Journal* 48, no. 1 (2005): 56–57.

29. Thayer, *Annals of the American Circus*, 3:189.

30. George Conklin, *The Ways of the Circus: Being the Memoires and Adventures of George Conklin, Tamer of Lions* (New York: Harper & Bros., 1921), 114.

31. Some researchers have recently studied Asian elephants with respect to "maintaining of subordinate behavior" to ask if alternatives to the dominance training still used in many captive contexts produce different elephant behavior and experience. They examined the use of operant conditioning through positive rewards, flexible training and performance schedules, or no schedule at all, and a pragmatic approach that privileges the performing animal's ability to choose to perform or not. By measuring the levels of the stress hormone cortisol in the blood in and the behaviour of elephants undergoing such training, they determined that those methods produce elephants that are more cooperative and less dangerous over the long term. Janine L. Brown, Nadja Wielebnowski, and Jacob V. Cheeran, "Pain, Stress, and Suffering in Elephants," in *Elephants and Ethics: Toward a Morality of Coexistence*, ed. Chris Wemmer and Catherine A. Christen (Baltimore: Johns Hopkins Univ. Press, 2008), 132–34; Gail Laule, "The Role of Fear in Abnormal Behavior and Animal Welfare," *Proceedings of the Seventh International Conference on Environmental Enrichment, New York, New York* (New York: Wildlife Conservation Society, 2005); Gail Ellen Laule, "Positive Reinforcement Training and Environmental Enrichment: Enhancing Animal Well-Being," *Journal of the American Veterinary Medical Association* 223, no. 7 (2003): 969–72.

32. "Romeo, 7980, Asiatic Male," Box 1, John "Chang" Reynolds Papers, CWM.

33. Ibid.; "American Elephant, Death of the Performing Elephant, Romeo, at Chicago," *Times* (London), July 4, 1872.

34. Sukumar, *Living Elephants*, 180–85.

35. Trevor B. Poole, "Principles in Training Animals," in *Animal Training: A Review and Commentary on Current Practice, Proceedings of a Symposium organized by the Universities Federation for Animal Welfare*, ed. UFAW (Potters Bar, UK: Universities Federation for Animal Welfare, 1990), 10.

36. "An Elephant and Lion Loose," *Georgia Weekly Telegraph* (Macon), Apr. 9, 1869.

37. Janet M. Davis, *The Circus Age: Culture and Society under the American Big Top* (Chapel Hill: Univ. of North Carolina Press, 2002), 42–52.

38. Gil Robinson, *Old Wagon Show Days* (Cincinnati: Brockwell, 1925), 81–82, 145–58.

39. "Elephant Billy," *Daily Public Ledger* (Maysfield, KY), Oct. 20, 1893.

40. Ibid.; "Chief, 270, Asiatic Male," Box 1, John "Chang" Reynolds Papers, CWM; "Execution of Vicious Elephants," *British Veterinary Journal* 28 (1889): 143.

41. Stuart Thayer, "One Sheet," *Bandwagon* 18, no. 5 (1974), 23.

42. Vicki Hearne, *Adam's Task: Calling Animals by Name* (New York: Knopf, 1982), 44–45.

43. Pamela Walker Laird, *Advertising Progress: American Business and the Rise of Consumer Marketing* (Baltimore: Johns Hopkins Univ. Press, 1998), 101–51; see also plate 6, opp. p. 256.

44. Davis, *Circus Age*, 10.

45. Joanna Latimer and Lynda Birke, "Natural Relations: Horses, Knowledge, Technology," *Sociological Review* 57, no. 1 (2009): 1–27.

46. "Niblo's Garden," *New York Times*, Mar. 14, 1859; "Van Amburgh's Mammoth Menagerie and Moral Exhibition," *New York Times*, Apr. 12, 1862; "The Northwest. Indiana. The Tusks," *Chicago Tribune*, Feb. 19, 1872, 4; "Tippo Saib, 916, Asiatic Male," Box 1, John "Chang" Reynolds Papers, CWM. Tippo Saib should not be confused with Tippoo Sultan, imported into the United States in 1821 and (probably) deceased sometime in the 1850s while owned by a Van Amburgh branded show. "The Museum," *Youth's Companion* 21, no. 43 (1848): 1; Richard W. Flint, "Entrepreneurial and Cultural Aspects of the Early-Nineteenth-Century Circus and Menagerie Business," in *Itinerancy in New England and New York: Dublin Seminar for New England Folklife: Annual Proceedings, 1984*, ed. Peter Benes (Boston: Boston University, 1986), 135n15; Thayer, "One Sheet," 23; "Tippoo Sultan, 917, Asiatic Male," Box 1, John "Chang" Reynolds Papers, CWM.

47. Janet Davis, "Spectacles of South Asia at the American Circus," *Visual Anthropology* 6, no. 2 (1993): 126. See, e.g., "Tippo Saib," *Time Piece and Literary Companion*, Nov. 1, 1797; "In India," *Philadelphia Monthly Magazine* 1 (Jan. 1798): 60; "Anecdotes of the Elephant," *Godey's Ladies Book* 51 (Oct. 1855): 300.

48. Kate Brittlebank, "Islamic Responses to the Fall of Srirangapattana and the Death of Tipu Sultan (1799)," in *Islam in History and Politics: Perspectives from South Asia*, ed. Asim Roy (New Delhi: Oxford Univ. Press, 2006), 88–92.

49. Thayer, "One Sheet," 23.

50. "Van Amburgh's & Co.'s Menageries and Great Moral Exhibition and Egyptian Caravan," *New York Times*, Mar. 3, 1864.

51. "Van Amburgh & Co's Mammoth Menagerie," *New York Times*, Nov. 24, 1863.

52. Dan Wylie, *Elephant* (London: Reaktion, 2008), 123.

53. "A Three Hours' Combat with an Elephant," *New York Times*, Nov. 11, 1867; "Subjugating an Elephant," *Scientific American* 17, no. 23 (1867): 355; Charlie Campbell, "Elephants, Good and Bad," *Circus Review* 5, no. 1 (1957): 2–3.

54. "A Three Hours' Combat with an Elephant," *New York Times*, Nov. 11, 1867. See also W. S. Adams, "Elephants and Their Keepers," *Wilkes' Spirit of the Times*, Mar. 18, 1865, 38.

55. See, e.g., W. S. Adams, "The Story of an Elephant," *Wilkes' Spirit of the Times*, June 10, 1865.

56. Garry Marvin, "A Passionate Pursuit: Foxhunting as Performance," in *Nature Performed: Environment, Culture, and Performance*, ed. Bronislaw Szerszynski, Wallace Heim, and Claire Waterton (Oxford: Wiley-Blackwell, 2003), 46–60.

57. Davis, *Circus Age*, 161–63.

58. David Carlyon, *Dan Rice: The Most Famous Man You've Never Heard Of* (New York: Public Affairs, 2001), 124.

59. "An Incident That Teaches," *Frank Leslie's Illustrated Newspaper*, Dec. 30, 1865, 237.

60. Louise E. Robbins, *Elephant Slaves and Pampered Parrots: Exotic Animals in Eighteenth-Century Paris* (Baltimore: Johns Hopkins Univ. Press, 2002), 97–99.

61. Rod Preece, *Awe for the Tiger, Love for the Lamb: A Chronicle of Sensibility to Animals* (Vancouver: Univ. of British Columbia Press, 2002), 265–69.

62. Ruth Miller Elson, *Guardians of Tradition: American Schoolbooks of the Nineteenth Century* (Lincoln: Univ. of Nebraska Press, 1964), 17; Erica Fudge, *Animal* (London: Reaktion, 2004), 21.

63. For some early examples of these arguments, see, e.g., "Animal Psychology," *Living Age* 17, no. 215 (1848): 596; "The Future Life of Animals," *Living Age* 2, no. 21 (1844): 555–56.

64. Alexander Murdoch Gow, *Good Morals and Gentle Manners: For Schools and Families* (Cincinnati: Van Antwerp, Bragg, 1873), 133–34.

65. Paul Semonin, *American Monster: How the Nation's First Prehistoric Creature became a Symbol of National Identity* (New York: New York Univ. Press, 2000), 400–404.

66. Diana Donald, "Pangs Watched in Perpetuity: Sir Edwin Landseer's Pictures of Dying Deer and the Ethos of Victorian Sportsmanship," in *Killing Animals*, ed. Animal Studies Group (Urbana: Univ. of Illinois Press, 2006), 51.

67. Fudge, *Animal*, 7–8.

68. S. S. Smith, *History of Wild Animals and Leading Curiosities and Guide to P. T. Barnum's New and Greatest Show on Earth* (New York: Wynkoop & Hallenbeck, 1878), 8.

69. "Training Elephants," *Harper's Weekly*, June 18, 1881, 393.

70. "Most Dangerous Business Is Handling of Elephants," *Billboard*, Dec. 21, 1907.

71. Carlyon, *Dan Rice*, 70.

72. "The Circus in America," *New York Times*, Mar. 12, 1882; "No Elephants Wanted," *New York Times*, July 22, 1886; "A Circus under the Hammer," *New York Times*, Feb. 18, 1887.

73. Conklin, *Ways of the Circus*, 137.

74. See, e.g., "How Much for the Elephant—A Circus under the Sheriff's Hammer," *New York Times*, Nov. 13, 1869; "A Bankrupt Circus Manager," *New York Times*, Aug. 8, 1877; "Coup's Circus Sold," *New York Times*, Sept. 20, 1882; "A Circus under the Hammer," *New York Times*, Feb. 18, 1887.

75. Brandy Parris, "Difficult Sympathy in the Reconstruction-Era Animal Stories of *Our Young Folks*," *Children's Literature* 31 (2003): 40.

76. Allegedly written by the elephant's well-known keeper and trainer, William Scott, the article was probably authored by a company ghostwriter and handed gratis to the magazine in the hopes child readers might nag their parents about visiting the circus. It is not clear that the subduing-an-elephant montage was supplied by Barnum & Bailey Company press agents inasmuch as it clashed sharply with the warm tone of the Scott story, which anthropomorphized Jumbo as a celebrity and friend to children.

77. Jennifer Mason, *Civilized Creatures: Urban Animals, Sentimental Culture, and American Literature, 1850–1900* (Baltimore: Johns Hopkins Univ. Press, 2005), 13.

78. Laird, *Advertising Progress*, 69, 93, 151; Teresa Mangum, "Dog Years, Human Fears," in *Representing Animals*, ed. Nigel Rothfels (Bloomington: Indiana Univ. Press, 2002), 35–47.

79. Mason, *Civilized Creatures*, 168.

80. Michael Denning, *Mechanic Accents: Dime Novels and Working-Class Culture in America* (New York: Verso, 1987), 29–30, 171–203; Michael Kimmel, *Manhood in America: A Cultural History* (New York: Free Press, 1996), 120–45; Mark A. Swiencicki, "Consuming Brotherhood: Men's Culture, Style and Recreation as Consumer Culture, 1880–1930," *Journal of Social History* 31 (summer 1998): 791–92.

81. Davis, *Circus Age*, 155; Nigel Rothfels, "Elephants, Ethics, and History," in Wemmer and Christen, *Elephants and Ethics*, 110–15.

82. Donald, "Pangs Watched in Perpetuity," 60; Wylie, *Elephant*, 83–84.

83. Marian Louise Scholtmeijer, *Animal Victims in Modern Fiction: From Sanctity to Sacrifice* (Toronto: Univ. of Toronto Press, 1993), 69.

84. Carlyon, *Dan Rice*, 60.

85. P. T. Barnum, *The Wild Beasts, Birds and Reptiles of the World: The Story of Their Capture* (Chicago: Werner, 1894); *Forest and Jungle; or, Thrilling Adventures in All Quarters of the Globe* (1896); P. T. Barnum, *Lion Jack: A Story of Perilous Adventures among Wild Men and the Capturing of Wild Beasts; Showing How Menageries are Made* (1876; New York: G. W. Dillingham, 1887); A. H. Saxon, *P. T. Barnum: The Legend and the Man* (New York: Columbia Univ. Press, 1989), 290.

86. "Anecdotes of the Elephant," *Godey's Ladies Book* 51 (Oct. 1855): 300.

87. John R. Corvell, "The Romance of a Menagerie," *St. Nicholas* 11, no. 12 (1884): 933.

88. Nicholas Ridout, "Animal Labour in the Theatrical Economy," *Theatre Research International* 29, no. 1 (2004): 60.

89. Parris, "Difficult Sympathy," 32–33.

90. Mason, *Civilized Creatures*, 12; Susan J. Pearson, *The Rights of the Defenseless: Protecting Animals and Children in Gilded Age America* (Chicago: Univ. of Chicago Press, 2011), 116–24.

91. Conklin, *Ways of the Circus*, 133–34; W. C. Coup, *Sawdust and Spangles: Stories and Secrets of the Circus* (Chicago: Herbert S. Stone, 1901), 186.

92. "Cruelty to Animals," *Saturday Evening Post*, Apr. 11, 1868.

93. April Louise Austin, "Illustrating Animals for the Working Classes: The Penny Magazine (1832–45)," *Anthrozoös* 23, no. 4 (2010): 365–82.

94. P. T. Barnum to Spencer Baird, Apr. 11, 1883, reprinted in A. H. Saxon, comp. and ed., *Selected Letters of P. T. Barnum* (New York: Columbia Univ. Press, 1983), 233.

95. "Snakes," and "Cruelty to Animals," *New York World*, Mar. 19, 1867.

96. Phineas T. Barnum, *Struggles and Triumphs; or, Sixty Years Recollection of P. T. Barnum* (Buffalo: Courier, 1889), 321–22.

97. "Henry Bergh and His Work," *Scribner's Monthly* 17, no. 6 (1879): 874, 879; Mason, *Civilized Creatures*, 17; Saxon, *P.T. Barnum*, 235–38.

98. James Turner, *Reckoning with the Beast: Animals, Pain, and Humanity in the Victorian Mind* (Baltimore: Johns Hopkins Univ. Press, 1980), 29.

99. Diane L. Beers, *For the Prevention of Cruelty: The History and Legacy of Animal Rights Activism* (Athens, OH: Swallow Press, 2006), 52–54; Katherine C. Grier, *Pets in America: A History* (Chapel Hill: Univ. of North Carolina Press, 2006), 128–43; Joel Tarr and Clay McShane, *Horse in the City: Living Machines in the Nineteenth Century* (Baltimore: Johns Hopkins Univ. Press, 2007), 78–82; Pearson, *Rights of the Defenseless*, 64.

100. Charles Nordhoff, "Peep at Elephant," *Harper's New Monthly Magazine* 20, no. 118 (1860): 460.

101. Nigel Rothfels, "Catching Animals," in *Animals in Human Histories: The Mirror of Nature and Culture*, ed. Mary J. Henninger-Voss (Rochester: Univ. of Rochester Press, 2002), 195, 199, 202–3.

102. "The Colonial Club," *New York Times*, Apr. 25, 1892; "American Authors' Guild," *New York Times*, May 12, 1896; "Against Foreign Penal Colonies," *New York Times*, Nov. 10, 1895; "Society," *New York Times*, Mar. 15, 1896; "Society," *New York Times*, Mar. 29, 1896.

103. "Dr. William Ballou, Writer on Science," *New York Times*, Dec. 1, 1937; David Raines

Wallace, *The Bonehunters' Revenge: Dinosaurs and Fate in the Gilded Age* (New York: Mariner Books / Houghton Mifflin, 1999), 210–11.

104. Alfred S. Romer, quoted in Wallace, *Bonehunters' Revenge*, 204–5.

105. Mason, *Civilized Creatures*, 20.

106. Harriet Ritvo, *The Platypus and the Mermaid: And Other Figments of the Classifying Imagination* (Cambridge, MA: Harvard Univ. Press, 1997), 1–49.

107. William Hosea Ballou, "Are the Lower Animals Approaching Man?" *North American Review* 145, no. 372 (1887): 516–23.

108. Parris, "Difficult Sympathy," 38.

109. "Execution of an Elephant," *New York Times*, July 29, 1866.

110. "All About Elephants," *Ohio Farmer*, Aug. 5, 1865.

111. Matt Cartmill, *A View to a Death in the Morning: Hunting and Nature through History* (Cambridge, MA: Harvard University Press, 1996), 156.

112. Kathleen Kete, "Animals and Human Empire," in *Cultural History of Animals in the Age of Empire*, ed. Kathleen Kete (London: Berg, 2009), 5–7.

113. "Jumbo's Successors," *New York Sun*, reprinted in *St. Paul Daily Globe* (MN), Dec. 26, 1885.

### Chapter 5 · Herd Management in the Gilded Age

1. George Wilbur Peck, *Peck's Bad Boy with the Circus* (Chicago: Stanton & Van Vliet, 1906), 72, 156–61.

2. "Barnum's Show in Winter Quarters," *Harper's Weekly*, 27, no. 1313 (1882): 106.

3. Lynda Birke, Joanna Hockenhull, and Emma Creighton, "The Horse's Tale: Narratives of Caring for/about Horses," *Society and Animals* 18, no. 4 (2010): 331–47.

4. Glenn Porter, *The Rise of Big Business: 1860–1920* (Wheeling, IL: Harlan Davidson, 2004), 6.

5. Janet M. Davis, *The Circus Age: Culture and Society under the American Big Top* (Chapel Hill: Univ. of North Carolina Press, 2002), 40–41.

6. Roger M. Olien and Diana Davids Olien, *Oil and Ideology: The Cultural Creation of the American Petroleum Industry* (Chapel Hill: Univ. of North Carolina Press, 2000), 15–16.

7. "Old Showman," *Spirit of the Times*, Aug. 10, 1850.

8. Pamela Walker Laird, *Advertising Progress: American Business and the Rise of Consumer Marketing* (Baltimore: Johns Hopkins Univ. Press, 1998), 44.

9. Earl Chapin May, *The Circus: From Rome to Ringling* (New York: Duffield & Green, 1932), 27–39.

10. Elbert Bowen, "The Circus in Early Rural Missouri," *Missouri Historical Review* 47 (1952): 1–17; George L. Chindahl, *A History of the Circus in America* (Caldwell, ID: Caxton, 1959), 22.

11. Chindahl, *History of the Circus*, 51.

12. Bowen, "Circus in Early Rural Missouri," 1–3; W. C. Coup, *Sawdust and Spangles: Stories and Secrets of the Circus* (Chicago: Herbert S. Stone, 1901), 63.

13. "With the Circus on the Road," newspaper clipping dated Dec. 14, 1890, "No. 8 Scrapbook. Barnum & Bailey, 1890–1891," box 24, McCaddon Collection of the Barnum and Bailey Circus, Manuscripts Division, Department of Rare Books and Special Collections, Princeton University Library.

14. Chindahl, *History of the Circus*, 51–60; May, *Circus*, 57, 239.

15. Coup, *Sawdust and Spangles*, 8.

16. Chindahl, *History of the Circus*, 17–18.

17. Ibid., 50–57.

18. Davis, *Circus Age*, 10, 15–25, 37, 41–52, 77–79.

19. Alan Trachtenberg, *The Incorporation of America: Culture and Society in the Gilded Age* (New York: Hill & Wang, 1982), 74.

20. Alfred D. Chandler, *The Visible Hand: The Managerial Revolution in American Business* (Cambridge, MA: Belknap Press of Harvard Univ. Press, 1977), 81–121.

21. Jim Cullen, *Art of Democracy: A Concise History of Popular Culture in the United States*, 2nd ed. (New York: Monthly Review Press, 2002), 102.

22. Davis, *Circus Age*, 77–79.

23. See, e.g., "Grand Entrance of the Asiatic Caravan, Museum, and Menagerie of P. T. Barnum," *Farmer's Cabinet*, July 31, 1851; "P. T. Barnum's Grand Colossal Museum and Menagerie," *Daily Quincy Whig* (IL), Aug. 19, 1853.

24. David Carlyon, *Dan Rice: The Most Famous Man You've Never Heard Of* (New York: Public Affairs, 2001), 162.

25. Charles H. Day, "The Elephant as Advertisement," *Billboard*, Mar. 23, 1901.

26. "The Circus in America," *New York Times*, Mar. 12, 1882.

27. "No. 8 Scrapbook. Barnum & Bailey, 1890–1891," box 24, McCaddon Collection.

28. Coup, *Sawdust and Spangles*, 124.

29. "Forepaugh's Wild Beasts," *New York Times*, June 3, 1882. See also "The Circus in America," *New York Times*, Mar. 12, 1882; "Amusements: P. T. Barnum's Hippodrome," *New York Times*, Apr. 6, 1882.

30. Olive Logan, *The Mimic World, and Public Exhibitions: Their History, Their Morals, and Effects* (Philadelphia: New-World Publishing, 1871), 348–49.

31. Eric Ames, *Carl Hagenbeck's Empire of Entertainments* (Seattle: Univ. of Washington Press, 2008), 29.

32. Nigel Rothfels, *Savages and Beasts: The Birth of the Modern Zoo* (Baltimore: Johns Hopkins Univ. Press, 2002), 85.

33. See, e.g., "Prices of Wild Beasts," *Chicago Daily Tribune*, Apr. 24, 1885; "A Circus under the Hammer," *New York Times*, Feb. 18, 1887.

34. "Wild Animals," *Billboard*, Nov. 9, 1907. See also Ames, *Carl Hagenbeck's Empire*, 28–30; Coup, *Sawdust and Spangles*, 20–22.

35. Carl Hagenbeck, *Beasts and Men, Being Carl Hagenbeck's Experiences for Half a Century among Wild Animals*, abridged and trans. Hugh S. R. Elliot and A. G. Thacker (London: Longmans, Green, 1909), 29; Nigel Rothfels, "Catching Animals," in *Animals in Human Histories: The Mirror of Nature and Culture*, ed. Mary Henninger-Voss (Rochester: Univ. of Rochester Press, 2002), 195.

36. Charles Nordhoff, "Peep at Elephant," *Harper's Monthly Magazine* 20, no. 118 (1860): 457.

37. "Jumbo's Successors," *New York Sun*, reprinted in St. Paul Daily Globe (MN), Dec. 26, 1885.

38. Ibid.; Bernhard Gissibl, "The Nature of Colonialism: Being an Elephant in German East Africa," paper presented at the "Animals in History: Examining the Not So Human Past" Conference, German Historical Institute, Köln, Germany, May 2005; John Frederick Walker, *Ivory's Ghosts: The White Gold of History and the Fate of Elephants* (New York: Atlantic Monthly Press, 2009), 83–135.

39. *The Great International! A Hand-Book of Interest, Illustrating some of the Great Features, to*

*be seen in the Leviathan Show of the World!* (New York: Torrey Bros., n.d.), 15–16, box 48, folder 1, McCaddon Collection.

40. *The Great International! A Hand-Book of Interest, Illustrating some of the Great Features, to be seen in the Leviathan Show of the World!* (New York: Torrey Bros., n.d.), 15–16, box 48, folder 1, McCaddon Collection.

41. "Barnum's Show in Winter Quarters," *Harper's Weekly*, 27, no. 1313 (1882): 106.

42. "Barnum & Bailey Limited Minute Book, No. 2," box 44, "Miscellaneous Materials: Minute Books," McCaddon Collection.

43. Davis, *Circus Age*, 15–25, 37, 41–52, 77–79.

44. "In the Shop of a Circus," *The Press* (New York), Feb. 15, 1891.

45. Ann Norton Greene, *Horses at Work: Harnessing Power in Industrial America* (Cambridge, MA: Harvard Univ. Press, 2008), 199; Jason Hribal, "'Animals Are Part of the Working Class': A Challenge to Labor History," *Labor History* 44, no. 4 (2003): 436.

46. William T. Hornaday, *Minds and Manners of Wild Animals* (1922), quoted in Davis, *Circus Age*, 154. See also Carol Joys, *The Wild Animal Trainer in America* (Boulder, CO: Pruett Publishing, 1983), xiii.

47. "Barnum's Baby Elephant," *New York Times*, Feb. 4, 1882.

48. Paul Eipper, *Circus: Men, Beasts, and Joys of the Road*, trans. Frederick H. Martens (New York: Viking, 1931), 44–45.

49. Dan Wylie, *Elephant* (London: Reaktion, 2008), 59.

50. Barbara Noske, *Beyond Boundaries: Humans and Animals* (Montreal: Black Rose Books, 1997), 15.

51. "Vagaries of Animals," *New York Times*, May 3, 1896.

52. "In a Circus Menagerie," *New York Times*, Apr. 30, 1882.

53. Ibid.

54. Garrard Tyrwhitt-Drake, *Beasts and Circuses* (Bristol: Arrowsmith, 1936), 105–6.

55. Ibid. In the last ten years, a handful of studies have shown elephants demonstrating an understanding of the immediate consequences of their actions in effecting the behavior of other beings, if not necessarily understanding the full nature of the state of mind in those other beings. For instance, elephants know that a living thing they squash under one foot will cease to move any longer, although they do not necessarily intend to "kill" him or her. Zoo elephants have been found to deceive one another by imitating human behaviors unknown to a fellow elephant. Robert W. Mitchell and Nicholas S. Thompson, "Ethological and Psychological Perspectives," in *Deception: Perspectives on Human and Nonhuman Deceit*, ed. Robert W. Mitchell and Nicholas S. Thompson (Albany, NY: SUNY Press, 1986), 149.

56. George Conklin, *The Ways of the Circus: Being the Memoires and Adventures of George Conklin, Tamer of Lions* (New York: Harper & Bros, 1921), 132.

57. "In a Circus Menagerie," *New York Times*, Apr. 30, 1882.

58. "Training Elephants," *Harper's Weekly*, June 18, 1881, 393.

59. "Managing an Elephant," *Daily Alta California*, Nov. 3, 1887.

60. John Eck, "Most Dangerous Business Is Handling of Elephants," *Billboard*, Dec. 21, 1907.

61. "The Elephant Was Wise," *Salt Lake Herald* (UT), Aug. 20, 1906.

62. See, e.g., "Vagaries of Animals," *New York Times*, May 3, 1896; Frank C. Bostock, *The Training of Wild Animals*, ed. Ellen Velvin (1903; repr. New York: Century, 1920), 149; Coup, *Sawdust and Spangles*, 160–62.

63. "Hebe's Wonderful Baby," *New York Sun*, Mar. 21, 1880.

64. Reports of this behavior are diverse. See, e.g., Caitlin O'Connel, *The Elephant's Secret Sense: The Hidden Life of the Wild Herds of Africa* (Chicago: Univ. of Chicago Press, 2007), 109–10; William Henry Giles Kingston, *Stories of Animal Sagacity* (Edinburgh: T. Nelson & Sons, 1874), 168–69; Joshua M. Plotnik, Richard Lair, Wirot Suphachoksahakun, and Frans B. M. de Waal, "Elephants Know When They Need a Helping Trunk in Cooperative Task," *PNAS (Proceedings of the National Academy of Sciences of the United States of America)* 108, no. 12 (2011): 5116–21.

65. J. Y. Henderson, *Circus Doctor: As Told to Richard Taplinger* (New York: Bantam Books, 1952), 55.

66. "Elephants and Their Keepers," *Wilkes' Spirit of the Times*, Mar. 18, 1865. See also, e.g., "The Story of an Elephant," *Wilkes' Spirit of the Times*, June 10, 1865; "An Elephant in Love," *Frank Leslie's Budge of Fun*, July 1, 1865.

67. Late-twentieth-century studies of the effects of picket lines and chaining on elephants have found evidence of "abnormal social behavior" and "social isolation" in individuals, even when stored together in large groups, since they cannot choose how and when to interact. Equally, elephants may display apparent irritation with and aggression against adjacent individuals. These effects are intensified in elephants chained for most of the day while arranged in picket lines by sex and height, that is, according to human sensibilities of aesthetic appeal or logistical convenience. That lack of choice in spatial arrangement counters the social development of elephants: "wild elephants always organize themselves in a typical spatial pattern in which neonates and infants stay close to their mothers and allomothers—females who give maternal care to the offspring of other mothers—and juvenile females stay close to neonates and infants." Fred Kurt, Khyne U. Mar, and Marion E. Garaï, "Giants in Chains: History, Biology, and Preservation of Asian Elephants in Captivity," in *Elephants and Ethics: Toward a Morality of Coexistence*, ed. Chris Wemmer and Catherine A. Cristen (Baltimore: Johns Hopkins Univ. Press, 2008), 340.

68. Cleveland Moffat, "Elephants in a Rage," *Los Angeles Times*, July 7, 1895.

69. Joys, *Wild Animal Trainer*, 12–14. See also, e.g., "Managing an Elephant," *Daily Alta California*, Nov. 3, 1887.

70. Thomas K. McGraw, *American Business, 1920–2000: How It Worked* (Wheeling, IL: Harlan Davidson, 2000), 7–8.

71. "No. 8 Scrapbook. Barnum & Bailey, 1890–1891," box 24, McCaddon Collection.

72. Conklin, *Ways of the Circus*, 126–30, 136–37.

73. "Dead elephants are good elephants. Tours 1902," photo # 356, "No. 11 Scrapbook. Photograph Album, Barnum & Bailey Foreign Tour, 1897–1902," box 27, McCaddon Collection.

74. Alan Roocroft and James Oosterhuis, "Foot Care for Captive Elephants," in *The Elephant's Foot: Prevention and Care of Foot Conditions in Captive Asian and African Elephants*, ed. Blair Csuti, Eva L. Sargent, Ursula S. Bechert (Ames: Iowa State Univ. Press, 2001), 21–22.

75. Joseph T. McCaddon to Editor, *Spectator*, Jan. 5, 1931, box 41, folder 7, McCaddon Collection.

76. McShane and Tarr, *Horse in the City*, 51.

77. *Twenty-Seventh Annual Report of the Pennsylvania Society for the Prevention of Cruelty to Animals* (Philadelphia: Burk & McFetridge, 1895), 14–15, 31, box 1, folder 15, Miscellaneous Circus Collection, Special Collections and Rare Books, Princeton University Library.

78. "Bergh on Barnum's Dead Elephant," *Atlanta Constitution*, Apr. 8, 1883. On Bergh's philosophies on animal training, see Susan J. Pearson, *The Rights of the Defenseless: Protecting Animals and Children in Gilded Age America* (Chicago: Univ. of Chicago Press, 2011), 44–51.

79. "Editor's Easy Chair," *Harper's New Monthly Magazine* 67 (June–Nov. 1883): 146–47.

80. Greene, *Horses at Work*, 22.

81. Coup, *Sawdust and Spangles*, 82.

82. Bostock, *Training of Wild Animals*, 73; John J. Jennings, *Theatrical and Circus Life; or, Secrets of the Stage, Green-Room, and Sawdust Arena* (St. Louis: Dan Linahan, 1882), 605; Logan, *Mimic World*, 297.

83. Conklin, *Ways of the Circus*, 132.

84. "Elephants at Sea—*Calcutta Englishman*," *Essex County Republican* (Keeseville, NY), May 21, 1874.

85. Ibid.; "The Elephant," *Farmer's Cabinet*, Apr. 24, 1835; "The Female Elephant," *Mackenzie's British, Irish and Canadian Gazette*, June 9, 1838; "Disembarkation of Elephants at Calcutta from Burmah," *Littell's Living Age*, Mar. 27, 1858.

86. Comte de Buffon, quoted in Sujit Sivasundaram, "Trading Knowledge: The East India Company's Elephants in India and Britain," *Historical Journal* 48, no. 1 (2005): 54.

87. "Sagacity of Elephants," *Phelps County New Era* (MO), Oct. 14, 1876; "Hebe's Wonderful Baby," *New York Sun*, Mar. 21, 1880.

88. Ralph Helfer, *Beauty of the Beasts: Tales of Hollywood's Wild Animal Stars* (n.p.: E-Reads, 1999), 104.

89. "Vagaries of Animals," *New York Times*, May 3, 1896.

90. "Angry Circus Elephant Crushes a Man to Death," *Brooklyn Daily Eagle*, May 28, 1902.

91. Ellen Velvin, *Rataplan: A Rogue Elephant and other Stories* (Philadelphia: H. Altemus, 1902), 1–3.

92. See, e.g., Charles A. Day, *The Animal Kingdom Illustrated and Sketches Descriptive of the Wild Beasts Contained in Forepaugh's Menagerie* (Philadelphia: Forepaugh Amusement Enterprises, 1876), 6–10, vol. 12, item 10, Leonidas Westerveld Circus Collection, New-York Historical Society, New York; S. S. Smith, *History of Wild Animals and Leading Curiosities and Guide to P. T. Barnum's New and Greatest Show on Earth* (New York: Wynkoop & Hallenbeck, 1878), 6–9.

93. Ros Clubb and Georgia Mason, *A Review of the Welfare of Zoo Elephants in Europe: A Report Commissioned by the RSPCA* (Oxford: University of Oxford, Animal Behaviour Research Group, 2002), 222–30; Georgia J. Mason, "Stereotypies: A Critical Review," *Animal Behaviour* 41 (1991): 1015–37; Georgia J. Mason, "Stereotypies and Suffering," *Behavioural Processes* 25 (1991): 103–15. See also Janine L. Brown, Nadja Wielebnowski, and Jacob V. Cheeran, "Pain, Stress, and Suffering in Elephants," in Wemmer and Christen, *Elephants and Ethics*, 125–26.

94. Researchers have, for instance, compared turn-of-the-twenty-first-century Asian elephants in circuses and zoos kept in an electrified paddock—a tool not available to nineteenth-century circuses—to those kept tethered with heavy chains around one or more ankles. They found that groups of elephants unshackled in paddocks chose to move around within the paddock and exhibited more social and play behaviors and in general displayed more species-typical behaviors, as well as no or radically reduced stereotypic movements, depending on the individual in question. When shackled the same elephants showed very frequent weaving. Andrzej Elzanowski and Agnieszka Sergiel, "Stereotypic Behavior of a Female Asiatic Elephant (*Elephas maximus*) in a Zoo," *Journal of Applied Animal Welfare Science* 9, no. 3 (2006): 223–32; T. H. Friend, "Behavior of Picketed Circus Elephants," *Applied Animal Behaviour Science* 62 (1999): 73–88; T. H. Friend and M. L. Parker, "The Effect of Penning versus Picketing on Stereotypic Behavior of Circus Elephants," *Applied Animal Behaviour Science* 64 (1999): 213–55; J. Schmid, "Keeping Circus Elephants Temporarily in Paddocks—The Effects on Their Be-

havior," *Animal Welfare* 4 (1995): 87–101. Nor have researchers observed such abnormal repetitive behaviors among elephants at large in rural Asia. Kurt, Mar, and Garaï, "Giants in Chains," 337.

95. Clubb and Mason explain the reasoning thusly: "Stereotypics are often used as indicators of poor welfare for three main reasons. Firstly, animals that are housed in environments we would consider to be poor, such as barren, cramped conditions, tend to develop more stereotypies than similar animals in large, enriched, surroundings. . . . Secondly, they tend to develop in animals that are frustrated or thwarted from performing highly motivated behaviours, such as feeding and foraging . . . , escape . . . , or nest-building. . . . Thirdly, they are sometimes associated with other indicators of poor welfare, such as injury . . . , adrenal hypertrophy and adrenocortical hyperactivity. . . . They are thus signs that animals are motivated to perform natural behaviors that cannot be performed naturally—exacerbated if this motivations is high and/or if the animal has much free time to fill. . . . Many studies therefore report reductions in stereotypies through providing the captive animal with more, or better preferred, activities to do." Clubb and Mason, *Review*, 223. We should also exercise caution, nonetheless. There is some recent evidence that stereotypic elephants will survive better long term because they use repetitive behavior to manage stress, so it is not clear that weaving elephants were more likely to become unresponsive or dangerous to people. Ibid., 222–30. See also Brown, Wielebnowski, and Cheeran, "Pain, Stress and Suffering in Elephants," 125–26.

96. "Hebe's Wonderful Baby," *New York Sun*, Mar. 21, 1880.

97. *Boston Journal*, quoted in "An Elephant's Bath in Boston Common," *Chicago Daily Tribune*, June 23, 1882.

98. "Elephants Cause Trouble in Troy," *New York Times*, Aug. 3, 1882; "Four Elephants at Large," *Washington Post*, Aug. 3, 1882; "An Elephant in a Rolling Mill," *Los Angeles Times*, Aug. 4, 1882; "Elephantine Pranks," *Chicago Daily Tribune*, Aug. 6, 1882; "Waifs and Strays," *Harper's Weekly* 26, no. 1339 (1882): 526.

99. Eric Scigliano, *Love, War, and Circuses: The Age-Old Relationship Between Elephants and Humans* (New York: Houghton Mifflin, 2002), 259. On the behavioral effects of confinement, see Vicki Hearne, *Adam's Task: Calling Animals by Name* (New York: Knopf, 1982), 21; Sue Savage-Rumbaugh, Kanzi Wamba, Panbanisha Wamba, and Nyota Wamba, "Welfare of Apes in Captive Environments: Comments on, and by, a Specific Group of Apes," *Journal of Applied Animal Welfare Science* 10, no. 1 (2007): 7–19.

100. "Hebe's Wonderful Baby," *New York Sun*, Mar. 21, 1880.

101. "Wild Animals Burned Alive," *New York Times*, May 23, 1879.

102. Coup, *Sawdust and Spangles*, 76.

## Chapter 6 · *Going Off Script*

1. August H. Kober, *Circus Nights and Circus Days*, trans. Claud W. Sykes (London: Sampson Low, Marston, 1928), 20.

2. John Eck, "Most Dangerous Business Is Handling of Elephants," *Billboard*, Dec. 21, 1907.

3. Garrard Tyrwhitt-Drake, *Beasts and Circuses* (Bristol: Arrowsmith, 1936), 27. See also Gil Robinson, *Old Wagon Show Days* (Cincinnati: Brockwell, 1925), 81.

4. See, e.g., "News of the Day," *New York Times*, May 7, 1867; "Nearly a Panic," *New York Times*, June 9, 1883; "Elephants in a Cellar," *New York Times*, Mar. 25, 1884; Paul Eipper, *Circus: Men, Beasts, and Joys of the Road*, trans. Frederick H. Martens (New York: Viking, 1931), 206–7; Robinson, *Old Wagon Show Days*, 102.

5. See, e.g., "How an Elephant Crossed a Bridge," *New York Times*, May 24, 1874; "A Huge Passenger," *New York Times*, Apr. 7, 1882; "In a Circus Menagerie," *New York Times*, Apr. 30, 1882; "Elephants in a Cellar," *New York Times*, Mar. 25, 1884; "Elephants Scared by Little Things," *New York Times*, July 22, 1886; Tyrwhitt-Drake, *Beasts and Circuses*, 103–4.

6. "Nearly a Panic," *New York Times*, June 9, 1883.

7. "Trouble in Barnum's Show," *New York Times*, June 8, 1883; "A Panic Narrowly Escaped," *New York Tribune*, June 8, 1883; "Trouble for Barnum," *Quincy Daily Herald* (IL), June 8, 1883.

8. "Nearly a Panic," *New York Times*, June 9, 1883.

9. "Only a Gleeful Baby Elephant," *New York Times*, June 10, 1883.

10. On the difficulty in tracking elephants who were sold or renamed frequently, William (Buckles) Woodcock notes, "Listing elephants can literally become a maddening hobby. Just about the time you think you have it figured out some obscure article disrupts everything." William (Buckles) Woodcock, "From Buckles #2," *Buckles Blog*, Oct. 2, 2010, http://bucklesw.blog spot.com/2010/10/from-buckles-2_02.html. Likewise, the "Chang" Reynolds elephant biographies at the Circus Historical Society in Baraboo, Wisconsin, show evidence of frequent corrections and updates as he pieced together the life stories of all the circus elephants in the United States up to the 1970s. Box 1, John "Chang" Reynolds Papers, Robert L. Parkinson Library and Research Center, Circus World Museum, Baraboo, Wisconsin (hereafter CWM).

11. Walter Lippmann, *Public Opinion* (1922; repr. New York: Free Press, 1997), 217, 221.

12. "Forepaugh and His Prize Beauty," *New York Times*, Mar. 25, 1883; See also John J. Jennings, *Theatrical and Circus Life; or, Secrets of the Stage, Green-Room, and Sawdust Arena* (St. Louis: Dan Linahan, 1882), 515.

13. "Circus Notes," *New York Dramatic News*, Nov. 14, 1882.

14. J. Clarence Hyde, "If You Would Join the Big Circus," *Daily Continent* (New York), Sunday, Apr. 19, 1891.

15. "Death of the Elephant Empress," *New York Times*, Aug. 6, 1875; "Mangled by an Elephant," *New York Times*, Sept. 1, 1885.

16. "Attacked by an Elephant," *New York Times*, July 19, 1885.

17. "Crushed by an Elephant," *New York Times*, Aug. 28, 1883. See also Robinson, *Old Wagon Show Days*, 81–82; Tyrwhitt-Drake, *Beasts and Circuses*, 105.

18. "Crushed by an Elephant," *New York Times*, Aug. 28, 1883.

19. Connie Clausen, *I Love You Honey, But the Season's Over* (New York: Holt, Rinehart & Winston, 1961), 9, 15.

20. "Topsy, 947, Asian Female," box 1, John "Chang" Reynolds Papers, CWM.

21. "Angry Circus Elephant Crushes a Man to Death," *Brooklyn Daily Eagle*, May 28, 1902.

22. Ibid.

23. W. C. Coup, *Sawdust and Spangles: Stories and Secrets of the Circus* (Chicago: Herbert S. Stone, 1901), 11.

24. David A. H. Wilson, "Politics, Press and the Performing Animals Controversy in Early Twentieth-Century Britain," *Anthrozoös* 21, no. 4 (2008): 319.

25. Eric Scigliano, *Love, War, and Circuses: The Age-Old Relationship between Elephants and Humans* (New York: Houghton Mifflin, 2002), 196–205.

26. See also Jonathan Burt, *Animals in Film* (London: Reaktion, 2002), 113; Sue Coe and Kim Stallwood, "An Elephant Never Forgets," *Blab* 18 (autumn 2007): 81–90; Akira Mizuta Lippit, *Electric Animal: Toward a Rhetoric of Wildlife* (Minneapolis: Univ. of Minnesota Press, 2000), 197, 248.

27. "Elephant Forgets Not His Murderous Act," *New York Times*, Apr. 28, 1901.

28. "Death of the Elephant Empress," *New York Times*, Aug. 6, 1875; "The Seventh," *Daily Public Ledger* (Maysfield, KY), Mar. 26, 1896.

29. "Vagaries of Animals," *New York Times*, May 3, 1896.

30. Ibid.

31. "Empress, 356, Indian Female," and "Gypsy, 412, Asiatic Female," box 1, John "Chang" Reynolds Papers, CWM.

32. "Killed by an Elephant," *New York Times*, Mar. 26, 1896.

33. See, e.g., ibid.; "Circus Elephant on Rampage," *San Francisco Call*, Mar. 26, 1896; "The Seventh," *Daily Public Ledger* (Maysfield, KY), Mar. 26, 1896; "Big Elephant Ran Amuck," *Omaha Daily Bee*, March 26, 1896; "Third Man Killed," *Salt Lake Herald*, March 26, 1896; "Elephant on a Rampage," *New York Sun*, Mar. 26, 1896; "Kills Her Keeper," *Richmond Dispatch* (VA), Mar. 26, 1896; "Mad Elephant," *Marietta Daily Leader* (OH), Mar. 27, 1896.

34. "Elephant Kills Him," *Chicago Tribune*, Mar. 26, 1896.

35. Ibid.

36. "Freaks of a Circus Trick Elephant, From the *Chicago Inter-Ocean*," *New York Times*, Sept. 1, 1895.

37. "Empress, 356, Indian Female," and "Gypsy, 412, Asiatic Female," box 1, John "Chang" Reynolds Papers, CWM.

38. "Elephant Kills Him," *Chicago Tribune*, Mar. 26, 1896.

39. "Used to Live Here," *Sterling Standard* (IL), Apr. 2, 1896.

40. George "Slim" Lewis, *I Loved Rogues* (Seattle, WA: Superior Publishing, 1978), 4–5.

41. Gregory J. Renoff, *The Big Tent: The Traveling Circus in Georgia, 1820–1930* (Athens: Univ. of Georgia Press, 2008), 106; Tom Parkinson, "Nickel Plate Harris and His Circus," *Bandwagon* 52, no. 3 (2008): 3–26.

42. George Wilbur Peck, *Peck's Bad Boy with the Circus* (Chicago: Stanton & Van Vliet, 1906), 65–75.

43. Renoff, *Big Tent*, 107.

44. See, e.g., "Jumbo's Successors," *McCook Tribune* (NE), Oct. 29, 1885.

45. Edward Paysen Evans, *The Criminal Prosecution and Capital Punishment of Animals: The Lost History of Europe's Animal Trials* (1906; London: Faber & Faber, 1987).

46. Ken Gonzales-Day, *Lynching in the West, 1850–1935* (Durham, NC: Duke Univ. Press, 2006); Stephen J. Leonard, *Lynching in Colorado, 1859–1919* (Denver: Univ. Press of Colorado, 2002), 123–52; Amy Louise Wood, *Lynching and Spectacle: Witnessing Racial Violence in America, 1890–1940* (Chapel Hill: Univ. of North Carolina Press, 2009), 4–26.

47. See, e.g., Evans, *Criminal Prosecution of Animals*, 161, 171, 176.

48. Mark V. Barrow, *A Passion for Birds: American Ornithology after Audubon* (Princeton, NJ: Princeton Univ. Press, 1998), 146–52; Mark V. Barrow, *Nature's Ghosts: Confronting Extinction from the Age of Jefferson to the Age of Ecology* (Chicago: Univ. of Chicago Press, 2009), 346–47.

49. Andrew C. Isenberg, "The Wild and the Tamed: Indians, Euroamericans, and the Destruction of the Bison," in *Animals in Human Histories: The Mirror of Nature and Culture*, ed. Mary Henninger-Voss (Rochester: Univ. of Rochester Press, 2002), 131.

50. Dan Wylie, *Elephant* (London: Reaktion, 2008), 53.

51. John Eck, "Most Dangerous Business Is Handling of Elephants," *Billboard*, Dec. 21, 1907.

52. Ibid.

53. Jill Howard Church, "The Elephants' Graveyard: Life in Captivity," in *A Primer on Animal*

*Rights*, ed. Kim W. Stallwood (Brooklyn, NY: Lantern Books, 2002), 132; see also Janine L. Brown, Nadja Wielebnowski, and Jacob V. Cheeran, "Pain, Stress, and Suffering in Elephants," in *Elephants and Ethics: Toward a Morality of Coexistence*, ed. Chris Wemmer and Catherine A. Christen (Baltimore: Johns Hopkins Univ. Press, 2008), 130–34.

54. G. A. Bradshaw, *Elephants on the Edge: What Elephants Teach Us about Humanity* (New Haven, CT: Yale Univ. Press, 2009), 14, 85; "Hebe's Wonderful Baby," *New York Sun*, Mar. 21, 1880.

55. G. A. Bradshaw, "Inside Looking Out: Neuroethological Compromise Effects on Elephants in Captivity," in *Elephant in the Room*, ed. Debra L. Forthman, Lisa F. Kane, David Hancocks, and Paul F. Waldau (North Grafton, MA: Tufts Center for Animals and Public Policy, 2009), 55–68; G. A. Bradshaw, Allan N. Schore, Janine L. Brown, Joyce H. Poole, and Cynthia J. Moss, "Elephant Breakdown," *Nature* 433 (Feb. 24, 2005): 807.

56. G. A. Bradshaw and Allan N. Schore, "How Elephants Are Opening Doors: Developmental Neuroethology, Attachment, and Social Context," *Ethology* 113 (2007): 426.

57. Sue Savage-Rumbaugh, Kanzi Wamba, Panbanisha Wamba, and Nyota Wamba, "Welfare of Apes in Captive Environments: Comments On, and By, a Specific Group of Apes," *Journal of Applied Animal Welfare Science* 10, no. 1 (2007): 16.

58. Lewis, *I Loved Rogues*, 4–5.

59. "Elephant Forgets Not His Murderous Act," *New York Times*, Apr. 28, 1901.

60. Hilda Kean, *Animal Rights: Political and Social Change in Britain since 1800* (London: Reaktion, 1998), 31.

61. Jennings, *Theatrical and Circus Life*, 604.

62. Albert F. McLean Jr., *American Vaudeville as Ritual* (Lexington: Univ. of Kentucky Press, 1965), 148–49.

63. Matt Cartmill, *A View to a Death in the Morning: Hunting and Nature through History* (Cambridge, MA: Harvard Univ. Press, 1996), 154. See also Ralph H. Lutts, "The Wild Animal Story: Animals and Ideas," in *The Wild Animal Story*, ed. Ralph H. Lutts (Philadelphia: Temple Univ. Press, 1998), 1–2.

64. McLean, *American Vaudeville*, 141–42.

65. Nigel Rothfels, *Savages and Beasts: The Birth of the Modern Zoo* (Baltimore: Johns Hopkins Univ. Press, 2002), 155–57.

66. Jesse Haney, *Haney's Art of Training Animals* (New York: J. Haney, 1869), 14–15.

67. Carol Joys, *The Wild Animal Trainer in America* (Boulder, CO: Pruett, 1983), 7.

68. Michael Peterson, "The Animal Apparatus: From a Theory of Animal Acting to an Ethics of Animal Acts," *TDR: The Drama Review* 51, no. 1 (2007): 35.

69. Harold J. Shiestone, "The Scientific Training of Wild Animals," *Scientific American*, Oct. 1902, 260.

70. Joys, *Wild Animal Trainer*, 21; John Stokes, " 'Lion Griefs': The Wild Animal Act as Theatre," *New Theatre Quarterly* 20, no. 2 (2004): 147.

71. "Hagenbeck's trained elephant 'shooting the chutes,' The Pike, World's Fair, St. Louis, Mo." Louisiana Purchase Exposition, St. Louis, 1904. Robert N. Dennis Collection of Stereoscopic Views, New York Public Library, New York.

72. "Hagenbeck Greater Shows Launched in Cincinnati," *Billboard*, Apr. 14, 1902.

73. Rothfels, *Savages and Beasts*, 160–61, 200; Stokes, "Lion Griefs," 146–47. For some twentieth- and twenty-first-century examples of this rhetoric of bad trainers and "kind" trainers, see

Frank C. Bostock, *The Training of Wild Animals*, ed. Ellen Velvin (1903; repr. New York: Century, 1920), 4–5; Joys, *Wild Animal Trainer*, 17–21; Mim Eichler Rivas, *Beautiful Jim Key: The Lost History of a Horse and Man Who Changed the World* (New York: Morrow, 2005), 25–26, 39–40, 56–58.

74. Harold J. Shiestone, "The Scientific Training of Wild Animals," *Scientific American*, Oct. 1902, 260.

75. Eric Ames, *Carl Hagenbeck's Empire of Entertainments* (Seattle: Univ. of Washington Press, 2008), 179–80; Joys, *Wild Animal Trainer*, 23–24.

76. August H. Kober, *Circus Nights and Circus Days*, trans. Claud W. Sykes (London: Sampson Low, Marston, 1928), 18.

77. Bostock, *Training of Wild Animals*, 29–33, 183–85; Stokes, "Lion Griefs," 140.

78. Jennifer Mason, *Civilized Creatures: Urban Animals, Sentimental Culture, and American Literature, 1850–1900* (Baltimore: Johns Hopkins Univ. Press, 2005), 47–49.

79. Ibid., 160. See also Gail Bederman, *Manliness and Civilization: A Cultural History of Gender and Race in the United States, 1880–1917* (Chicago: Univ. of Chicago Press, 1996).

80. "His Friends the Animals," *New York Sun*, Jan. 19, 1902.

81. Joys, *Wild Animal Trainer*, 19.

82. See, e.g., Bostock, *Training of Wild Animals*, 233; Coup, *Sawdust and Spangles*, 183–85.

83. McLean, *American Vaudeville*, 147.

84. William Johnson, *The Rose-Tinted Menagerie* (London: Heretipptic, 1990), 52–53.

85. Joys, *Wild Animal Trainer*, vii, 25.

86. "Miller's Elephants Now Playing Vaudeville," *Billboard*, Nov. 4, 1905.

87. "His Friends the Animals," *New York Sun*, Jan. 19, 1902.

88. Tanja Schwalm, "'No Circus without Animals?' Animal Acts and Ideology in the Virtual Circus," in *Knowing Animals*, ed. Laurence Simmons and Philip Armstrong, Brill Human Animal Studies (Leiden: Brill, 2007), 87.

89. Horace Townsend, "Animals and Their Trainers," *Frank Leslie's Popular Monthly* 26, no. 6 (1888): 733; see also, e.g., McLean, *American Vaudeville*, 148.

90. Jonathan Safran Foer, *Eating Animals* (New York: Little, Brown, 2009), 99.

## Chapter 7 · *Animal Cultures Lost in the Circus, Then and Now*

1. "California Enterprise," *Daily Alta California*, May 21, 1859.

2. "Jumbo's Successors," *New York Sun*, reprinted in *St. Paul Daily Globe* (MN), Dec. 26, 1885.

3. "Hebe's Wonderful Baby," *New York Sun*, Mar. 21, 1880.

4. "Townsend Walsh Collection Circus Press Releases," box 5, Townsend Walsh Collection, Rare Books Division, New York Public Library, New York.

5. "A Baby Elephant," *Harper's Weekly*, Apr. 3, 1880, 219; see also "Hebe's Wonderful Baby," *New York Sun*, Mar. 21, 1880.

6. "A Baby Elephant," *Harper's Weekly*, Apr. 3, 1880, 219.

7. Ibid.

8. Raymond Sukumar, *The Living Elephants: Evolutionary Ecology, Behaviour, and Conservation* (New York: Oxford Univ. Press, 2003), 137–40.

9. Fred Kurt, Khyne U. Mar, and Marion E. Garaï, "Giants in Chains: History, Biology, and Preservation of Asian Elephants in Captivity," in *Elephants and Ethics: Toward a Morality of Coexistence*, ed. Chris Wemmer and Catherine A. Cristen (Baltimore: Johns Hopkins Univ. Press, 2008), 340.

10. Ros Clubb, Marcus Rowcliffe, Phyllis Lee, Khyne U. Mar, Cynthia Moss, and Georgia J.

Mason, "Compromised Survivorship in Zoo Elephants," *Science* 322, no. 598 (2008): 1649; Kurt, Mar, and Garaï, "Giants in Chains," 340.

11. "A Baby Elephant," *Harper's Weekly*, Apr. 3, 1880, 219; see also "The Mode of Suckling of the Elephant Calf," *Science* 1, no. 1 (1880): 10.

12. "How Jumbo Fares in Winter Quarters," *New York Tribune*, Feb. 18, 1883.

13. Shana Alexander has questioned whether the mating of Hebe and Mandarin took place in the United States, suggesting that Hebe was imported less than two years before Columbia's birth and came into the country pregnant, although this view may emanate from later, incorrect newspaper items. Alexander, *The Astonishing Elephant* (New York: Random House, 2000), 195–97; "Columbia, 283, Asiatic Female," box 1, John "Chang" Reynolds Papers, Robert L. Parkinson Library and Research Center, Circus World Museum, Baraboo, WI.

14. "Editorial," *Science* 1, no. 17 (1880): 202.

15. "Hebe's Wonderful Baby," *New York Sun*, Mar. 21, 1880.

16. Barbara Noske, *Beyond Boundaries: Humans and Animals* (Montreal: Black Rose Books, 1997), 16–17.

17. "Route Book of Cooper & Bailey's Great London Circus and Allied Shows of the Season of 1880. Compiled by W. G. Crowley. Philadelphia: Merrihew & Son, 135 N. Third St. 1880," box 46 "Miscellaneous Materials: Route Books," McCaddon Collection of the Barnum and Bailey Circus, Manuscripts Division, Department of Rare Books and Special Collections, Princeton University Library.

18. Janet M. Davis, *The Circus Age: Culture and Society under the American Big Top* (Chapel Hill: Univ. of North Carolina Press, 2002), 39–45; Earl Chapin May, *The Circus: From Rome to Ringling* (New York: Duffield & Green, 1932), 121.

19. "Barnum's Baby Elephant," *New York Times*, Feb. 4, 1882.

20. *Forepaugh's Sacred White Elephant* (Buffalo: n.p., n.d.), box 48, folder 7, McCaddon Collection; W. C. Coup, *Sawdust and Spangles: Stories and Secrets of the Circus* (Chicago: Herbert S. Stone, 1901), 39–43; Davis, *Circus Age*, 53, 252n58.

21. Davis, *Circus Age*, 44–45.

22. "Barnum and London," *Brooklyn Daily Eagle*, May 15, 1884.

23. *New York Sun*, reprinted in "Baby Bridgeport's Birthday," *Daily Standard* (Bridgeport, CT), Saturday Evening, Feb. 3, 1883.

24. "Barnum's Baby Elephant," *New York Times*, Feb. 4, 1882.

25. "In a Circus Menagerie," *New York Times*, Apr. 30, 1882.

26. "Vagaries of Animals," *New York Times*, May 3, 1896.

27. "Editorial," *Science* 1, no. 17 (1880): 202.

28. "In a Circus Menagerie," *New York Times*, Apr. 30, 1882.

29. "Barnum's Baby Elephant," *New York Times*, Feb. 4, 1882.

30. "Murderous Circus Elephant," *New York Sun*, Mar. 23, 1903.

31. "Saved from Elephant," *New York Tribune*, Mar. 23, 1903.

32. "Townsend Walsh Collection Circus Press Releases," box 5, Townsend Walsh Collection.

33. "Choke Elephant to Death," *New York Times*, Nov. 9, 1907. See also, e.g., "Vicious Elephant Is Executed by Owners," *San Francisco Call*, Nov. 10, 1907; "Columbia Became Ugly," *Salt Lake Herald*, Nov. 10, 1907; "Keepers Execute Elephant," *Daily Sun* (Gainesville, FL), Nov. 12, 1907; "First Native American Elephant Is Executed," *Los Angeles Herald*, Nov. 10, 1907; "Elephant Strangled by Showmen," *New York Tribune*, Nov. 9, 1907.

34. George Conklin, *The Ways of the Circus: Being the Memoires and Adventures of George Conklin, Tamer of Lions* (New York: Harper & Bros., 1921), 130–31, 136.

35. See, e.g., "Jumbo's Successors," *New York Sun*, reprinted in *St. Paul Daily Globe* (MN), Dec. 26, 1885; Charlie Campbell, "Elephants, Good and Bad," *Circus Review* 5, no. 1 (1957): 2–3; Jay Teel, *The Crime and Execution of Black Diamond, the Insane Elephant* (Ansted, WV: Petland Press, 1930), folder 15, vol. 8, Leonidas Westervelt Circus Collection, New-York Historical Society, New York; Charles Edwin Price, *The Day They Hung the Elephant* (Johnson City, TN: Overmountain Press, 1992).

36. Ian R. Tyrrell, *Transnational Nation: United States History in Global Perspective since 1789* (New York: Palgrave Macmillan, 2007), 8.

37. Mark St. Leon, "Yankee Circus to the Fabled Land: The Australian-American Circus Connection," *Journal of Popular Culture* 33, no. 1 (1999): 77–89.

38. Harvey L. Watkins, *Four Years in Europe: The Barnum & Bailey Greatest Show on Earth in the Old World* (n.p.: Henry L. Watkins, 1907), 22–23.

39. "Scrapbook 2: Barnum & Bailey Ltd., London," box 18, McCaddon Collection.

40. Barney Nelson, *The Wild and the Domestic: Animal Representation, Ecocriticism, and Western American Literature* (Reno: Univ. of Nevada Press, 2000), 1, 4.

41. Virginia DeJohn Anderson, *Creatures of Empire: How Domestic Animals Transformed Early America* (Oxford: Oxford Univ. Press, 2004), 96.

42. Gil Robinson, *Old Wagon Show Days* (Cincinnati: Brockwell, 1925), 33.

43. Mason, pers. corr., Aug. 30, 2008; Clubb et al., "Compromised Survivorship," 1649.

44. "In a Circus Menagerie," *New York Times*, Apr. 30, 1882.

45. Lucia Zora, *Sawdust and Solitude* (Boston: Little, Brown, 1928), 77.

46. Ann Norton Greene, *Horses at Work: Harnessing Power in Industrial America* (Cambridge, MA: Harvard Univ. Press, 2008), 39.

47. Connie Clausen, *I Love You Honey, But the Season's Over* (New York: Holt, Rinehart & Winston, 1961), 197.

48. Christine Geraghty, "Re-examining Stardom: Questions of Texts, Bodies and Performance," in *Stardom and Celebrity: A Reader*, ed. Sean Redmond and Su Holmes (London: Sage, 2007), 98.

49. Albert F. McLean Jr., *American Vaudeville as Ritual* (Lexington: Univ. of Kentucky Press, 1965), 150–51; John Berger, "Why Look at Animals?" in *About Looking* (New York: Pantheon Books, 1980), 1–26.

50. May, *Circus*, 223–237; George L. Chindahl, *A History of the Circus in America* (Caldwell, ID: Caxton, 1959), 118–57; Marcello Truzzi, "The American Circus as a Source of Folklore: An Introduction," *Southern Folklore Quarterly* 30, no. 4 (1966): 290.

51. "Elephant Census," *Billboard*, Apr. 12, 1952; "It Couldn't Be They Weren't at Home," *Billboard*, Apr. 26, 1952.

52. Eric Scigliano, *Love, War, and Circuses: The Age-Old Relationship between Elephants and Humans* (New York: Houghton Mifflin, 2002), 205.

53. Gregg Mitman, "Pachyderm Personalities: The Media of Science, Politics, and Conservation," in *Thinking with Animals: New Perspectives in Anthropomorphism*, ed. Loraine Daston and Gregg Mitman (New York: Columbia Univ. Press, 2008), 176.

54. Alan Beardsworth and Alan Bryman, "The Wild Animal in Late Modernity," *Tourist Studies* 1, no. 1 (2001): 90–91.

55. David Hammarstrom, quoted in Fred D. Pfening Jr., "The Circus Year in Review," *Bandwagon* 54, no. 2 (2010): 7.

56. Michael Peterson, "The Animal Apparatus: From a Theory of Animal Acting to an Ethics of Animal Acts," *TDR: The Drama Review* 51, no. 1 (2007): 41; see, e.g., Humane Society of the United States, "Animal-Free Circuses," www.hsus.org/wildlife/issues_facing_wildlife/circuses/ animalfree_circuses_and_entertainment/a_list_of_animalfree_circuses.html; People for the Ethical Treatment of Animals, "Animal-Free Circuses," www.circuses.com.

57. Richard Corliss, "That Old Feeling IV, A Tale of Two Circuses," *Time Magazine*, Apr. 20, 2001, www.time.com/time/columnist/corliss/article/0,9565,107192-1,00.html.

58. "About the Center for Elephant Conservation," www.elephantcenter.com/?id=3624, accessed, Apr. 15, 2010; "It's a Boy! Ringling Bros. First Calf Born from Artificial Insemination," Center for Elephant Conservation Press Release, Jan. 27, 2010, www.elephantcenter.com/?id= 5080.

59. Staff at North American zoos, at Carsen & Barnes "Endangered Ark Foundation" in Hugo, OK, and at Ringling Brother's Florida "Elephant Conservation Center" breeding / winter quarters facilities have recently resorted to artificial insemination to try to produce more elephants for entertainment and show since elephants are notoriously reticent to breed naturally once removed from life at large in Asia and Africa. It is not clear if these attempts at breeding, especially by impregnating juvenile cows (since later in life captive females develop infections and other health problems that inhibit oestrus and impregnation) will succeed in producing self-sustaining populations of elephants with the behaviors and knowledge required to survive outside captivity. Moreover, so far captive-born elephants have usually died before age five from disease, inexpert care from socially deprived or juvenile mothers, and other factors produced by dominant modes of elephant captivity. Shana Alexander, *Astonishing Elephant*, 255; G. A. Bradshaw, *Elephants in Circuses: Analysis of Practice, Policy, and Future* (Ann Arbor, MI: Animals and Society Institute, 2007), 23n10; Clubb et al., "Compromised Survivorship," 1649; Sukumar, *Living Elephants*, 388–89.

60. Barack Photo Gallery, www.elephantcenter.com/photogallery.aspx#.

61. "About the Center for Elephant Conservation," www.elephantcenter.com/?id=3624.

62. Shana Alexander, *Astonishing Elephant*, 175; Scigliano, *Love, War, and Circuses*, 263–64. For images of the facility, see websites by Ringling Brothers and PETA. www.elephantcenter .com/; www.ringlingbeatsanimals.com/bound-babies.asp.

63. Schwalm, "No Circus without Animals?" 84–85.

64. Kenneth Feld, quoted in "The Circus Is Just One of His Acts," *New York Times*, Mar. 24, 1993; Mark Albright, "For Kenneth Feld, the 141st Edition of Ringling Bros. and Barnum & Bailey Circus Embraces a New Generation," *St. Petersburg Times* (FL), Dec. 28, 2010.

65. Bob Brooke, "Step Right Up," *History Magazine*, Oct.–Nov. 2001, www.history-magazine. com/circuses.html.

66. Ibid.; John H. McConnell, "Shrine Circus—R.I.P.?," *Bandwagon* 53, no. 3 (2009): 12–18.

67. McConnell, "Shrine Circus—R.I.P.?," 12–18.

68. "Baby Circus Elephant Recovering from Virus," *United Press International*, Feb. 10, 2010, www.upi.com/Odd_News/2010/02/10/Baby-circus-elephant-recovering-from-virus/UPI-24571 265841420/.

69. "Sick Young Ringling Elephant Pulled from Circus," *Miami Herald*, Feb. 6, 2010, accessed Feb. 9, 2010; Eloísa Ruano González, "Ringling Bros.' Baby Elephant, Barack, Fighting Deadly Virus," *Orlando Sentinel*, Feb. 5, 2010, http://articles.orlandosentinel.com/2010-02-05/

news/os-elephant-herpes-20100205_1_center-for-elephant-conservation-second-elephant-herpes-virus; "Leading Elephant Expert Joins in Defense of Animals in Condemning St. Louis Zoo for Deadly Breeding Practices," IDA Press Release, Jan. 21, 2010, www.idanews.org/ida-breaking-news/leading-elephant-expert-01-21-2010/.

70. See, e.g., Scott Lawrence, "PETA Brings Circus Protest to School," KFDM News 6 (Beaumont, TX), www.kfdm.com/news/school-37424-peta-protest.html; Michelle Kim, "PETA Plans Protest at Pine Hills Elementary with Bloody Elephant Costume," WRBG CBS 6 (Albany, NY), www.cbs6albany.com/news/elephant-1272964-bloody-animals.html.

71. Scigliano, *Love, War, and Circuses*, 255–56.

72. This group also represents those food producers that dispute the health dangers of obesity as a "myth." "An Epidemic of Obesity Myths," www.obesitymyths.com; www.consumerfreedom.com/.

73. USDA officials and veterinarians, as well meaning as some are, have acted to censure and fine the circuses only in the most egregious and public cases—elephants found dead in the back of circus company box vans or touring circus elephants killed because denied veterinary care by their owners. Scigliano, *Love, War, and Circuses*, 265–70.

74. Australia has recently seen its last circus elephant retired. Bolivia, Norway, and many individual cities and municipalities already ban animal-act circuses. Similar legislation may be forthcoming in Britain. A March 2010 survey by the Department for Environment, Food and Rural Affairs of ten thousand Britons showed that 95.5 percent of respondents answered no to the question: "Do you think that there are any species of wild animal which it is acceptable to use in travelling circuses?" Department of Food, Environment and Rural Affairs, "Initial Summary of the Responses to the DEFRA Public Consultation Exercise on the Use of Wild Animals in Circuses," Mar. 2010, www.defra.gov.uk/foodfarm/farmanimal/welfare/act/secondary-legis/circus.htm.

75. People for the Ethical Treatment of Animals, "Elephants in Circuses: Training and Tragedy," www.petatv.com/tvpopup/video.asp?video=training_tragedy&Player=qt; Will Hoover, "Slain Elephant Left Tenuous Legacy in Animal Rights," *Honolulu Advertiser*, Aug. 20, 2004, http://the.honoluluadvertiser.com/article/2004/Aug/20/ln/ln19a.html.

76. See, e.g., Lisa F. Kane, "Contemporary Zoo Elephant Management: Captive to a 19th-Century Vision," in *An Elephant in the Room: The Science and Well-being of Elephants in Captivity*, ed. Debra L. Forthman, Lisa F. Kane, David Hancocks, and Paul F. Waldau (North Grafton, MA: Tufts Center for Animals and Public Policy, 2009), 87–96.

77. Association of Zoos and Aquariums, "AZA Standards of Care for Elephants" (2003), www.aza.org/elephant-standards-care/. See also John Lehnhardt, "Elephant Handling: A Problem of Risk Management and Resource Allocation," *American Association of Zoological Parks and Aquariums Annual Conference Proceedings* (1991): 569–75; Margaret Whittaker and Gail Laule, "Protected Contact and Elephant Welfare," in Forthman et al., *An Elephant in the Room*, 181–88.

78. For various views on the Oregon zoo episode, see, e.g., In Defense of Animals, "Oregon Zoo Fact Sheet," www.helpelephants.com/oregon_zoo_fact_sheet.html; "Oregon DA Refuses to Prosecute for Elephant Abuse," *Animals' Advocate: Quarterly Newsletter of the Animal Legal Defense Fund* 87, no. 4 (2000): 4; Scigliano, *Love, War, and Circuses*, 279.

79. American Zoological Association, "Accredited Zoos and Aquariums," www.aza.org/current-accreditation-list/.

### Primary Sources

Although there is an abundance of historical evidence documenting circuses and animals in the United States, like many primary source bases, it offers great detail with respect to some favored topics, people, and practices while obscuring many of the workaday realities, experiences of people, and actions of animals that were equally crucial to what the circuses were. Specifically, the robust archival collections of circus history around the nation—and there are at least half a dozen—tend to contain abundant "circus paper," that is, advertising materials in the form of broadsides, handbills, newspaper clippings, and souvenir or promotional photographs that tell us much about the public face of these businesses. Early-twentieth-century celebrity memoirs of circus life are even easier to locate in many libraries, and these expose different aspects of circus life, although always with a proindustry agenda. Less common are accounts by working-class circus people, candid behind-the-scenes photos (especially for the nineteenth century, before the birth of consumer photography), or business records documenting labor relations or the caste-system subcultures of the circuses.

Nevertheless, several archival collections were crucial to this work, and I found that they held far more riches than I could possibly fit into this little book. The Joseph T. McCaddon Collection of the Barnum and Bailey Circus, Manuscripts Division, Department of Rare Books and Special Collections, Princeton University Library is a unique archive documenting the late-nineteenth-century circus business, with its telegram receipts, loan and engagement contracts, personal photographs, and systematically compiled surveys of the press agent's performance by way of day-by-day clippings of planted newspaper pieces assembled in scrapbooks. I also made happy use of the Leonidas Westervelt Circus Collection at the New-York Historical Society, which was particularly helpful for the early history of the animal trade, P. T. Barnum's entry into the circus trade, elephant biographies, and fan newsletters. The small P. T. Barnum Collection of letters between Barnum and his agents, as well as press agent Townsend Walsh's papers in the Rare Books Division of the New York Public Library, helped with an exploration of the "there's no such thing as bad press" philosophy pioneered by P. T. Barnum and others with respect to animal exhibitions. Additionally, the Tibbals Digital Collection of circus advertising and broadsides, available through the John and Mabel Ringling Museum of Art in Sarasota, Florida, emerged and expanded during the writing of this book to become the foremost online source for circus advertising art—and circus paper truly should be taken as a pre-eminent art form of the nineteenth- and early-twentieth-century United States. The Tibbals collection allows for surveys of advertising trends over decades that would be very difficult and potentially damaging to these fragile items if performed in person. Of course, the king of circus collections is housed at the Robert L. Parkinson Library and Research Center at the Circus World Museum in Baraboo, Wisconsin. Although of somewhat limited use for this project because its business records emphasize twentieth-century circuses like Ringling Brothers, it probably holds every book every published on circuses, and it has a particularly dedicated staff.

This project also benefited from a survey of circus and animal show materials in the Early American Imprints Series and the American Broadsides and Ephemera Series (both drawn from the American Antiquarian Society collections in Worcester, New York), which provide most of the early advertising supporting the first several chapters of this book. For the entire century, the Early American Newspapers and the American Periodical Series searchable online newspaper archives offer an overwhelming avalanche of independent news coverage and planted circus press notices. Taken in combination, these two forms—circus-authored pseudo news and local reporting on circuses—contain more information than any other source on the trade. In combination, they also draw our attention to the struggles of circus press men to control the messages the public received about elephant captivity, animal training, circus labor relations, and more.

To get a sense for how elephants behaved and the ways circus people talked about elephants (as both blessing and curse in the business), I did exploit a number of celebrity-authored memoirs and exposés, which revealed the cultures of traveling shows, events in the lives of particular elephants, as well as the nature of animal acquisition, training, and disposition: P. T. Barnum, *Struggles and Triumphs; or, Forty Years' Recollections of P. T. Barnum* (1871; Buffalo, NY: Warren, Johnson, 1873); Frank C. Bostock, *The Training of Wild Animals*, ed. Ellen Velvin (1903; repr. New York: Century, 1920); Connie Clausen, *I Love You Honey, But the Season's Over* (New York: Holt, Rinehart & Winston, 1961); George Conklin, *The Ways of the Circus: Being the Memoires and Adventures of George Conklin, Tamer of Lions* (New York: Harper & Bros., 1921); W. C. Coup, *Sawdust and Spangles: Stories and Secrets of the Circus* (Chicago: Herbert S. Stone, 1901); Paul Eipper, *Circus: Men, Beasts, and Joys of the Road*, trans. Frederick H. Martens (New York: Viking, 1931); Jesse Haney, *Haney's Art of Training Animals* (New York: J. Haney, 1869); J. Y. Henderson, *Circus Doctor: As Told to Richard Taplinger* (New York: Bantam Books, 1952); John J. Jennings, *Theatrical and Circus Life; or, Secrets of the Stage, Green-Room and Sawdust Arena* (St. Louis: Dan Linahan, 1882); August H. Kober, *Circus Nights and Circus Days*, trans. Claud W. Sykes (London: Sampson Low, Marston, 1928); Olive Logan, *The Mimic World, and Public Exhibitions: Their History, Their Morals, and Effects* (Philadelphia: New-World Publishing, 1871); George A. Palmer, *Secrets and Mysteries of Educating Animals: A Practical Experience of Twenty Years* (Portland, IN: Sun Job Department, 1890); Gil Robinson, *Old Wagon Show*

*Days* (Cincinnati: Brockwell, 1925); Garrard Tyrwhitt-Drake, *Beasts and Circuses* (Bristol: Arrowsmith, 1936); Lucia Zora, *Sawdust and Solitude* (Boston: Little, Brown, 1928).

Like newspapers and memoirs, circus trade journals, fan newsletters, and history magazines can be either a gold mine or a problem to the researcher, depending upon one's perspective. The *New York Clipper* provides plenty of entertainment news for the mid-nineteenth century but is surprisingly quiet about the traveling show trade. The classic industry rag, the *Billboard*, is the go-to source here, although it must be handled with care since its editors and writers were not involved in investigative reporting. Rather, they were entrusted with providing news, gossip, and advertising designed to facilitate the circus trade and help circus folk find work and locate equipment and animals, among other things. Much of the *Billboard*'s reportage on the nineteenth century are twentieth-century "wisdom of an old circus hand"–type memory pieces that convey a sense of circus folklore, rather than plain-spoken critique or complete accuracy with respect to the who, what, when, and where of circus history. Fan magazines like *White Tops*, the *Circus Review*, and the circus history magazine *Bandwagon* similarly show evidence of the same struggle to accurately represent both circus history and the historical memory of circus people—two overlapping but distinct and often conflicting mandates. In fact-checking accounts of particular elephants' lives for this project, I found that much of what these publications contain is often historically inaccurate. They do usefully convey a sense of the pride with which many circus people did their work under very difficult working conditions for limited pay, although with an uncritical eye that takes worker injuries, problematic animal welfare, and audience criticism of circus companies—issues many circus people did worry and argue about—as "bad news" best avoided for fear of tarnishing the reputation and earning power of the industry. Mining their pages, serious researchers must be selective.

### Circus Histories

Looking through the contemporary circus-history magazine *Bandwagon*, sometimes a valuable secondary source, I found the work of Richard W. Flint and Stuart Thayer to be indispensable. See, e.g., Richard W. Flint, "Origin of the Circus in America," *Bandwagon* 25, no. 2 (1981): 18; Stuart Thayer, "One Sheet," *Bandwagon* 18, no. 5 (1974): 23; Stuart Thayer, "The Keeper Will Enter the Cage: Early American Wild Animal Trainers," *Bandwagon* 26, no. 6 (1982): 38; Stuart Thayer, "The Elephant in America Before 1840," *Bandwagon* 31, no. 1 (1987): 20–26; Stuart Thayer, "The Elephant in America, 1840–1860," *Bandwagon* 35, no. 5 (1991): 34–37. *Bandwagon* is indexed, with many articles online, at www.circushistory.org/Bandwagon/Bandwagon Index.htm.

Scholarly circus history is abundant and reaches back at least to the mid-twentieth century. There is far too much of this work to list here, but especially useful for this project for the early British and American animal shows, horse circuses, and menageries are the Flint and Thayer pieces noted above, as well as: Elbert Bowen, "The Circus in Early Rural Missouri," *Missouri Historical Review* 47 (1952): 1–17; Maurice

Willson Disher, *Greatest Show on Earth* (London: Bell & Sons, 1937); Richard W. Flint, "Entrepreneurial and Cultural Aspects of the Early-Nineteenth-Century Circus and Menagerie Business," in *Itinerancy in New England and New York: Dublin Seminar for New England Folklife: Annual Proceedings, 1984,* ed. Peter Benes (Boston: Boston University, 1986), 131–49; R. J. Hoage and William A. Deiss, eds., *New Worlds, New Animals: From Menagerie to Zoological Park in the Nineteenth Century* (Baltimore: Johns Hopkins Univ. Press, 1996); Penelope M. Leavitt and James S. Moy, "Spalding and Rogers' Floating Palace, 1852–1859," *Theatre Survey* 25, no. 1 (1984): 15–27; Brett Mizelle, "Contested Exhibitions: The Debate over Proper Animal Sights in Post-Revolutionary America," *Worldviews: Environment, Culture, Religion* 9, no. 2 (2005): 219–35; Brett Mizelle, "'Man Cannot Behold It without Contemplating Himself': Monkeys, Apes, and Human Identity in the Early American Republic," *Explorations in Early American Culture: A Supplemental Issue of Pennsylvania History* 66 (1999): 144–73; Brett Mizelle, "'I Have Brought My Pig to a Fine Market': Animals, Their Exhibitors, and Market Culture in the Early Republic," in *Cultural Change and the Market Revolution in America, 1789–1860,* ed. Scott C. Martin (Lanham, MD: Rowman & Littlefield, 2005), 182–207; James S. Moy, "Entertainments at John B. Ricketts' Circus, 1793–1800," *Educational Theatre Journal* 30, no. 2 (1978): 186–202; A. H. Saxon, *Enter Foot and Horse: A History of Hippodrama in England and France* (New Haven, CT: Yale Univ. Press, 1968); A. H. Saxon, *The Life and Art of Andrew Ducrow and the Romantic Age of the English Circus* (Hamden, CT: Archon Books, 1978); R. W. G. Vail, "This Way to the Big Top," *New-York Historical Society Quarterly Bulletin* 29, no. 3 (1945): 138–42. And one cannot overlook Stuart Thayer's monumental three-volume documentation of the antebellum menagerie and circus trades, *Annals of the American Circus: 1793–1829,* 3 vols. (Manchester, MI: Rymark Printing, 1976–92).

For the later nineteenth-century and twentieth-century combined modern circuses, prominent studies include those by: George L. Chindahl, *A History of the Circus in America* (Caldwell, ID: Caxton, 1959); Rupert Croft-Cooke and Peter Cotes, *Circus: A World History* (London: Elek, 1976); Janet M. Davis, *The Circus Age: Culture and Society under the American Big Top* (Chapel Hill: Univ. of North Carolina Press, 2002); Earl Chapin May, *The Circus: From Rome to Ringling* (New York: Duffield & Green, 1932); Marcello Truzzi, "The American Circus as a Source of Folklore: An Introduction," *Southern Folklore Quarterly* 30, no. 4 (1966): 289–300; Gregory J. Renoff, *The Big Tent: The Traveling Circus in Georgia, 1820–1930* (Athens: Univ. of Georgia Press, 2008); George Speaight, *History of the Circus* (London: Tantivy Press, 1980). More theoretical approaches to circus and history can be found in Paul Bouissac, *Circus and Culture: A Semiotic Approach* (Bloomington: Indiana Univ. Press, 1976), and Helen Stoddart, *Rings of Desire: Circus History and Representation* (Manchester: Manchester Univ. Press, 2000).

### Business and Cultural History of Nineteenth-Century Entertainment

The labor, management, and business histories of entertainment in the United States before the 1910s are helpful, but their authors struggled with the same problems I faced in knowing about working-class and animal actors, internal company disputes, and audience reception, namely, correspondence not saved by their authors (P. T. Barnum intentionally destroyed his personal papers before his death), a paucity of working-class or audience-authored accounts, and the fact that so much of what went on was oral or ephemeral and customary and so produced no paper record. Still, in order to think about early celebrity and the trials of a traveling show, one can consult Louis S. Warren, *Buffalo Bill's America: William Cody and the Wild West Show* (New York: Knopf, 2005), and David Carlyon's exhaustive study, *Dan Rice: The Most Famous Man You've Never Heard Of* (New York: Public Affairs, 2001). The scholarly literature on P. T. Barnum is a genre in its own right; see, e.g., Bluford Adams, *E Pluribus Barnum: The Great Showman and the Making of U.S. Popular Culture* (Minneapolis: Univ. of Minnesota Press, 1997); John Rickards Betts, "P. T. Barnum and the Popularization of Natural History," *Journal of the History of Ideas* 20, no. 3 (1959): 353–68; James W. Cook, *The Arts of Deception: Playing with Fraud in the Age of Barnum* (Cambridge, MA: Harvard Univ. Press, 2001); Neil Harris, *Humbug: The Art of P. T. Barnum* (Chicago: Univ. of Chicago Press, 1973); A. H. Saxon, *P. T. Barnum: The Legend and the Man* (New York: Columbia Univ. Press, 1995).

## Elephants and History

With respect to specific examinations of circus elephants, elephant biographies, which constitute an important part of the work written over the twentieth century, indicate the degree to which Americans and others have long been interested in the stories of individual elephants as historical figures. Prominent among these is the Reynolds elephant biographies housed at the Richard L. Parkinson Library and Research Center, Circus World Museum in Baraboo, Wisconsin. Although they could do with some updating—they are now about fifty years old—"Chang" Reynolds's extensive records provide the most comprehensive early survey of elephant life in the United States: John "Chang" Reynolds Papers, Robert L. Parkinson Library and Research Center, Circus World Museum, Baraboo, WI. Today the online database www
.elephant.se holds even more of this kind of information on elephant births, deaths, sales, and more on a global scale; it was also helpful in understanding elephants as historical beings. A few scholarly accounts of individual elephants were carefully researched in the twentieth century. Shettel, McClung and McClung, and Goodwin were the first authors to do serious archival research on the first elephant imported into the United States. Their articles and the Thayer pieces in *Bandwagon*, mentioned above, represent sincere attempts to write Asian elephants into American history as individuals, depicting a given elephant as a "she" or "he," rather than an "it." I am particularly indebted to their labors in innumerable small community archives and old newspapers before the age of digital reproductions and search engines. George G. Goodwin, "The Crowninshield Elephant: The Surprising Story of Old Bet," *Natural History: the Journal of the American Museum* 60 (Oct. 1951): 357–39; George G. Goodwin, "The First Living Elephant in America," *Journal of Mammology* 6, no. 4 (1925): 256–63; Robert McClung and Gale McClung, "Captain Crowninshield Brings Home an Elephant," *American Neptune: A Quarterly Journal of Maritime History* 18, no. 2 (1958): 137–41; Robert M. McClung and Gale S. McClung, "America's First Elephant,"

*Nature Magazine* 50 (Oct. 1957): 403; James W. Shettel, "The First Elephant in the United States," *Circus Scrapbook*, July 1929, 7–8.

The bulk of elephant biographies and "census" information, however, is found in newspapers, fan-magazines, self-published books and pamphlets, and countless newspaper stories of varying accuracy. Some of the more prominent magazine and book-length biographies include: "Elephant Census," *Billboard*, Apr. 12, 1952; "It Couldn't Be They Weren't at Home," *Billboard*, 26 Apr. 1952; Charlie Campbell, "Elephants, Good and Bad," *Circus Review* 5, no. 1 (1957): 2–3; Ann Colver, *Old Bet* (New York: Knopf, 1957); Charles Edwin Price, *The Day They Hung the Elephant* (Johnson City, TN: Overmountain Press, 1992); Morton Smith, "Elephant Census of the United States," *Hobbies: The Magazine for Collectors* (Mar. 1942): 33, item #6, box (vol.) 6, Leonidas Westervelt Circus Collection, New-York Historical Society, New York; Phil Stong, *A Beast Called an Elephant* (New York: Dodd, Mead, 1955); Jay Teel, "True Facts and Pictures: The Crime and Execution of Diamond the Insane Elephant," (1930), item #15, box 14 "Pamphlets and minor books, [18–]–c1947," Leonidas Westervelt Circus Collection, New-York Historical Society, New York; *Murder of the Elephant: An Accurate Account of the Death of the Noble Animal* (Boston: Coverley, 1816), 8. See also the sources in Chap. 1, n2.

Beyond elephant biographies, there exist only a few historical works on circus animals, most of which do not problematize elephant experience or elephant captivity and the difficulties it imposed on circus workers: Pamela Alberts Abraham, "The Circus Elephant: A History of America's Most Outstanding Performer," M.A. thesis, University of California, Los Angeles, 1971; Bob Cline, *America's Elephants* (n.p.: T'Belle LLC Productions, 2009); William Johnson, *The Rose-Tinted Menagerie* (London: Heretic, 1990); Jennifer L. Mosier, "The Big Attraction: The Circus Elephant and American Culture," *Journal of American Culture* 22, no. 2 (1999), 9. The genre of Jumbo/Jumbo-mania stories, some rather fictionalized (even excluding the many, many children's books written about Jumbo) also indicates our need to take elephants (and other animals) as individuals and historically contingent figures to better reflect *human* experience. It includes: Jan Bondesin, *The Feejee Mermaid and Other Essays in Natural and Unnatural History* (Ithaca, NY: Cornell Univ. Press, 1999), chap. 5; Paul Chambers, *Jumbo: This Being the True Story of the Greatest Elephant in the World* (Hannover, NH: Steerforth Press, 2009); Les Harding, *Elephant Story: Jumbo and P. T. Barnum under the Big Top* (Jefferson, NC: McFarland, 2000); Richard W. Flint, "Jumbo Recycled," *Bandwagon* 2, no. 4 (1979): 17; James L. Haley, "Jumbo: The Colossus of his Kind," *American Heritage* 24, no. 5 (1973): 62–68, 82–85; Bill Kelly, "P. T. Barnum's Biggest Star," *American History* 32, no. 6 (1998): 37–40, 59; Eric Mathieson, *The True Story of Jumbo the Elephant* (New York: Coward-McCann, 1964); Theodore James Jr., "World Went Mad When Jumbo Came to Town," *Smithsonian* 13, no. 2 (1982): 134–52; W. P. Jolly, *Jumbo* (London: Constable Press, 1976).

Numerous scholars have demonstrated that people and elephants have been in close interaction for millennia, a process documented by natural histories and human cultural histories of elephants: Shana Alexander, *The Astonishing Elephant* (New York: Random House, 2000); Stephen Alter, *Elephas Maximus: A Portrait of the Indian Ele-*

*phant* (New York: Houghton Mifflin Harcourt, 2004); Silvio A. Bedini, *The Pope's Elephant: An Elephant's Journey from Deep in India to the Heart of Rome* (New York: Penguin, 1997); Richard Carrington, *Elephants: Their Natural History, Evolution, and Influence on Mankind* (New York: Basic Books, 1958); J. C. Daniel, *The Asian Elephant: A Natural History* (1998; repr. Dehra Dun: Natraj, 2009); Oliver Goldsmith, *The Asian Elephant: A Natural History* (1955); John M. Kistler, *War Elephants* (Lincoln: Univ. of Nebraska Press, 2007); Donald F. Lach, "Asian Elephants in Renaissance Europe," *Journal of Asian History* 1 (1967): 133–76; Ivan T. Sanderson, *The Dynasty of Abu: A History and Natural History of the Elephants and Their Relatives, Past and Present* (New York: Knopf, 1962); Eric Scigliano, *Love, War, and Circuses: The Age-Old Relationship between Elephants and Humans* (New York: Houghton Mifflin, 2002); H. H. Scullard, *The Elephant in the Greek and Roman World* (London: Thames & Hudson, 1974); Sujit Sivasundaram, "Trading Knowledge: The East India Company's Elephants in India and Britain," *Historical Journal* 48, no. 1 (2005): 27–63; Dan Wylie, *Elephant* (London: Reaktion, 2008).

### Ethological / Ethical / Animal Welfare Research on Elephants

Any study of animals as historical figures and factors of causation in the past requires a knowledge of the literature on contemporary animals that can be used to interpret how historical humans reported on other species. The difficulty here is that, due to scientific convention, ethological, behavioral, animal welfare, and veterinary research is usually published in concise, narrow studies that isolate one, two, or three variables. Thus, in scientific journals from *Science* to *Animal Behavior* to *Pachyderm*, humanities and social science researchers may find isolated segments of research findings, the context of which are difficult to know. On the other hand, much of this ethological and animal welfare science research on elephants (and most animals) is still under way, and scientific journals are able to publish such findings years before they might appear in syntheses or trade books. Another issue to keep in mind is that many animal science journals carry particular perspectives that should inform our reading of them. A publication like *Zoo Biology* is, very generally speaking, supportive of the continued existence of zoos, so it tends to offer research arguing how zoo animal captivity benefits animals or should be reformed to perpetuate zoos and support in-zoo breeding programs. Meanwhile, the *Journal of Applied Animal Welfare* and *Animal Welfare* support the "five freedoms" of animals, and their research papers tend to privilege animal experience and the quest to find ways to measure it with the idea that all animal welfare problems are caused by people. Certainly, scientists are not alone in this predilection: women's studies journals tend to have a broadly feminist perspective, labor history journals likewise support workers' right to collective bargaining, and business and management journals tend not to publish research that calls for the abolition of capitalism. So, all research fields carry their own assumptions as starting points for asking questions about the world.

In any event, this is where synthetic accounts, episodic works, and scientist's memoirs save the day. By digesting all the smaller research studies into narratives that put contemporary elephants and others in political, environmental, institutional, and commercial context, they provide a kind of theory by which to ask questions about historical elephants and people: G. A. Bradshaw, *Elephants in Circuses: Analysis of Practice, Policy and Future, Animals and Society Institute Policy Paper* (Ann Arbor, MI: Animals and Society Institute, 2007); G. A. Bradshaw, *Elephants on the Edge: What Elephants Teach Us about Humanity* (New Haven, CT: Yale Univ. Press, 2009); Ros Clubb and Georgia Mason, *A Review of the Welfare of Zoo Elephants in Europe: A Report Commissioned by the RSPCA* (Oxford: University of Oxford, Animal Behaviour Research Group, 2002), 4–7; Debra L. Forthman, Lisa F. Kane, David Hancocks and Paul F. Waldau, eds., *An Elephant in the Room: The Science and Well-Being of Elephants in Captivity* (North Grafton, MA: Tufts Center for Animals and Public Policy, 2009); Cynthia J. Moss, Harvey Croze and Phyllis C. Lee, *The Amboseli Elephants: A Long-Term Perspective on a Long-Lived Mammal* (Chicago: Univ. of Chicago Press, 2011); H. M. Schwammer, T. J. Foose, M. Fouraker, and D. Olson, *A Research Update on Elephants and Rhinos: Proceedings of the International Elephant and Rhino Research Symposium, Vienna, June 7–11, 2001* (Münster: Schüling Verlag, 2002); Chris Wemmer and Catherine A. Christen, eds., *Elephants and Ethics: Toward a Morality of Coexistence* (Baltimore: Johns Hopkins Univ. Press, 2008); Raman Sukumar, *Elephant Days and Nights: Ten Years with the Indian Elephant* (New York: Oxford Univ. Press, 1996); Raymond Sukumar, *The Living Elephants: Evolutionary Ecology, Behaviour, and Conservation* (New York: Oxford Univ. Press, 2003).

For a primer on the animal welfare approach to animal science, see the various publications of the Universities Federation for Animal Welfare (UFAW), including: David Fraser, *Understanding Animal Welfare: The Science in Its Cultural Context* (London: Wiley-Blackwell, 2008); UFAW, ed., *Animal Training: A Review and Commentary on Current Practice, Proceedings of a Symposium organized by the Universities Federation for Animal Welfare* (Potters Bar, UK: Universities Federation for Animal Welfare, 1990); John Webster, *Animal Welfare: Limping towards Eden* (London: Blackwell, 2005). See also Temple Grandin, ed., *Improving Animal Welfare: A Practical Approach* (Cambridge, MA: CAB International, 2010); Richard D. Ryder, "Measuring Animal Welfare," *Journal of Applied Animal Welfare Science* 1, no. 1 (1998): 75–80.

## Animal Histories

The number of scholarly books and articles that strive to document the lives and actions of historical animals are few and far between. Nonetheless, there are a few authors who do so, if not as the central task of their work, at least with respect to particular historical questions. This is a literature in the spirit of what Harriet Ritvo notes is an "increasingly inclusive or democratic trend (sometimes called 'history from the bottom up') within the historical profession." Harriet Ritvo, "History and Animal Studies," *Society and Animals* 10, no. 4 (2002): 404. It emerged from the revolutions in social, cultural, and environmental history in the later twentieth century and has begun to destabilize the idea that any human has been an entirely independent actor in the past. As historians of the environment or nonliterate human societies have found ways to document elements of the past created by forces or people that did not leave behind textual sources, these historians of human-animal interactions have found ways to document how some animal behavior—on a species-wide or regional level at least—has shaped human life. They do so by employing research on ecology, health, or behavior of other species in varying degrees of intensity in order to examine the behaviors, migrations, and numbers of large wild populations (e.g., bison wolves) or, in lieu of individuals, across species in large geographic regions (e.g., cattle, pigs, horses). See, e.g., Virginia DeJohn Anderson, *Creatures of Empire: How Domestic Animals Transformed Early America* (Oxford: Oxford Univ. Press, 2004); Richard W. Bulliet, *Hunters, Herders, and Hamburgers: The Past and Future of Human-Animal Relationships* (New York: Columbia Univ. Press, 2005); Jon T. Coleman, *Vicious: Wolves and Men in America* (New Haven, CT: Yale Univ. Press, 2004); Andrew C. Isenberg, *The Destruction of the Bison: An Environmental History, 1750–1920* (Cambridge: Cambridge Univ. Press, 2000); Robert Malcolmson and Stephanos Mastoris, *The English Pig: A History* (London: Hambleton & London, 2001); Sarah E. McFarland and Ryan Hediger, eds., *Animals and Agency: An Interdisciplinary Exploration* (Leiden: Brill, 2009); Helena Pycior, "Together in War and Memory: Fala and President Franklin Delano Roosevelt," paper presented at the Organization of American Historians Annual Meeting, Mar. 29, 2008; Clay McShane and Joel Tarr, *The Horse in the City: Living Machines in the Nineteenth Century* (Baltimore: Johns Hopkins Univ. Press, 2007); Sandra Swart, *Riding High: Horses, Humans, and History in South Africa* (Johannesburg: Wits Univ. Press, 2010). On African elephants who "might have actively influenced their history," see Bernhard Gissibl, "The Nature of Colonialism: Being an Elephant in German East Africa," paper presented at the "Animals in History: Examining the Not So Human Past" Conference, German Historical Institute, Cologne, Germany, May 2005.

### *Animal Studies / Histories of Animal Representation and Use*

The emerging body of work on human representations and uses of other species is critical in answering questions about how nonhuman animals have served as metaphors or sites of conflict or social change or how people have used them to define what it was to be human, masculine, white, middle-class, "of the circus," and more. This literature has become too large to discuss completely here, so for works documenting the eighteenth to early twentieth century or the United States, see, e.g., April Louise Austin, "Illustrating Animals for the Working Classes: *The Penny Magazine* (1832–45)," *Anthrozoös* 23, no. 4 (2010): 365–82; Steve Baker, *Picturing the Beast: Animals, Identity, Representation* (Manchester: Manchester Univ. Press, 1993); Peter Benes, ed., *New England's Creatures, 1400–1900, Dublin Seminar for New England Folklife Annual Proceedings* 18 (Boston: Boston University, 1995); Diane L. Beers, *For the Prevention of Cruelty: The History and Legacy of Animal Rights Activism in the United States* (Athens: Swallow Press of Ohio Univ. Press, 2006); Derek Bousé, *Wildlife Films* (Philadelphia: Univ. of Pennsylvania Press, 2000); Jonathan Burt, *Animals in Films* (London: Reaktion, 2002); Rebecca Cassidy, *Horse People* (Baltimore: Johns Hopkins Univ. Press, 2008); Cynthia Chris, *Watching Wildlife* (Minneapolis: Univ. of Minnesota Press, 2006); Eileen Crist, *Images of Animals* (Philadelphia: Temple Univ. Press, 2000); Lorraine Daston and Gregg Mitman, eds., *Thinking with Animals: New Perspec-*

*tives on Anthropomorphism* (New York: Columbia Univ. Press, 2006); Susan Davis, *Spectacular Nature: Corporate Culture and the Sea World Experience* (Berkeley: Univ. of California Press, 1997); Diana Donald, "Pangs Watched in Perpetuity: Sir Edwin Landseer's Pictures of Dying Deer and the Ethos of Victorian Sportsmanship," in *Killing Animals*, ed. Animal Studies Group (Urbana: Univ. of Illinois Press, 2006), 50–68; Erica Fudge, *Animal* (London: Reaktion, 2004), and her other edited editions and monographs; Donna Haraway, *Primate Visions: Gender, Race, and Nature in the World of Modern Science* (New York: Routledge, 1989); Mary Henninger-Voss, ed., *Animals in Human Histories: The Mirror of Nature and Culture* (Rochester: Univ. of Rochester Press, 2002); R. J. Hoage, ed., *Perceptions of Animals in American Culture* (Washington, DC: Smithsonian Institution Press, 1989); Walter Hogan, *Animals in Young Adult Fiction* (Lanham, MD: Scarecrow Press, 2008); Linda Kalof, *Looking at Animals in Human Histories* (London: Reaktion, 2007); Susan D. Jones, *Valuing Animals: Veterinarians and Their Patients in Modern America* (Baltimore: Johns Hopkins Univ. Press, 2003); Kathleen Kete, ed., *A Cultural History of Animals in the Age of Empire* (London: Berg, 2009); Margaret J. King, "The Audience in the Wilderness," *Journal of Popular Film and Television* 24, no. 2 (1996): 60–68; Randy Malamud, ed., *A Cultural History of Animals in the Modern Age* (London: Berg Publishers, 2007), 1–26; Ralph H. Lutts, ed., *The Wild Animal Story* (Philadelphia: Temple Univ. Press, 1998), 1–2; Ralph H. Lutts, *The Nature Fakers: Wildlife, Science, and Sentiment* (Charlottesville: Univ. Press of Virginia, 1990); Jennifer Mason, *Civilized Creatures: Urban Animals, Sentimental Culture, and American Literature, 1850–1900* (Baltimore: Johns Hopkins Univ. Press, 2005); Gregg Mitman, *Reel Nature: America's Romance with Wildlife on Film* (Seattle: Univ. of Washington Press, 1999); Barney Nelson, *The Wild and the Domestic: Animal Representation, Ecocriticism, and Western American Literature* (Reno: Univ. of Nevada Press, 2000); Barbara Noske, *Beyond Boundaries: Humans and Animals* (Montreal: Black Rose Books, 1997); Brandy Parris, "Difficult Sympathy in the Reconstruction-Era Animal Stories of Our Young Folks," *Children's Literature* 31 (2003): 25–49; Susan J. Pearson, *The Rights of the Defenseless: Protecting Animals and Children in Gilded Age America* (Chicago: Univ. of Chicago Press, 2011); Louise E. Robbins, *Elephant Slaves and Pampered Parrots: Exotic Animals in Eighteenth-Century Paris* (Baltimore: Johns Hopkins Univ. Press, 2002); Harriet Ritvo, *The Animal Estate: The English and Other Creatures in the Victorian Age* (Cambridge, MA: Harvard Univ. Press, 1987); Harriet Ritvo, The *Platypus and the Mermaid: And Other Figments of the Classifying Imagination* (Cambridge, MA: Harvard Univ. Press, 1997); Nigel Rothfels, *Savages and Beasts: The Birth of the Modern Zoo* (Baltimore: Johns Hopkins Univ. Press, 2002); Nigel Rothfels, ed., *Representing Animals* (Bloomington: Indiana Univ. Press, 2002); Marian Louise Scholtmeijer, *Animal Victims in Modern Fiction: From Sanctity to Sacrifice* (Toronto: Univ. of Toronto Press, 1993); James Turner, *Reckoning with the Beast: Animals, Pain, and Humanity in the Victorian Mind* (Baltimore: Johns Hopkins Univ. Press, 1980).

### Animals and Capitalism

Some of the most important insights about how elephants and others could function as pet, villain, asset, and speculation for circus people and fans comes from historians of breeding, technology, and the market: Margaret Derry, *Bred for Perfection: Shorthorn Cattle, Collies, and Arabian Horses since 1800* (Baltimore: Johns Hopkins Univ. Press, 2003); Margaret Derry, *Horses in Society: A Story of Animal Breeding and Marketing Culture, 1800–1920* (Toronto: Univ. of Toronto Press, 2006); Ann Norton Greene, *Horses at Work: Harnessing Power in Industrial America* (Cambridge, MA: Harvard Univ. Press, 2008); Katherine C. Grier, "Buying Your Friends: The Pet Business and American Consumer Culture," in *Commodifying Everything: Relationship of the Market,* ed. Susan Strasser (New York: Routledge, 2003); Katherine C. Grier, *Pets in America: A History* (Chapel Hill: Univ. of North Carolina Press, 2006); Jason Hribal, " 'Animals Are Part of the Working Class': A Challenge to Labor History," *Labor History* 44, no. 4 (2003): 435–36; Nicole Shukin, *Animal Capital: Rendering Life in Biopolitical Times* (Minneapolis: Univ. of Minnesota Press, 2009); Joel Tarr and Clay McShane, *Horse in the City: Living Machines in the Nineteenth Century* (Baltimore: Johns Hopkins Univ. Press, 2007); Richard White, "Animals and Enterprise," in *The Oxford History of the American West,* ed. Clyde A. Milner II, Carol A. O'Connor, and Martha A. Sandweiss (New York: Oxford Univ. Press, 1994). Two books, Susan R. Schrepfer and Philip Scranton, ed., *Industrializing Organisms: Introducing Evolutionary History* (New York: Routledge, 2004), and Sarah Franklin, *The Dolly Mixtures: The Remaking of Genealogy* (Durham, NC: Duke Univ. Press, 2007), add a conclusion to the circus man's dream of a native American captive elephant population by showing how human manipulation of animals for profit has proved particularly profitable (for some, at least) with respect to sheep, cattle, and horses, among others. Shana Alexander's volume, *The Astonishing Elephant* (New York: Random House, 2000), indeed discusses some of the more ambitious schemes afoot since the late twentieth century by which a handful of elephant men have labored to reproduce elephants by artificial insemination and selective breeding to create an easily managed pool of domesticated elephants designed to serve the specific needs of zoos and circuses.

### Animals on Stage

The history and theory of animals in performance is a new and exciting area of scholarly research that I intend this book to bolster and inform. In thinking at length about the idea of elephants as performers, "actors," and famous individuals, I drew great insight and inspiration from: Matthew Bliss, "Property or Performer: Animals on the Elizabethan Stage," *Theatre Studies* 39 (1994): 45–59; Yoram S. Carmeli, "The Sight of Cruelty: The Case of Circus Animal Acts," *Visual Anthropology* 10 (1997): 1–15; Yoram S. Carmeli, " 'Cruelty to Animals' and Nostalgic Totality: Performances of a Traveling Circus in Britain," *International Journal of Sociology and Social Policy* 22, no. 11/12 (2002): 73–88; Suzanne Laba Cataldi, "Animals and the Concept of Dignity: Critical Reflections on a Circus Performance," *Ethics and the Environment* 7, no. 2 (2002): 104–26; Michael Dobson, "A Dog at All Things: The Transformation of the Onstage Canine, 1550–1850," *Performance Research* 5, no. 2 (2000): 116–24; Michael Kirby, "On Acting and Non-Acting," *Drama Review* 16, no. 1 (1972): 3–15; James Knowles, " 'Can ye not tell a man from a marmoset': Apes and Others on the Early Modern Stage," in *Renaissance Beasts: Of Humans, Animals, and Other Wonderful Creatures,* ed. Erica

Fudge (Urbana: Univ. of Illinois Press, 2004), 138–62; Garry Marvin, "A Passionate Pursuit: Foxhunting as Performance," in *Nature Performed: Environment, Culture, and Performance*, ed. Bronislaw Szerszynski, Wallace Heim, and Claire Waterton (Oxford: Wiley-Blackwell, 2003), 46–60; Michael Peterson, "The Animal Apparatus: From a Theory of Animal Acting to an Ethics of Animal Acts," *TDR: The Drama Review* 51, no. 1 (2007): 33–48; Richard Schechner, *Performance Theory* (London: Routledge, 1988), 225, 248; Nicholas Ridout, "Animal Labour in the Theatrical Economy," *Theatre Research International* 29, no. 1 (2004): 57–65; John Stokes, "'Lion Griefs': The Wild Animal Act as Theatre," *New Theatre Quarterly* 20, no. 2 (2004): 138–54; David Williams, "The Right Horse, The Animal Eye: Bartabas and Théâtre Zingaro," *Performance Research* 5, no. 2 (2000): 29–40. A less-critical account nonetheless containing broadly valuable information on the use of animals in shows at zoos, circuses, and beyond is provided by Carol Joys, *The Wild Animal Trainer in America* (Boulder, CO: Pruett, 1983).

breeding, 46, 78, 145, 158, 220; and American
power, 209; captive elephant, 208–13,
226–27, 232
Bridgeport (elephant), 209, 214
British East India Company, 18, 44
bull elephants, 39, 105, 112, 125–27, 136–37,
159–60, 170; assaults by, on circus workers,
107–11, 115, 121, 162, 182, 187–90; assaults by,
on members of the public, 116; and musth,
79, 105–6, 109, 111–14, 125–26; as rogues,
105, 109, 117, 126, 201

California, 67, 143, 208
capitalism: and domesticated animals, 46, 124,
144–45, 220; and domestic ethic of kindness,
117–18, 128, 134–35; and ideology of animal
"improvement," 46, 78; and ideology of
human supremacy, 5, 62–64, 123–25, 131,
135, 154–55, 194, 200; and speculation in
wild animals, 43–44, 46
captivity, dysfunction of animals in, 18, 26–27,
140–41, 169, 176–77, 196–97, 220
Carsen & Barnes Circus, 228
celebrity, animal, 38, 42, 45–46, 60–62, 69, 121,
186, 208, 215–16, 222–23; defined, 4–5
Center for Consumer Freedom, 230
Center for Elephant Conservation, 226
Chief (elephant), 116–17, 193
Chieftain (elephant), 214
children's literature, 11, 61, 73–74, 78, 85, 124,
127–31, 180
Christian dominionism, 63–65, 74–75, 124–25
Chuny (elephant), 79–80, 112, 168
circuses: animal management in, 151–73; criti-
cism of, 99–100, 127–36, 164, 180, 195, 213;
historical development of, 1–3, 13–14, 67,
75–76, 119, 141–44, 223, 226, 227–28; histori-
ography of, 6, 7; industrialization of, 106,
146, 170; paradox of, 14, 175
Cirque du Soleil, 226
Clark, M. L., 187
class conflict and racism, 140, 164, 193–94
Clausen, Connie, 183, 222
Clever Hans (horse), 80–81
Clyde Beatty Circus, 223
Columbia (elephant), 162, 209–20, 213
Columbia Museum, Boston, 32
Columbus (elephant), 45, 109–10, 121–22

communication technologies: electronic
media, 222; national development of, 20,
43, 65, 119–20
Comte de Buffon (George Louis Leclerc), 21, 31, 168
Conklin, George, 77, 88, 91, 93, 113, 131, 156,
162–65, 219
conservationism, 194–95
consumer society: and animal performance, 5,
64–65, 68, 221, 223–24; historical development of, 3–4; as sign of American power, 5,
21, 48–49, 144, 148
Cooper, James E., 88, 218
Cooper, Thomas Apthorp, 56
Cooper & Bailey Circus, 209–10, 214–15
Corvell, John, 130
Coup, William Charles, 54, 77, 131, 143, 148, 173
cow elephants: as allomothers, 18, 210–12,
218–19; assaults by, on circus workers,
182–83, 187, 191, 217; assaults by, on
members of the public, 185, 190; as "mad,"
176–207, 217; and reproduction, 210–12
Craven, Stuart, 88, 113, 158, 170–71, 172–73, 196,
212, 214
Crowninshield, Jacob, 15, 17, 19, 20
Crumb, William, 109
Cummings, Gordon, 134
Curtis, Mr., 39, 40

Dampier, Calvin, 192–93
Dan Rice Circus Company, 67
Darwin, Charles, 72, 99, 124, 136
Davis, John, 35
Disney, 5, 176–77, 225
dogs, 28–29, 53, 55, 58–59, 85, 95–96, 108, 178,
222; on stage, 33, 56–58, 65, 67, 71, 74, 102,
199–202, 222; training of, 78, 85, 91, 96–99,
199–200, 204
Douglas-Hamilton, Iain, 225
Driesbach, Herr, 110
Drinker, Elizabeth, 28, 33
*Dumbo* (film), 176–77, 223
Dunlap, William, 56

Eck, John, 195
Edison, Thomas, 185
*Electrocuting an Elephant* (film), 185
"Elephant, The," 17–36, 30, 56, 57; as metaphor
for early republic, 16; as naturalist's curiosity,